Asia Conference on Power and Electrical Engineering 2016 (ACPEE 2016)

MATEC Web of Conferences Volume 55 (2016)

Bangkok, Thailand
20 - 22 March 2016

Editors:

Tek-Tjing Lie
Moustafa Eissa
Emanuele Calabro

ISBN: 978-1-5108-2333-4

Printed from e-media with permission by:

Curran Associates, Inc.
57 Morehouse Lane
Red Hook, NY 12571

Some format issues inherent in the e-media version may also appear in this print version.

Printed by Curran Associates, Inc. (2016)

For additional information, please contact EDP Sciences – Web of Conferences at the address below.

EDP Sciences – Web of Conferences
17, Avenue du Hoggar
Parc d'Activité de Courtabœuf
BP 112
F-91944 Les Ulis Cedex A
France

Phone: +33 (0) 1 69 18 75 75
Fax: +33 (0) 1 69 28 84 91

contact-edps@webofconferences.org

Additional copies of this publication are available from:

Curran Associates, Inc.
57 Morehouse Lane
Red Hook, NY 12571 USA
Phone: 845-758-0400
Fax: 845-758-2633
Email: curran@proceedings.com
Web: www.proceedings.com

TABLE OF CONTENTS

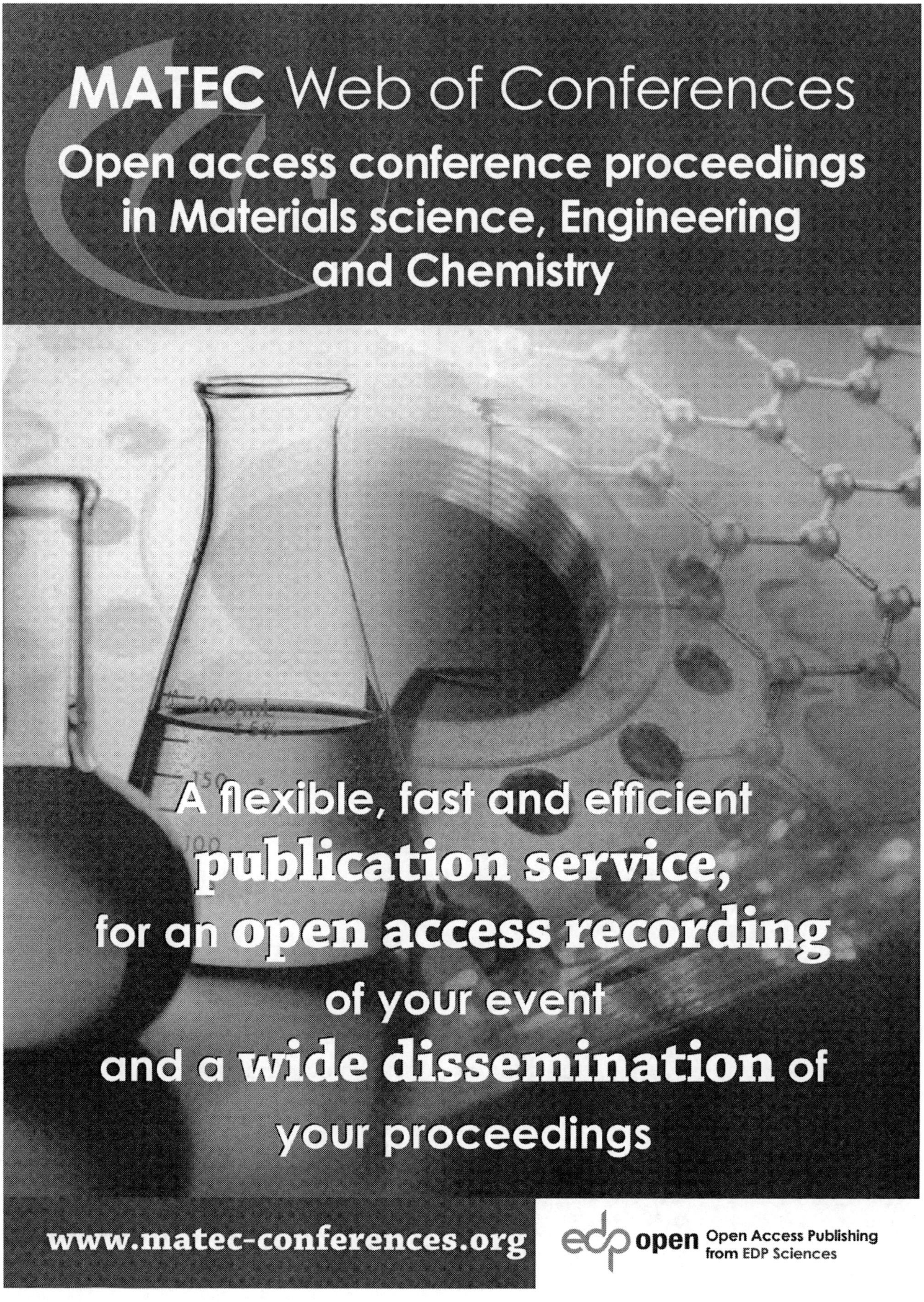
MATEC Web of Conferences
Open access conference proceedings in Materials science, Engineering and Chemistry
A flexible, fast and efficient
publication service,
for an open access recording
of your event
and a wide dissemination of
your proceedings
www.matec-conferences.org
edp open Open Access Publishing from EDP Sciences

PREFACE

The APCEE 2016 conference on Power and Electrical Engineering was successfully held at Thammasat University Rangsit Campus during March 20-21, 2016.

APCEE provided a forum for accessing to the most recent research and experience in the field of Power and Electrical Engineering from academics, industrials, professional with best practice. The conference proceeding will be a real good documents in the power and electrical engineering that can help other researchers, academics, professional engineers working in this field.

The outstanding scientific program provided potential and distinguished keynote materials and also convincing presentations by leaders in various topics of Power and Electrical Engineering. The authors in this field have contributed by valuable works and long outcomes from research in different topics covering "power electronics, control strategy, power system protection, electrical machines, renewable resources, dynamic modeling and high voltage". All papers including in this proceedings had undergone very extensive peer reviewing via experts before acceptance.

The published papers have knowledge and experiences that will be very valuable for others. By sharing this proceeding will be of great benefits for all. We do believe that ACPEE 2016 conference has the success to be very memorable professional and scientific experience in a unique Power and Electrical Engineering environment.

We would like to thank keynote speakers, invited speakers, authors of this proceeding for their valuable contributions. We would also like to thank all referees for their constructive comments on the papers, all the thanks for the organizing committee for the effort done before and after the conference.

Prof. M M Eissa

Faculty of Engineering, Helwan University at Helwan

Committees

International Advisory Board

Prof. Wei SHYY , Hong Kong University of Science and Technology, Hong Kong
Prof. Miroslaw Wyszynski, University of Birmingham, UK

Conference Committee

Dr. Nopbhorn Leeprechanon, Thammasat University, Thailand
Prof. Issa Batarseh, University of Central Florida, USA
Prof. Chan Siew Hwa, Nanyang Technology University, Singapore
Prof. Tek-Tjing Lie, AUT University, Auckland, New Zealand
Prof. Emanuele Calabrò, Institute of Industrial Technology, Italy

Program Committee

Prof. Somchart Chantasiriwan, Thammasat University, Thailand
Prof. Zhe Chen, Aalborg University, Denmark
Dr. Yaser Qudaih, Kyushu Institute of Technology, JAPAN

International Technical Committee

Prof. Moustafa Eissa, Helwan university, Egypt
Prof. Chandima Gomes, Universiti Putra Malaysia, Malaysia
Prof. Mohan Kolhe, University of Agder, Norway
Prof. Dimitris Labridis, Aristotle University of Thessaloniki, Greece
Dr. Shaifulazuar Bin Rozali, University of Malaya, Malaysia
Dr. Shuheng Chen, University of Electronic Science and Technology of China, China
Dr. Taib Iskandar Mohamad, Universiti Kebangsaan Malaysia, Malaysia
Dr. Zulfiqar Khan, Bournemouth University, UK
Dr. Khoa Dang Hoang, University of Sheffield, UK
Dr. Michael Gschwendtner, AUT University, Auckland, New Zealand
Dr. Ahmed Zobaa, Brunel University London, UK
Dr. Khaled Al-Soufi, King Fahd University of Petroleum & Minerals, Saudi Arabia
Prof. Alejandro A. Franco, Laboratoire de Réactivité et Chimie des Solides, France
Prof. Abraham Lomi, Institut Teknologi Nasional Malang, Indonesia

edp sciences

MATEC Web of Conferences

Statement of Peer review

In submitting conference proceedings to MATEC Web of Conferences, I certify to the Publisher that I adhere to the **Policy on Publishing Integrity** of the journal in order to safeguard good scientific practice in publishing.

1. All articles have been subjected to peer review administered by the proceedings editors.
2. Reviews have been conducted by expert referees, who have been requested to provide unbiased and constructive comments aimed, whenever possible, at improving the work.
3. Proceedings editors have taken all reasonable steps to ensure the quality of the materials they publish and their decision to accept or reject a paper for publication has been based only on the merits of the work and the relevance to the journal.

Title, date and place of the conference

2016 Asia Conference on Power and Electrical Engineering (ACPEE 2016)

March 20-22, 2016 Bangkok, Thailand

Proceedings editor(s):

Prof. Tek-Tjing Lie, Prof. Moustafa Eissa, Prof. Emanuele Calabrò

Date and editor's signature 30/03/2016

MATEC Web of Conferences **55**, 01001 (2016) DOI: 10.1051/matecconf/20165501001
ACPEE 2016

Research on Improved Control Strategy for STATCOM Based on Virtual Matrix Method

Xudong Wang[1,a], Shengtie Wang[1]

[1]*College of Electric Power, Inner Mongolia University of Technology, 010080 Hohhot, China*

Abstract. Fast and accurate detection of reactive current is the precondition for the realization of static synchronous compensator (STATCOM) reactive power compensation and harmonic suppression. Aiming at deviation and delay of the traditional reactive current detection algorithm with phase-locked loop (PLL) and low-pass filter (LPF) of STATCOM, a novel improved reactive current detection algorithm without PLL is proposed, in which the virtual matrix (VM) is built to replace the original PLL, and improved current average value filter is used to realize the function of LPF, so as to improve the real-time performance and robustness of reactive current detection. The realization process of VM detection method is derived in this paper, and improved control strategy for STATCOM is designed based on the VM detection method. Simulation analysis of the proposed detection algorithm and control strategy is conducted in Matlab platform so as to verify the correctness and effectiveness of the control strategy. The VM detection has the advantages of simple structure, fast response and easy for digital realization, which provides reference for the improvement of reactive power compensation precision for STATCOM.

1 Introduction

With the increase of wind power capacity and the development of micro-grid technology, the structure of power system is becoming more and more complex, which brings higher requirements for power quality of power network [1-4]. Reactive power compensation device is one of the effective ways to solve the problem of power quality. STATCOM is an important part of flexible AC transmission system[5], which is a hot spot in the field of reactive power compensation[6-8].

STATCOM control system is composed of reactive current detection module, control algorithm, pulse trigger module and so on. Fast and accurate detection of reactive current is the premise of realizing STATCOM reactive power compensation and harmonic suppression [9-11]. The p-q and i_p-i_q detection method based on instantaneous reactive power theory has the effect of voltage distortion and impact load, and the response speed and robustness of the system are reduced[12-14]. In [15], a software PLL is designed to replace the traditional PLL, but the error caused by the phase change of the grid voltage is not considered in this paper. In [16], the concept of "intelligent impedance" is introduced, and the function of the PLL is replaced by the intelligent impedance and proportional resonant controller. However, this method brings higher requirements for the design of parameters, which is not easy to achieve. In [17], synchronous rotation angular velocity is preset, so as to find the

[a] Corresponding author : wxd010353@imut.edu.cn

way to replace the PLL by sine/cosine function table lookup, but the error caused by the grid voltage phase change is not eliminated. In [18-19], the neural network controller is designed to replace the traditional PLL, but the response speed and the number of iterations are important factors that affect the performance of the algorithm. In [20], the fundamental positive sequence voltage extraction module is used for detecting the fundamental positive sequence voltage to replace the original PLL, but the introduction of two low pass filter reduces the real-time performance of the detection.

A new method for improving the reactive current detection based on VM is proposed in this paper. In this algorithm, the VM is used to replace the original PLL, and the current average value method is adopted to replace the LPF. The improved control strategy for STATCOM is designed based on VM reactive current detection algorithm. The effectiveness of the proposed algorithm and control strategy is verified by Matlab simulation.

2 Virtual Matrix Method

Assuming three-phase detection currents are:

$$\begin{bmatrix} i_a \\ i_b \\ i_c \end{bmatrix} = \sqrt{2} \begin{bmatrix} \sum_{m=1}^{\infty} I_m \sin(m\omega t + \theta_m) \\ \sum_{m=1}^{\infty} I_m \sin(m\omega t - \frac{2m\pi}{3} + \theta_m) \\ \sum_{m=1}^{\infty} I_m \sin(m\omega t + \frac{2m\pi}{3} + \theta_m) \end{bmatrix} \tag{1}$$

i_a, i_b and i_c are three-phase detected currents; I_m and θ_m are effective values of each measured harmonic current and the initial phase angle; m is the number of harmonics.
According to the traditional i_p-i_q detection method:

$$\begin{bmatrix} i_p \\ i_q \end{bmatrix} = \mathbf{C}\mathbf{C}_{32} \begin{bmatrix} i_a \\ i_b \\ i_c \end{bmatrix} = \sqrt{\frac{2}{3}} \mathbf{C} \begin{bmatrix} 1 & -1/2 & -1/2 \\ 0 & \sqrt{3}/2 & -\sqrt{3}/2 \end{bmatrix} \begin{bmatrix} i_a \\ i_b \\ i_c \end{bmatrix} \tag{2}$$

C is the coefficient matrix obtained by the PLL circuit.

In order to simplify the detection circuit, the virtual matrix $\mathbf{C}_v$ is introduced to replace matrix **C**. The expression of $\mathbf{C}_v$ is:

$$\mathbf{C}_v = \begin{bmatrix} \sin(\omega t + \theta) & -\cos(\omega t + \theta) \\ -\cos(\omega t + \theta) & -\sin(\omega t + \theta) \end{bmatrix} \tag{3}$$

θ is an arbitrary initial phase angle.
The matrix $\mathbf{C}_v$ is substituted into the expression (2), then:

$$\begin{bmatrix} i_p \\ i_q \end{bmatrix} = \mathbf{C}_v\mathbf{C}_{32} \begin{bmatrix} i_a \\ i_b \\ i_c \end{bmatrix} = \sqrt{3} \begin{bmatrix} \sum_{n=1}^{\infty} I_n \cos[(n \pm 1)\omega t + \theta_n - \theta] \\ -\sum_{n=1}^{\infty} I_n \sin[(n \pm 1)\omega t + \theta_n - \theta] \end{bmatrix} \tag{4}$$

When n=1, the DC components of i_p and i_q are:

$$\begin{bmatrix} i_{\mathrm{pm}} \\ \mathrm{i}_{\mathrm{qm}} \end{bmatrix} = \sqrt{3}\begin{bmatrix} I_1\cos(\theta_1-\theta) \\ -I_1\sin(\theta_1-\theta) \end{bmatrix} \tag{5}$$

Three-phase fundamental currents are:

$$\begin{bmatrix} i_{\mathrm{af}} \\ i_{\mathrm{bf}} \\ i_{\mathrm{cf}} \end{bmatrix} = \mathbf{C}_{23}\mathbf{C}_v^{-1}\begin{bmatrix} i_{\mathrm{pm}} \\ i_{\mathrm{qm}} \end{bmatrix} = \sqrt{2}\begin{bmatrix} I_1\sin(\omega t+\theta_1) \\ I_1\sin(\omega t-\frac{2\pi}{3}+\theta_1) \\ I_1\sin(\omega t+\frac{2\pi}{3}+\theta_1) \end{bmatrix} \tag{6}$$

$$\begin{aligned} \begin{bmatrix} i_{\mathrm{afp}} \\ i_{\mathrm{bfp}} \\ i_{\mathrm{cfp}} \end{bmatrix} &= \sqrt{2}\begin{bmatrix} I_1\sin(\omega t+\theta) \\ I_1\sin(\omega t-\frac{2\pi}{3}+\theta) \\ I_1\sin(\omega t+\frac{2\pi}{3}+\theta) \end{bmatrix}\cos(\theta_1-\theta) \\ \begin{bmatrix} i_{\mathrm{afq}} \\ i_{\mathrm{bfq}} \\ i_{\mathrm{cfq}} \end{bmatrix} &= \sqrt{2}\begin{bmatrix} I_1\cos(\omega t+\theta) \\ I_1\cos(\omega t-\frac{2\pi}{3}+\theta) \\ I_1\cos(\omega t+\frac{2\pi}{3}+\theta) \end{bmatrix}\sin(\theta_1-\theta) \end{aligned} \tag{7}$$

i_{afp}, i_{bfp} and i_{cfp} are active components of three-phase fundamental currents; i_{afq}, i_{bfq} and i_{cfq} are reactive components of three-phase fundamental currents.

According to the expression (6)-(7), it is still capable of detecting the fundamental active and reactive components of the detected current without errors. Compensation instruction current can be expressed as follows:

$$\begin{bmatrix} i_{\mathrm{aref}} \\ i_{\mathrm{bref}} \\ i_{\mathrm{cref}} \end{bmatrix} = \begin{bmatrix} i_{\mathrm{a}} \\ i_{\mathrm{b}} \\ i_{\mathrm{c}} \end{bmatrix} - \begin{bmatrix} i_{\mathrm{afp}} \\ i_{\mathrm{bfp}} \\ i_{\mathrm{cfp}} \end{bmatrix} \tag{8}$$

Similarly, assuming the three-phase supply voltage are:

$$\begin{bmatrix} e_{\mathrm{a}} \\ e_{\mathrm{b}} \\ e_{\mathrm{c}} \end{bmatrix} = \sqrt{2}\begin{bmatrix} \sum_{n=1}^{\infty}E_{1n}\sin(n\omega t)+\sum_{n=1}^{\infty}E_{2n}\sin(n\omega t+\theta_{2n}) \\ \sum_{n=1}^{\infty}E_{1n}\sin(n\omega t-\frac{2\pi}{3})+\sum_{n=1}^{\infty}E_{2n}\sin(n\omega t+\frac{2\pi}{3}+\theta_{2n}) \\ \sum_{n=1}^{\infty}E_{1n}\sin(n\omega t+\frac{2\pi}{3})+\sum_{n=1}^{\infty}E_{2n}\sin(n\omega t-\frac{2\pi}{3}+\theta_{2n}) \end{bmatrix} \tag{9}$$

e_{a}, e_{b} and e_{c} are three-phase voltages; $E_{1\mathrm{n}}$ and $E_{2\mathrm{n}}$ are the effective values of positive and negative sequence components of supply voltage harmonics.

According to the traditional i_p-i_q detection method:

$$\begin{bmatrix} e_p \\ e_q \end{bmatrix} = \sqrt{3}\begin{bmatrix} \sum_{n=1}^{\infty} E_{1n}\cos\left[(n-1)\omega t-\theta\right]+\sum_{n=1}^{\infty} E_{2n}\cos\left[(n+1)\omega t+\theta_{2n}+\theta\right] \\ -\sum_{n=1}^{\infty} E_{1n}\sin\left[(n-1)\omega t-\theta\right]-\sum_{n=1}^{\infty} E_{2n}\sin\left[(n+1)\omega t+\theta_{2n}+\theta\right] \end{bmatrix} \tag{10}$$

When n=1, the DC components of e_p and e_q are:

$$\begin{bmatrix} e_{pm} \\ e_{qm} \end{bmatrix} = \sqrt{3}\begin{bmatrix} E_1\cos\theta \\ E_1\sin\theta \end{bmatrix} \tag{11}$$

Fundamental positive sequence voltages are:

$$\begin{bmatrix} e_{af} \\ e_{bf} \\ e_{cf} \end{bmatrix} = \mathbf{C}_{23}\mathbf{C}_v^{-1}\begin{bmatrix} e_{pm} \\ e_{qm} \end{bmatrix} = \sqrt{2}\begin{bmatrix} E_1\sin(\omega t) \\ E_1\sin(\omega t-\frac{2\pi}{3}) \\ E_1\sin(\omega t+\frac{2\pi}{3}) \end{bmatrix} \tag{12}$$

According to the expression (12), it is still capable of detecting the fundamental active and reactive components of the detected power supply voltage without errors.

Three-phase instantaneous active power is:

$$p = \begin{bmatrix} e_{pm} \\ e_{qm} \end{bmatrix}^T \begin{bmatrix} i_{pm} \\ i_{qm} \end{bmatrix} = e_{pm}i_{pm} + e_{qm}i_{qm} \tag{13}$$

Active components of three-phase fundamental currents are:

$$\begin{bmatrix} i_{afp} \\ i_{bfp} \\ i_{cfp} \end{bmatrix} = \frac{p}{e_{pm}^2 + e_{qm}^2}\mathbf{C}_{23}\mathbf{C}_v^{-1}\begin{bmatrix} e_{pm} \\ e_{qm} \end{bmatrix} \tag{14}$$

The simplified schematic diagram of the virtual matrix detection method is shown in Fig. 1.

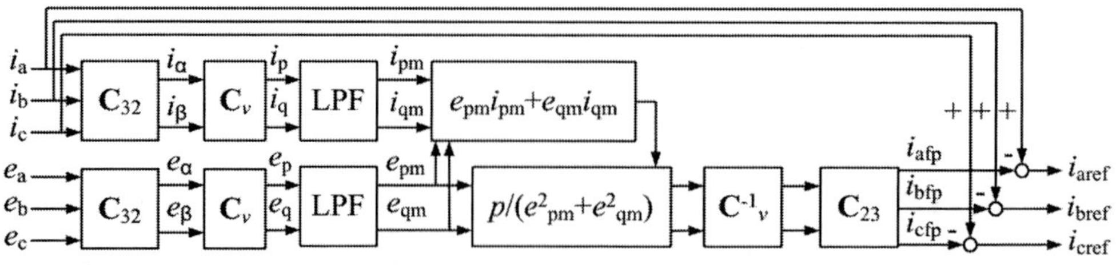

Figure 1. Schematic diagram of VM detection method.

3 Improved Current Average Value Filter

As shown in Fig. 1, the LPF is used to filter out the AC component of the detected current or voltage signal, and then, the fundamental positive sequence active current or voltage is obtained. Commonly used filters are FIR (Finite Impulse Response) and IIR (Infinite Impulse Response). But both of them have the problems such as complex parameter design, extension of time and so on. In [21], The integral period of traditional current average value method is changed from T/3 to the least common multiple of various order AC components, so as to filter out all AC components.

According to the expression (4) and (10), if a current or voltage signal does not contain even harmonics, the positive and negative sequence harmonic are $6k$ and $6k+6$ (k=0,1,2,...), the minimum common multiple is 6. The integral period of the current average value filtering method is T/6, that is, the DC component can be obtained by the 1/6 integral cycle time delay. The filter structure is simple and easy to be realized. The schematic diagram of the current average value filter is shown in Fig. 2.

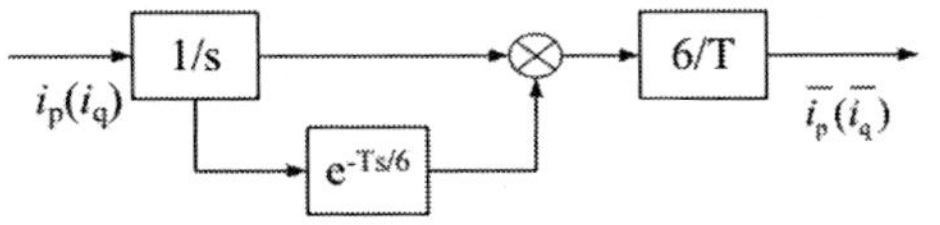

Figure 2. Schematic diagram of current average value filter.

4 Performance Comparison of Detection Method

In order to compare the advantages and disadvantages of the proposed method with the traditional p-q detection method and i_p-i_q detection method, the simulation models of three detection methods are built in Matlab/Simulink platform. In the simulation, it is assumed that the effective value of the power supply voltage is 311V, the effective value of the fifth and the seventh harmonic are respectively 4% and 3% of the effective value of the fundamental wave. The fundamental active value of the measured current is 20A, while the effective value of the fifth and the seventh harmonic are respectively 4% and 3% of the effective value of the fundamental wave. Simulation results are shown in Fig. 3.

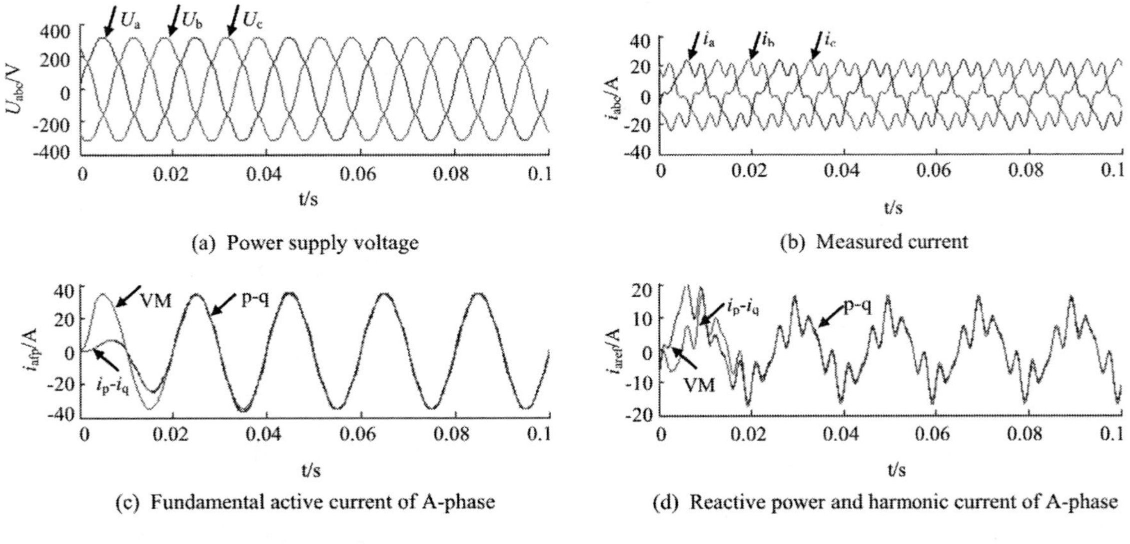

Figure 3. Simulation waveforms of three detection methods.

According to the results: 1) There is a certain degree of distortion in the power supply voltage and the measured current, and they contain fifth and seventh harmonic component respectively; 2) when the power supply voltage is distorted, the active current component can not be extracted accurately by p-q method, and the active component of the current can be extracted correctly by VM and i_p-i_q method; 3) both p-q and i_p-i_q methods require a power cycle to complete the correct acquisition of active current components, but the VM method based on current average value filter can achieve the above functions only in 1/6 power cycle; 4) the accuracy and speed of the VM method for reactive power and harmonic current extraction is better than p-q method and i_p-i_q method.

Fast Fourier Transformation (FFT) analysis is performed respectively on the current active component of the three detection methods in Fig. 3. The results are shown in Table 1.

Table 1. The results of FFT analysis.

Method	Effective value of fundamental current /A	THD/(%)
p-q	18.89	5.26
i_p-i_q	18.73	2.15
VM	19.92	0.4

According to the results in Table 1, the fundamental current effective value error of the VM method is 0.4%, while the errors of the other two methods are 5.6% and 6.35% respectively. Therefore, the total harmonic distortion rate of the reactive current obtained by the VM method is the smallest, the detection value is the most accurate, and the detection effect is better than the other two detection methods.

5 Simulation research on STATCOM improved control strategy

5.1 Improved control strategy

According to the schematic diagram of the instruction current detection based on the VM detection method, the improved control strategy for STATCOM is shown in Fig. 4.

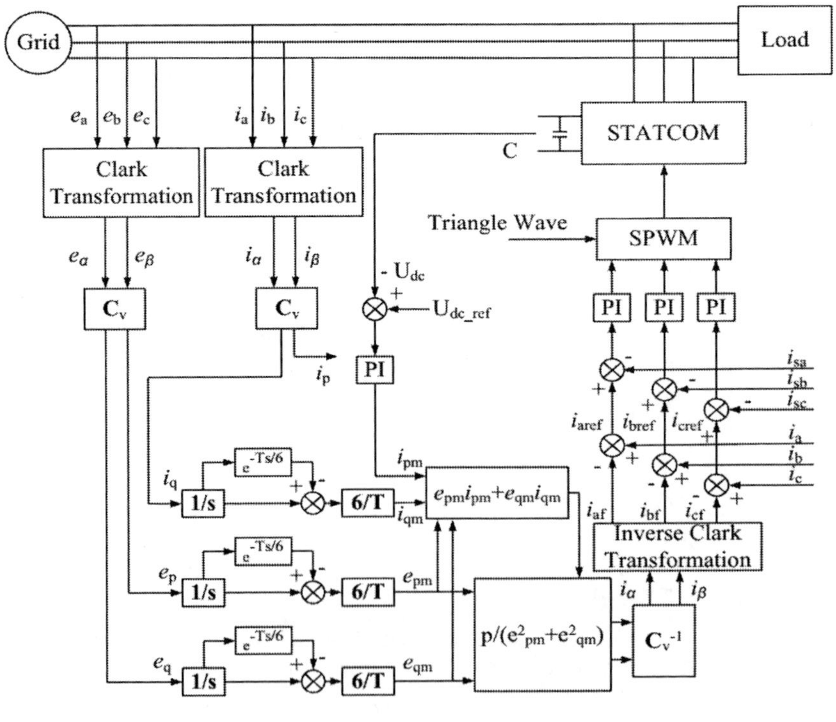

Figure 4. Block diagram of improved control strategy for STATCOM.

The control strategy is based on double closed-loop decoupling control of voltage outer loop and current inner loop. The instantaneous values of three-phase voltage in the grid side are e_a, e_b and e_c respectively, the instantaneous value of three-phase current are i_a, i_b and i_c respectively. The voltage and current are transformed to the two-phase orthogonal coordinate system, the instantaneous voltage e_α, e_β and the instantaneous current i_α, i_β are obtained respectively. After the $\mathbf{C}_v$ operation, the instantaneous active component e_p and the instantaneous reactive component e_q of voltage, the instantaneous active component i_p and the instantaneous reactive component i_q of current are obtained. The DC component of the instantaneous reactive current i_{qm} is obtained by the current average value filter. Similarly, the DC component of the instantaneous voltage e_{pm} and e_{qm} can be obtained. The DC component of active current i_{pm} is adjusted by PI controller after the STATCOM DC side capacitor voltage U_{dc} is subtracted from a given value U_{dc_ref}. Based on the expression (13)-(14) and the inverse matrix $\mathbf{C}_v^{-1}$ of the virtual matrix, the component of the compensation instruction in two phase orthogonal coordinate system $i_{\alpha f}$, $i_{\beta f}$ are obtained. After α-β/abc coordinate transformation, the compensation current i_{aref}, i_{bref} and i_{cref} are obtained. The STATCOM output current i_{sa}, i_{sb} and i_{sc} are subtracted from the compensation current instruction signal respectively. The signals are fed into a sinusoidal pulse width modulator by the respective PI regulator, and the inverter bridge circuit is triggered after the triangular carrier signal modulation.

5.2 STATCOM parameter design

In this paper, the capacity of STATCOM Q=1Mvar; the DC side voltage U_{dc}=2400V; the parameters of transformer T: the capacity S_N=2Mvar, the primary side voltage U_{1N}=35kV, the secondary side voltage U_{2N}=690V.

5.2.1 DC side capacitor selection

The effective value of reactive current for STATCOM is:

$$I_e = \frac{Q}{3U_{2N}} = \frac{1\times10^6}{3\times690} = 483(\mathrm{A}) \tag{15}$$

In this paper, the DC side capacitance value is calculated according to the following empirical formula:

$$C = \frac{0.2I_e}{\omega U_{dc}K} = \frac{0.2I_e}{2\pi f U_{dc}K} \tag{16}$$

K is the DC side voltage fluctuation coefficient; f is the grid frequency. Select K=0.01, the value of the DC side capacitor estimated by the equation (16) is 12.8mF.

5.2.2 Connection reactor selection

In this paper, the reactor parameter is calculated according to the following empirical formula:

$$L \le \frac{3U_{2N}(0.353m_{max}U_{dc} - U_{2N})}{314Q} \tag{17}$$

U_{2N} is the effective value of the secondary side voltage; U_{dc} is the DC side voltage of STATCOM; m_{max} is the maximum modulation ratio. The SPWM modulation method is used in this paper, so m_{max}=0.9. The reactor parameter calculated by the equation (17) is 0.4mH.

DOI: 10.1051/matecconf/20165501001

5.3 Simulation modelling and analysis

According to the STATCOM improved control strategy diagram shown in Fig. 4, the simulation model of STATCOM is built in Matlab/Simulink environment, as shown in Fig. 5.

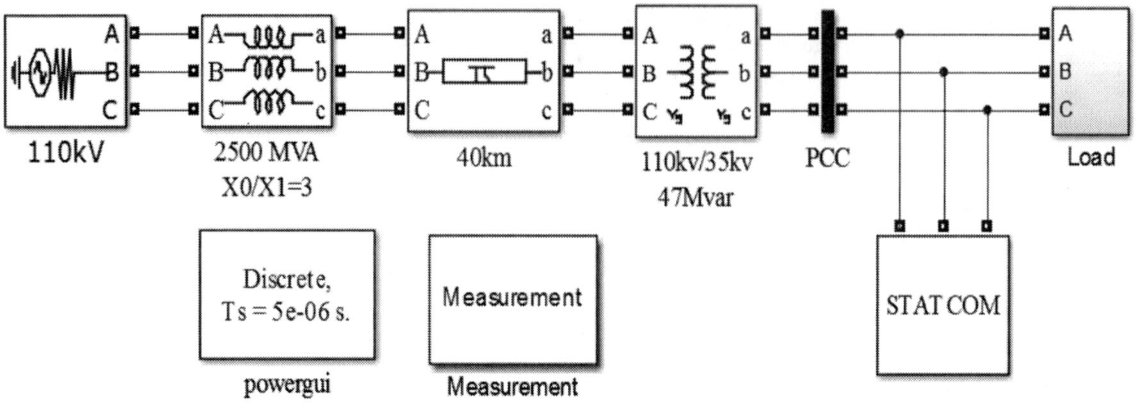

Figure 5. Simulation model of STATCOM.

5.3.1 Inductive load

The working conditions are as follows: the perceptual load capacity is 0.4MW during 0~0.2s, 1MW during 0.2~0.3s, 0.6MW during 0.3~0.4s; at 0.4s, STATCOM is broken away from network. Simulation results are shown in Fig. 6.

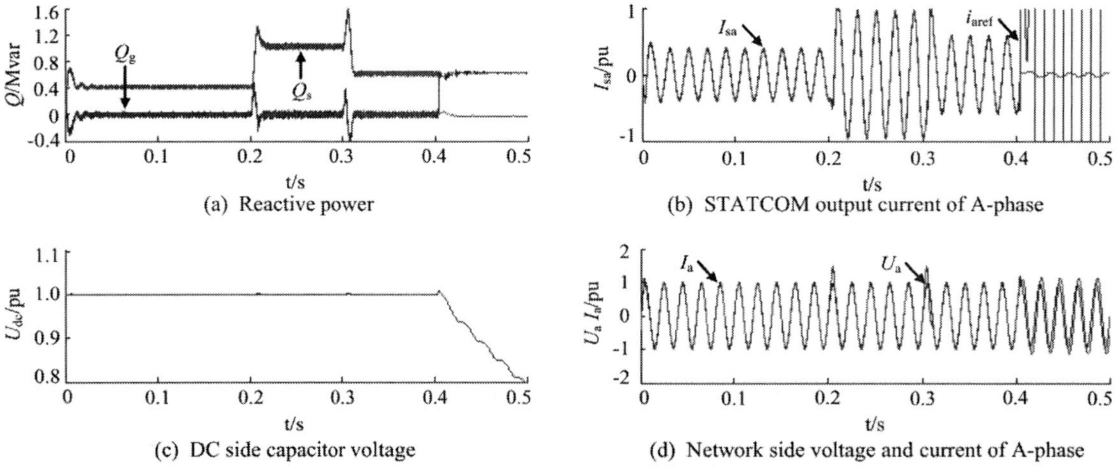

Figure 6. Simulation waveforms of inductive load.

Fig. (a) shows the reactive power of the network side output and the STATCOM output. During 0~0.4s, the required reactive power of the load is supplied by STATCOM, while the reactive power of the STATCOM output is compensated with the reactive power demand. The reactive power of the load is supplied by the power grid after 0.4s. The reactive power compensation effect of STATCOM is verified. Fig. (b) shows the A-phase of reactive power compensation reference current I_{aref} and STATCOM output current I_{sa}. I_{sa} can track the change of I_{aref} during 0~0.4s, and I_{sa} is reduced to 0 at 0.4s, the reactive power compensation instruction current increases with the load change. Fig. (c) shows the DC side capacitor voltage U_{dc}=1pu during 0~0.4s, the capacitor can provide reliable DC voltage support. Fig. (d) shows the A-phase of network side voltage U_a and current I_a. U_a and I_a are in the same phase during 0~0.4s. The current I_a lags behind voltage U_a after 0.4s, as the SATATCOM is broken away from network.

5.3.2 Capacitive load

The working conditions are as follows: the capacitive load capacity is 0.4MW during 0~0.2s, 1MW during 0.2~0.3s, 0.6MW during 0.3~0.4s; at 0.4s, STATCOM is broken away from network. Simulation results are shown in Fig. 7.

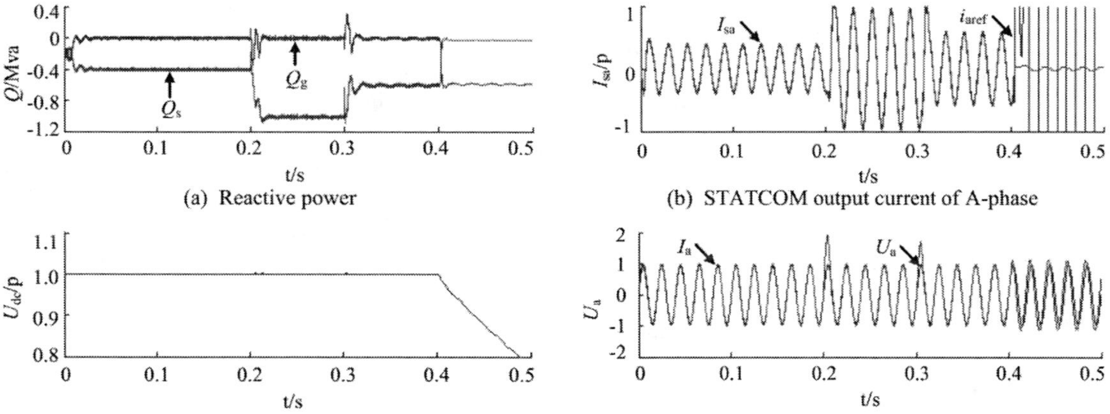

(a) Reactive power (b) STATCOM output current of A-phase

(c) DC side capacitor voltage (d) Network side voltage and current of A-phase

Figure 7. Simulation waveforms of capacitive load.

Fig. (a) shows the reactive power of the network side output and the STATCOM output. During 0~0.4s, the required reactive power of the load is supplied by STATCOM. The reactive power of the load is supplied by the network after 0.4s. Fig. (b) shows the A-phase of reactive power compensation reference current I_{aref} and STATCOM output current I_{sa}. I_{sa} can track the change of I_{aref} during 0~0.4s, and I_{sa} is reduced to 0 at 0.4s, the system no longer has the ability to compensate for reactive power. Fig. (c) shows the DC side capacitor voltage U_{dc}=1pu during 0~0.4s, the capacitor can provide reliable DC voltage support. Fig. (d) shows the A-phase of network side voltage U_a and current I_a. U_a and I_a are in the same phase during 0~0.4s. The current I_a ahead of voltage U_a after 0.4s, as the SATATCOM is broken away from network.

6 Conclusion

Aiming at deviation and delay of the traditional reactive current detection algorithm with PLL and LPF of STATCOM, a novel improved reactive current detection algorithm without PLL is proposed, in which the VM is built to replace the original PLL, and improved current average value filter is used to realize the function of LPF. The improved control strategy for STATCOM is designed based on the VM detection method. Based on Matlab/Simulink platform, simulation analysis of the detection algorithm and control strategy is conducted so as to verify the correctness and effectiveness of the control strategy. The work done in this paper lays the good theoretical foundation for further study of STATCOM.

Acknowledgement

This work was supported by National Natural Science Foundation of China (51267015) and Specialized Research Fund for the Doctoral Program of Higher Education of China (20121514110005).

References

1. Y. B. Guo, X. Zhou, X. H. Zhang, Electric Power Automation Equipment **32**, 50 (2012)
2. C. P. Wang, X. G. Yin, Q. Xiong, Power System Protection and Control **40**, 121 (2012)
3. D. Z. Li, B. Y. Chen, Electrical Measurement & Instrumentation **48**, 49 (2011)
4. X. L. Yao, L. F. Luo, J. Z. Xu, Advanced Technology of Electrical Engineering and Energy **32**, 76 (2013)
5. Z. Z. Song, X. P. Ma, Y. Wang, Electrical Measurement & Instrumentation **49**, 57 (2012)
6. S. Wang, L.X Li, Y. H. Zheng, Electric Power Automation Equipment **32**, 42 (2012)
7. X. R. Chang, H. Y. Wang, H. S. Zhang, Power System Technology **37**, 2819 (2013)
8. G. Q. Wang, H. S. Sun, X. Y. Zhu, Automation of Electric Power Systems **37**, 33 (2013)
9. X. R. Chang, H. Y. Wang, Y. J. Guo, Power System Protection and Control **42**, 117 (2014)
10. G.Y. Liu, S. P. Lig, Z. Q. Qin, Power System Technology **34**, 87 (2010)
11. Y. Wang, X. M. Mu, C. Qi, Power System Protection and Control **43**, 131 (2015)
12. K. Chatterjee, A. Chandra, K. A. Haddad, 36th IEEE Power Electronics Specialists Conference, Brazil (2005)
13. L. J. Su, B. C. Lu, C. C. Li, Power Sysytem Protection and Control **42**, 121 (2014)
14. M. K. Ghartemani, S. A. Khajehoddin, P. K. Jain, IEEE Transsactions on Power Electronics **27**, 1830 (2012)
15. F. Z. Wu, S. P. Pei, 2nd International Workshop on Intelligent Systems and Applications, China (2010)
16. R. B. Gonzatti, S. C. Ferreira, D. Silva, 28th Annual IEEE Applied Power Electronics Conference and Exposition, USA (2013)
17. H. Yang, R. X. Zhao, F. B. Cheng, Proceedings of the CSEE **28**, 78 (2008)
18. N. Gupta, S. P. Singh, S. P. Dubey, 5th IEEE Conference on Industrial Electronics and Applications, Taiwan (2010)
19. S. Q. Xue, J. D. Cai, Proceedings of the CSEE **32**, 91 (2012)
20. L. P. Shi, P. Liu, T. R. Xu, Electrical Measurement & Instrumentation **51**, 56 (2014)
21. L. Zhou, J. Meng, Q. Liu, Power Electronics **42**, 21 (2008)

MATEC Web of Conferences **55**, 01002 (2016) DOI: 10.1051/matecconf/20165501002
ACPEE 2016

Coordinated Voltage Control of Active Distribution Network

Jiang Xie[1,a], Wenxia Liu[1]and Huiting Xu[1],Xu Han[2],Shuya Niu[1]

[1] *School of Electrical and Electronic Engineering, North China Electric Power University, 102206 Beijing, China*
[2] *State Grid Baicheng Power Supply Company, 137000 Baicheng, China*

Abstract. This paper presents a centralized coordinated voltage control method for active distribution network to solve off-limit problem of voltage after incorporation of distributed generation (DG). The proposed method consists of two parts, it coordinated primal-dual interior point method-based voltage regulation schemes of DG reactive powers and capacitors with centralized on-load tap changer (OLTC) controlling method which utilizes system's maximum and minimum voltages, to improve the qualified rate of voltage and reduce the operation numbers of OLTC. The proposed coordination has considered the cost of capacitors. The method is tested using a radial edited IEEE-33 nodes distribution network which is modelled using MATLAB.

1 Introduction

Large-scale development of DG, will bring problem to traditional distribution network such as bidirectional power flow and voltage off-limit. Traditional distribution network is designed based on the assumption that power flow is unidirectional and the network does not contain any controlled resources. Voltage control is mainly through OLTC and shunt capacitors in substation. These devices are run locally in most distribution network, without any communication or coordination with each other [1] [2]. In [3], the author proposed a simple automatic voltage control method, which controls local reactive power output of DG, reducing active power output of DG as well. These local control are usually restricted to one bus voltage, relying only on local measurements.

In fact, information and controllable equipment of all network can be used basing on more and more convenient measurement and communication system. In [4], the author has proposed a coordinated voltage control method of substation OLTC and reactive output of DERs based on maximum and minimum voltage of the network. As the number of controllable components increase, the difficulty of determining control rules becomes increasingly complex. In this paper, a centralized coordinated voltage method has been proposed including coordination between primal-dual interior point method-based voltage regulation schemes of DG reactive powers and capacitors, and centralized on-load tap changer (OLTC) controlling method which utilizes the system's maximum and minimum voltages, to ensure a stable and economical distribution network voltage.

2 Proposed centralized coordinated voltage control

[a] Jiang Xie : xiejiang12334@163.com

The proposed method consists of two parts: primal-dual interior point method-based voltage regulation schemes of DG reactive powers and capacitors, centralized on-load tap changer (OLTC) controlling method which utilizes the system's maximum and minimum voltages proposed in [5].

2.1 Model designs for optimizing schemes

2.1.1 Objective function

$$M\text{in}\, I = I_v + I_{QDG} + I_C \tag{1}$$

Objective function consists of three parts,

1) Minimizing extreme value of voltage

Coordination between optimizing schemes and centralized OLTC control is also reflected by this sub-objective function:

$$I_v = \omega_v \frac{\Delta V_{Upper} + \Delta V_{Lower}}{\Delta V_{\max}} \tag{2}$$

where

$$\Delta V_{Upper} = \begin{cases} V_{\max}^{system} - V_{Upper} & V_{\max}^{system} > V_{Upper}, \\ 0 & V_{\max}^{system} \le V_{Upper} \end{cases}$$

$$\Delta V_{Lower} = \begin{cases} V_{Lower} - V_{\min}^{system} & V_{\min}^{system} < V_{Lower} \\ 0 & V_{\min}^{system} \ge V_{Lower} \end{cases}$$

$$\Delta V_{\max} = V_{Upper} - V_{Lower}$$

where ω_v is proportionality coefficient; V_{Upper} and V_{Lower} respectively are standard upper and lower limit of feeder voltage, eg 1.07 p.u. and 0.93 p.u.; $V_{\min}^{system}$ and $V_{\min}^{system}$ respectively are maximum and minimum voltage of the network measured.

2) Minimizing use amount of reactive power of DGs

DGs' reactive power are finite:

$$I_{QDG} = \omega_{QDG} \frac{\sum_{i=1}^{N_{DG}} |Q_{DG,i}|}{\sum_{i=1}^{N_{DG}} |Q_{DG,i}^{\max}|} \tag{3}$$

where ω_{QDG} is proportionality coefficient; N_{DG} is total number of DGs; Q_{DGi} is reactive power of DG i; $Q_{DGi}^{\max}$ is maximal reactive power of DG i, and $Q_{DGi}^{pf} = P_{DGi}\tan(\cos^{-1}(pf_{DGi}))$, $Q_{DGi}^{S} = \sqrt{S_{DGi}^2 - P_{DGi}^2}$, $Q_{DGi}^{v} = \sqrt{\left(\frac{V_{DGi}^{\max} V_g}{X}\right)^2 - P_{DGi}^2} - \frac{V_g^2}{X}$,where P_{DGi} is active power of DG; $V_{DGi}^{\max}$ is maximal output voltage of DG; pf_{DGi} is power factor of DG; S_{DGi} is apparent power of DG; V_g is voltage of system power supply; X represents the total reactance of the DG interfacing transformer and filter in case of inverter-based DG or it represents the synchronous reactance in case of synchronous machine-based DG. For inverter-based DG, $V_{DGi}^{\max}$ relies on the dc-link voltage, while for synchronous machine-based DG $V_{DGi}^{\max}$ depends on the field voltage.

3) Minimizing the cost of using shunt capacitors

Limiting the frequency of switching capacitor is intended to ensure the normal life of the compensating device and reduce the workload of operation and maintenance:

$$I_C = \omega_q \sum_{i=1}^{C_N} \Delta q_j \tag{4}$$

where c_N is the total number of nodes with shunt capacitor bank; ω_q is the cost factor of each switching operation of capacitor; Δq_j is the change in the number of compensation capacitor of node j before and after each optimization. In general, ω_v and ω_{QDG} are all set to 1.

2.1.2 Constraints

1) Power flow constraints

$$G(x,u) = 0 \tag{5}$$

where x and u respectively is state variable and control variable.

2) Constrains of reactive power of DGs

$$|Q_{DGi}| \le Q_{DGi}^{\max} \tag{6}$$

where Q_{DGi} is reactive power of DG i; $Q_{DGi}^{\max}$ is maximal reactive power of DG i.

3) Constrains of capacitors

$$0 \le q_j \le q_{j,\max} \quad (j = 1,2,\cdots,C_N) \tag{7}$$

where q_j is compensation capacity of shunt capacitor bank in node j; $q_{j,\max}$ is total capacity of shunt capacitor bank in node j; c_N is the total number of nodes with shunt capacitor bank.

4) Constrains of voltage magnitude

The node voltage amplitude must be controlled at rated voltage of 7% ~ 7%.

From a mathematical point of view, optimizing schemes are multi-objective nonlinear programming, and mixed-integer programming.

2.2 Algorithm

Primal-dual interior point method is used to deal with mixed integer optimization. Continuous control variables are reactive power output of DGs, discrete control variables is tap position of OLTC and numbers of groups of shunt capacitor banks switching on.

2.3 Evaluation Index

Comprehensive evaluation index is used to evaluate the proposed control strategy:

$$SI = \frac{1}{m \times n} \sum_{i=1}^{m} \sum_{j=1}^{n} |\Delta U_{ij}| \tag{8}$$

For further analysis and comparison, average offset dimension of voltage of each sample point is:

$$AVO_i = \frac{1}{n}\sum_{j=1}^{n}\left|\Delta U_{ij}\right| \tag{9}$$

where i is sampling point, j is node number, and m is total number of sampling point set as 96 for 15 minutes interval between i and i+1, n is total number of nodes, $\left|\Delta U_{ij}\right|$ is absolute value of offset dimension of voltage of node j at sampling point i.

3 Simulation result

The simulation takes modified IEEE-33 system. Rated voltage is 12.66kV. Intermittent solar energy can significantly change the system voltage profile and is usually located at the end of feeder. So 4 solar-based DGs are connected at different positions. The total number of tap positions of substation OLTC is 19. The time for voltage control is a typical 24-hour day, control interval is 15 minutes, and the total number of sampling is 96.The topological graph of IEEE-33 system is shown in figure 1. Curves of power of residential loads and solar-based DG during a day are shown in figure 2. Node 2, 15, 31 respectively are installed with shunt capacitor bank, each group of capacity is 200 kvar, and number of groups are 10, 3, and 3, respectively.

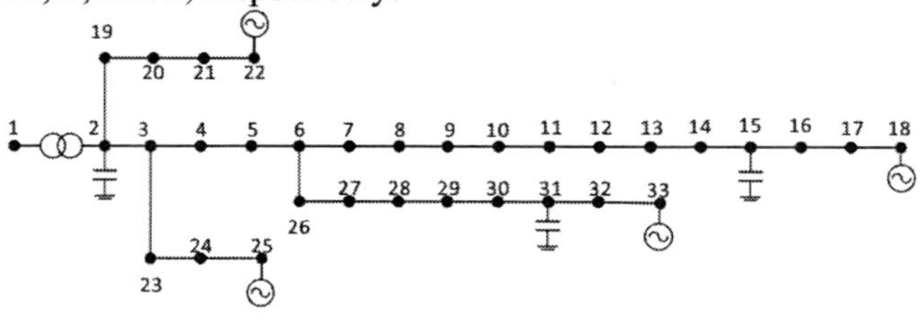

Figure 1. Modified IEEE-33 power system.

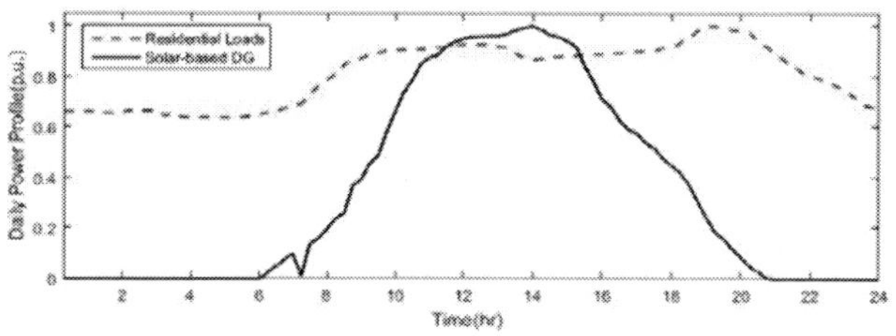

Figure 2. Power of residential loads and solar-based DG

The capacity of four DG is 3 MVA. Range of power factor are set to 0.95 lagging to 0.95 leading. To easily test effect of proposed coordinated voltage control scheme, assume the distributed photovoltaic DGs adopt decoupling control of active and reactive power. Figure 3 shows system's maximum and minimum voltages under two cases:

Case 1: Power factors of all DGs are unit, numbers of all shunt capacitor bank are zero. OLTC, shunt capacitor banks and reactive power of DGs are not controlled.

Case 2: Tap position, shunt capacitor bank and reactive power of DGs is all controlled using proposed coordinated voltage control scheme. Active power of DGs output is same as case 1. On the other hand, reactive power of DGs is adjustable according to constraints shown in formula (6).

Shown in figure 3 and 4, system's maximum voltage is off-limit of 1.07 p.u. during about 12:00 to 16:00 when DGs have greater active power output, system's minimum voltage is off-limit of 0.93 p.u. during about 20:00 to 22:00 without any control; Proposed coordinated control scheme have solved the off-limit problem of voltage successfully, remarkably reduced average offset dimension of voltage specially in both overvoltage and under-voltage.

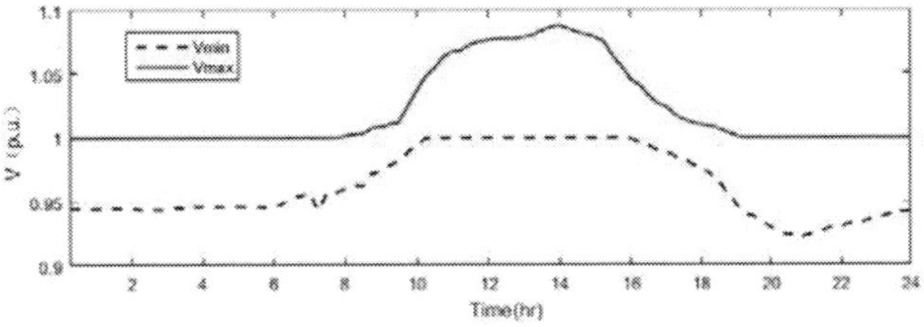

(a)

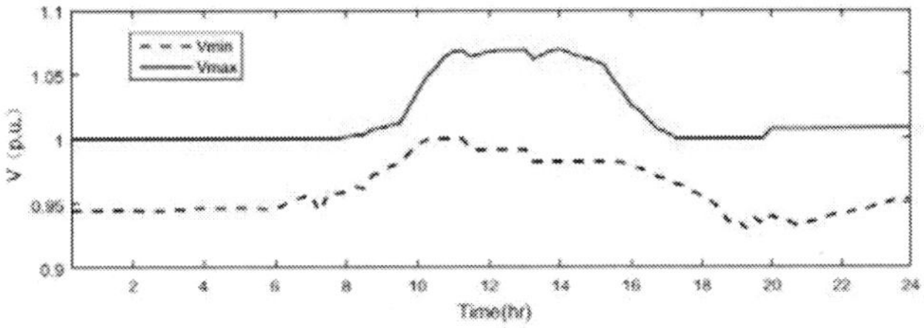

(b)

Figure 3. System's maximum and minimum voltages; (a) Without control, (b) Proposed coordinated control

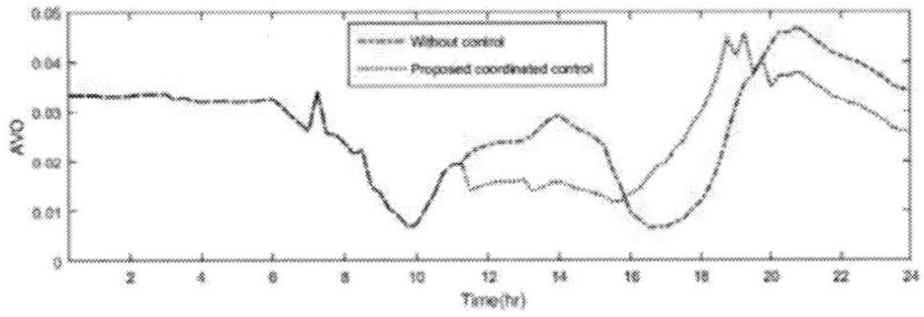

Figure 4. Average offset dimension of voltage (AVO)

Proposed method is contrasted by centralized voltage control method without capacitors. Proposed centralized coordinated voltage control scheme has considered the effect of shunt capacitor banks because shunt reactive power compensation can help improve under-voltage. Figure 5 shows the tap position of both scheme. The proposed scheme has decreased daily tap operation from 9 to 7 taps/day. Comprehensive evaluation index of all three cases respectively are 0.0270, 0.0260 and 0.0257 shown in table 1, both centralized and proposed centralized coordinated voltage control scheme can reduce offset dimension of system voltage, effect of proposed method this paper is better specially in under-voltage result from heavy load and light DG output.

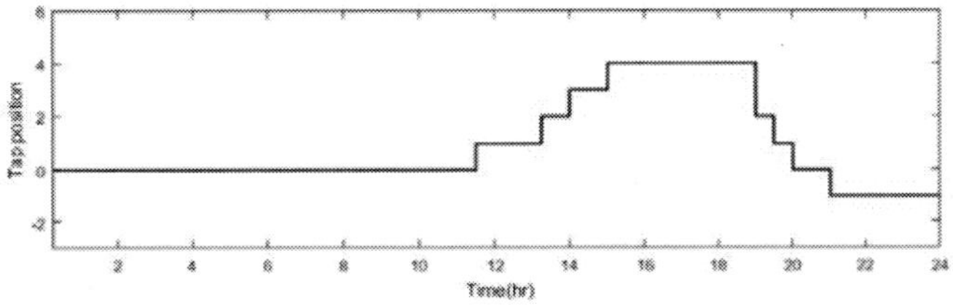

(a)

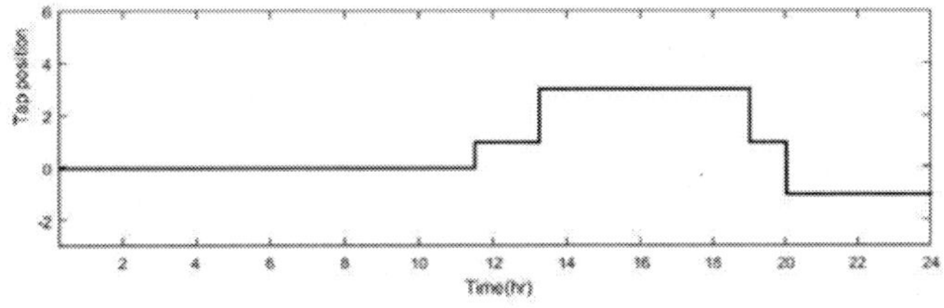

(b)

Figure 5. Tap position of OLTC; (a) Control without capacitors, (b) Proposed coordinated control.

Table 1. Comprehensive evaluation index (SI).

Condition / Element	Without control	Centralized control without capacitors	Proposed coordinated control
SI	0.0270	0.0260	0.0257

4 Conclusion

This paper presents a centralized coordinated voltage control scheme suited for active power distribution network. Once appears off-limit voltage, the reactive power of DG and shunt capacitor bank is optimized using primal-dual interior point method first of all, coordinated with substation OLTC by setting specific objective function then, to provide a more relaxed operation of the OLTC which can increase its life time; Especially in the optimization part, the proposed coordination has considered the cost of capacitor and incorporated it in whole coordinated control, reducing voltage excursion of active distribution network and decreased operation of OLTC facing heavy loading.

References

1. T.A.Short,Electric power distribution handbook Boca Raton,FL： CRC,(2004)
2. T.Gonen,Electric power distribution system New York: McGraw,(1986)
3. W.Freitas,J.C.M Vieira,A.Morelato,W.Xu,Influence of excitation system control modes on the allowable penetration level of distributed synchronous generators,IEEE Transactions on Energy Convers,2005,20(2)： 474 – 480
4. C.M.Hird,H.Leite,N.Jenkins,H.Li.Network voltage controller for distributed generation,IEEE Transactions on Generation,2004,151(2)： 150 – 156
5. Maher A. Azzouz and Ehab F. El-Saadany. Optimal coordinated Volt/Var control in active distribution networks[C]// PES General Meeting, 2014 IEEE. IEEE, 2014:1-5

MATEC Web of Conferences **55**, 01003 (2016) DOI: 10.1051/matecconf/20165501003
ACPEE 2016

Inductor Compensation in Three Phase PFC Control with Decoupling the Input Voltage and Bus Voltage

Guang Luo[1,a], Wu Ai[1] , Bing Chen[1] and Yi Liu[1]

[1]*Huazhong University of Science and Technology, School of Mechanical Science and Engineering, 430074 Wuhan , China*

Abstract. As the nonlinear structure of the boost ac/dc topology, the PFC is difficult to be controlled. The average state space theory has revealed that the choke current is determined by duty-ratio coupled with input voltage and bus voltage, which makes the high-performance sine-wave current track control more challenging. To remove disturbing variables and reduce the zero-crossing distortion, the decoupling control strategy is presented. Also, it is shown in this paper that the variable inductance has a great influence on PFC and THDI. As the inductor in boost converter varies with the current, a compensating coefficient of current is also proposed. The decoupling strategy and the inductor compensation strategy are implemented on a UPS with three phase input voltage. Numerical simulation and experimental results has indicated the high-performance of these control strategy.

1 Introduction

As a non-isolate ac/dc PFC converter, the boost converter is widely used in various industrial applications due to its simple circuit structure, high reliability, low cost and low current ripple in high frequency. Among the three current modes, CCM, DCM and BCM, CCM achieves the highest occupation ratio in medium and high power field [1]. In general, the sinusoidal wave generator is used for providing the set value of inner current loop. Its realization algorithm includes looking up table and following mains supply. Generally, the former is able to offer a better performance in terms of total harmonic distortion (THD), while the latter is able to offer a better performance in terms of power factor (PF). Besides, the look-up table needs a phase locked loop (PLL) accurately detect the zero crossing of sinusoidal wave, which will increase the complexity of the control system and load of the CPU [2].

The conventional analysis method is using the average state space theory under high-frequency carrier, and establishing controlled object [3-4]. When the input voltage is very low, boost IGBT works at an almost 100% duty ratio, which will lead to the phase lead of input current and thus results in zero distortion, especially under high ac line frequencies [5]. Therefore, the merely conventional PID correction is unable to reduce the effect of the leading phase of input current.

So, many new control technique have been proposed. In [6], a feedforward current control approach is proposed for a boost PFC converter. In this paper, an additional feedforward signal is shown to further reduce input current harmonic distortion, especially near the zero crossing of the

[a] Corresponding author : andylaw_lg@qq.mail

input voltage. In [7], a duty-ratio feedforward control approach is proposed. Essentially, both of the control techniques add an open-loop control method to the closed-loop control method. In [8], a new current detect method is proposed, but it will aggravate processing burden. In [9-11], various kinds of topology is proposed, but additional hardware may increase the cost.

In this paper, a decoupling control strategy is proposed on the basis of the conventional double-loop control strategy. After decoupling the input voltage and bus voltage, the input current is completely determined by duty-ratio. Then, the segment from duty-ratio to input current can be considered as a pure integrator model. Due to the pure integrator, a compensating coefficient could exactly offset the fluctuation of inductance value which plays a great role in the control.

2 The Small Signal Model of The Boost Topology

The averaged dynamics of the PFC over one switching cycle is given by

$$\begin{cases} \dfrac{di_L}{dt} = \dfrac{-(1-d(t))}{L} V_{BUS}(t) + \dfrac{V_{in}(t)}{L} \\ \dfrac{dV_{BUS}(t)}{dt} = \dfrac{(1-d(t))i_L(t)}{C} - \dfrac{V_{BUS}(t)}{CZ} \end{cases}. \tag{1}$$

The average value and its fluctuation is given by

$$\begin{cases} d(t) = D + \hat{d} \\ i_L(t) = i_L + \hat{i}_L \\ V_{BUS}(t) = V_{BUS} + \hat{V}_{BUS} \\ V_{in}(t) = V_{in} + \hat{V}_{in} \end{cases} \tag{2}$$

Thus, the balance equation and the small signal linear equation can be deduced as follows:

$$\begin{cases} V_{in} = (1-D)\cdot V_{BUS} \\ (1-D)\cdot i_L = V_{BUS}/Z \end{cases} \tag{3}$$

$$\begin{cases} \dfrac{d\hat{i}_L}{dt} = -\dfrac{1-D}{L}\hat{V}_{BUS} + \dfrac{\hat{d}}{L}V_{PBUS} + \dfrac{\hat{V}_{in}}{L} \\ \dfrac{d\hat{V}_{BUS}}{dt} = \dfrac{1-D}{C}\hat{i}_L - \dfrac{\hat{d}}{C}i_L - \dfrac{\hat{V}_{BUS}}{CZ} \end{cases}. \tag{4}$$

Deduced from these equations, an inner current controlled object from $\hat{d}$ to $\hat{i}_L$ is written as

$$\frac{\hat{i}_L(s)}{\hat{d}(s)} = G_d\left(1+\frac{s}{\omega_Z}\right)\Big/\left(\left(\frac{s}{\omega_0}\right)^2 + \frac{s}{Q\cdot\omega_0} + 1\right) \tag{5}$$

$$G_d = 2V_{BUS}/Z(1-D)^2, \omega_0 = (1-D)/\sqrt{LC},$$
$$\omega_Z = 2/CZ, Q = Z(1-D)\sqrt{C/L}.$$

And a outer voltage controlled object from $\hat{i}_L$ to $\hat{V}_{BUS}$ is written as

$$\frac{\hat{V}_{BUS}(s)}{\hat{i}_L(s)} = \left((1-D)^2 Z - Ls\right)\Big/\left((CZs+2)(1-D)\right). \tag{6}$$

3 The Control Stratagy

3. 1 Close Loop Control Strategy

The conventional control method is to establish the double loop model using PID regulator based on the two controlled objects as following (5), (6).

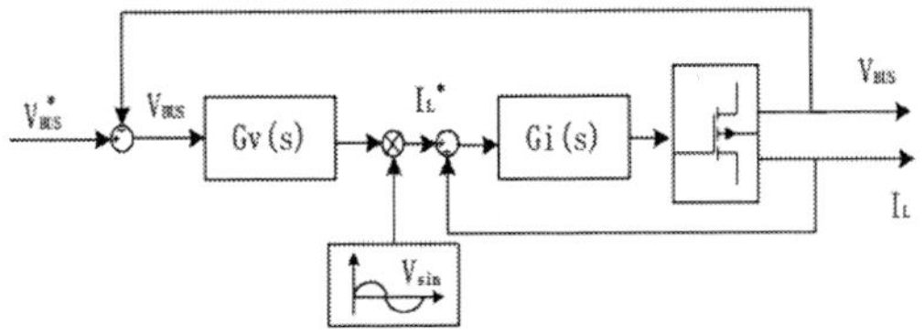

Figure 1. Conventional control block diagram

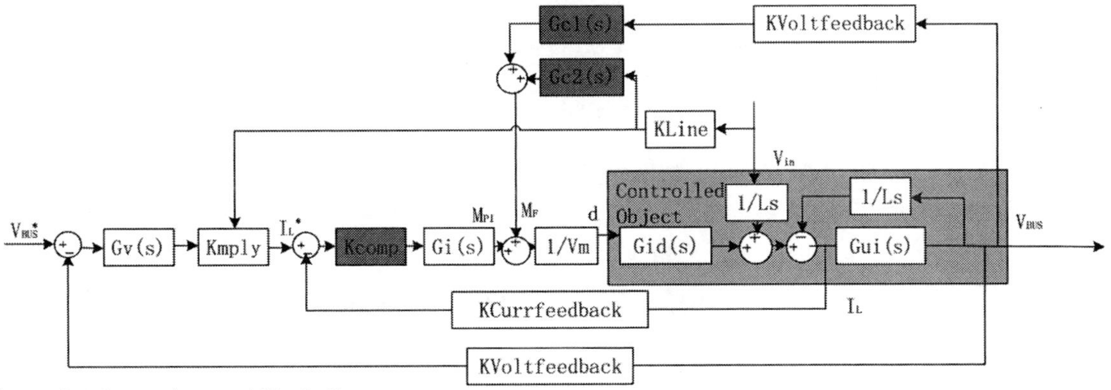

Figure 2. Advanced control block diagram

The voltage loop bandwidth is desired under 20 Hz to ensure the linear modulation of the 50Hz sine multiplier, while the current loop bandwidth is desired much higher to ensure a fast tracking of sinusoidal input reference signals.

In conventional algorithm analysis under high frequency carrier, equation (5) is simplified as a simply pure integral model, and equation (6) is simplified as an inertial model. In fact, however, they are not.

According to analysis of bode graph, the high-performance compensating controllers are designed.

3.2 Proposed Decoupling Strategy

The controlled model of the boost PFC converter given by (1), represents a nonlinear system owing to the coupling of input duty ratio $d(t)$ with $V_{in}(t)$ and $V_{BUS}(t)$. But the conventional controlled objects (5) and (6) are established without taking input voltage $V_{in}(t)$ and bus voltage $V_{BUS}(t)$ into consideration. In order to establish a linear system, a decoupling strategy is proposed.

Table 1. Control Parameters

Name	Value
` *Vm*	3906
KLine	10
KVoltfeedback	10
KCurrfeedback	100
Kmply	220

According to (1),

$$\frac{di_L}{dt}=\frac{d(t)}{L}V_{BUS}(t)+\frac{V_{in}(t)}{L}-\frac{V_{BUS}(t)}{L} \tag{7}$$

Under the high frequency, the change rate of $d(t)$ is much higher than that of $V_{BUS}(t)$.So, the $V_{BUS}(t)$ in $\frac{d(t)}{L}V_{BUS}(t)$ can be regarded as bus voltage set value V^*_{BUS}.

$$i_L(s)=Gid(s)*d(s)+\frac{V_{in}(s)}{Ls}-\frac{V_{BUS}(s)}{Ls} \tag{8}$$

where

$$Gid(s)=\frac{V^*_{BUS}}{Ls} \tag{9}$$

As is shown in Fig.2, $Gc1$ and $Gc2$ are respectively presented to offset the voltage disturbance, and dc bus voltage disturbance.
Then,

$$Gc1*KVoltfeedback*\frac{1}{Vm}*Gid-\frac{1}{Ls}=0 \tag{10}$$

And,

$$Gc2*KLine*\frac{1}{Vm}*Gid+\frac{1}{Ls}=0 \tag{11}$$

So,

$$\begin{cases} Gc1=V^*_{BUS}/(Vm*KVoltfeedback) \\ Gc2=-V^*_{BUS}/(Vm*Kline) \end{cases} \tag{12}$$

After decoupling the input voltage and bus voltage, equation (8) can be simplified to

$$i_L(s)=Gid(s)*d(s). \tag{13}$$

In contrast with equation (5), the inner current model is much briefer, owe to the decoupling strategy.

3.3 Proposed Inductor Compensation

According to hysteresis loop, the higher current is, the less the inductor value will become. The attenuation of the inductance value will make the inductor current rise sharply when IGBT is off, which will increasing current ripple in switch frequency when it is peak. Thus, the THDI will get worse. To achieve a high performance and a precise control of the current loop, the inductance value should be considered as constant. Therefore, the coefficient $Kcomp$ is needed.

In the experience, Fe-Si is selected for the inductance core. The character graphs of the inductance is expressed in Fig.3.

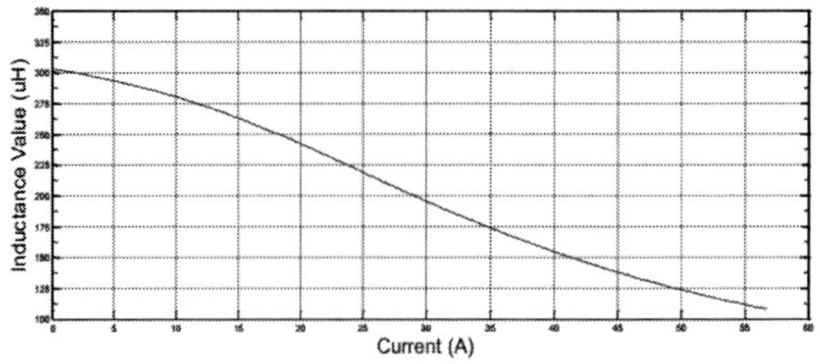

Figure 3. Decay curve of inductance

The coefficient is designed as follows:

$$Kcomp = L(i)/L_0, \tag{14}$$

Defining $L(i)$ the inductance value when current is i.

In the experience, a compensatory table should be firstly established. Then the new current controlled object can be improved to be as following:

$$\begin{aligned} Gid'(s) &= Gid(s) * Kcomp \\ &= \frac{V^*_{BUS}}{L(i)s} * \frac{L(i)}{L_0} = \frac{V^*_{BUS}}{L_0 s} \end{aligned} \tag{15}$$

This is the final controlled object without decoupling the input voltage, bus voltage and even the variable inductance.

4 Operation Result And Analysis

4.1 Introduced Experimental Platform

In order to verify the performance of the control strategy, it was implemented on a 20-kva ac/dc converter. It should be noted that, for such high-power applications, a three-phase boost converter is necessary to reduce the load of the converter on each phase. Fig.4 shows the topology of the converter, and Fig.5 shows the experimental prototype. The specifications of the converter are illustrated in Table 2.

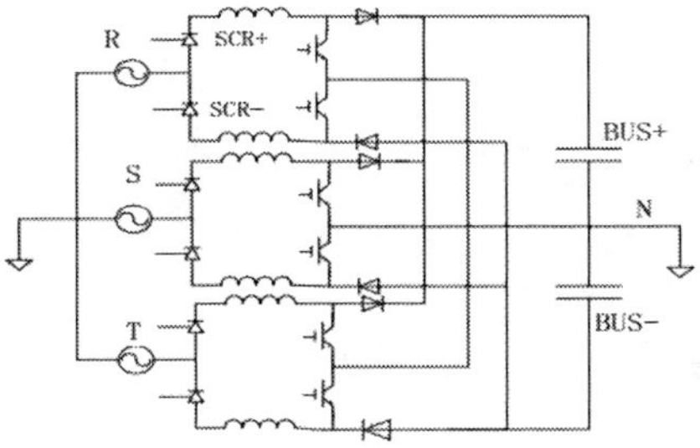

Figure 4. The Topology of the Converter

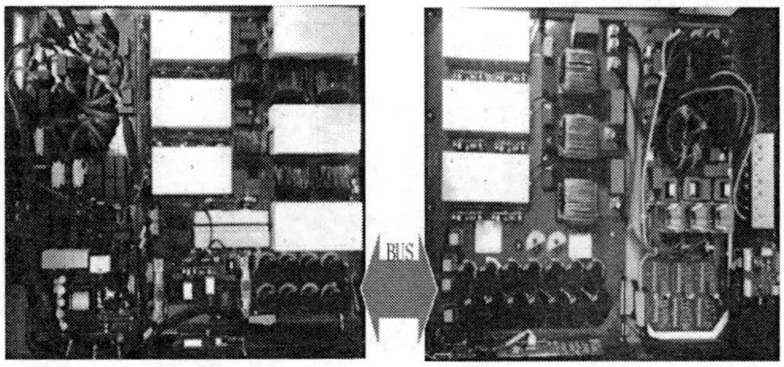

Figure 5. The Experimental Prototype

Table 2. Converter Specifications

Symbol	Parameter	Value
L	Boost PFC Inductor	300uH
C	BUS Capacitor	8160uF
V_{bus}	BUS Voltage	360V
P	Output Power	18KW
V_{in}	Grid Input Voltage	176-264V AC
η	Efficiency	0.92
PF	Power Factor	>0.98
f_s	Switch Frequency	19.2KHz
f_{cnt}	System Clock Frequency	150MHz

4.2 Dynamic Performance in Experiment

Fig.6 shows the dynamics in Matlab Simulation. In the simulation, the technology of design based on the model has been used. The C-language program code which is running on the DSP has been imported into the Matlab C-Sfunction model. Therefore, the performance of the simulation is exactly the same as that of real experiment.

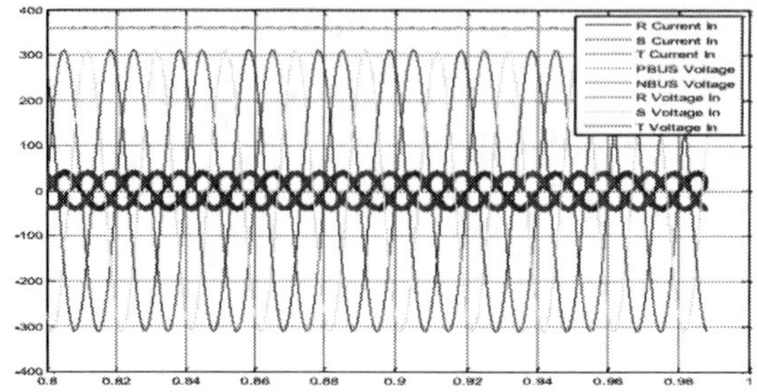

Figure 6. Matlab Simulation Result

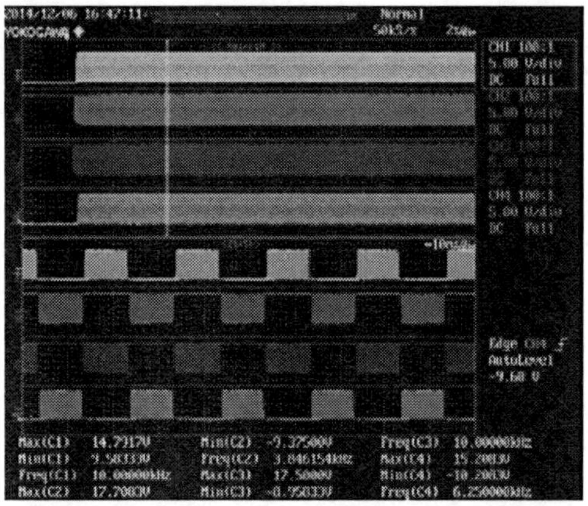

Figure 7. PWM duty-ratio of Phase R and Phase S

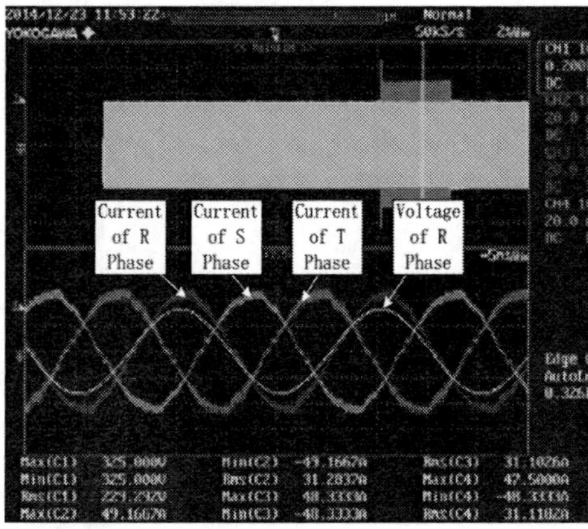

Figure 8. Three-phase input current and R phase input voltage

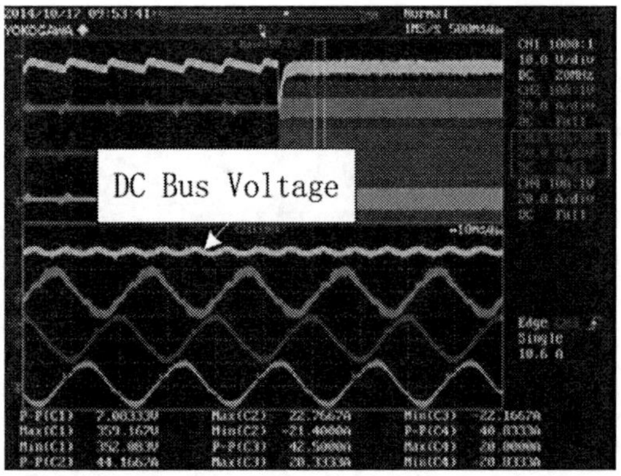

Figure 9. Response of the bus voltage for a 100% positive step load change

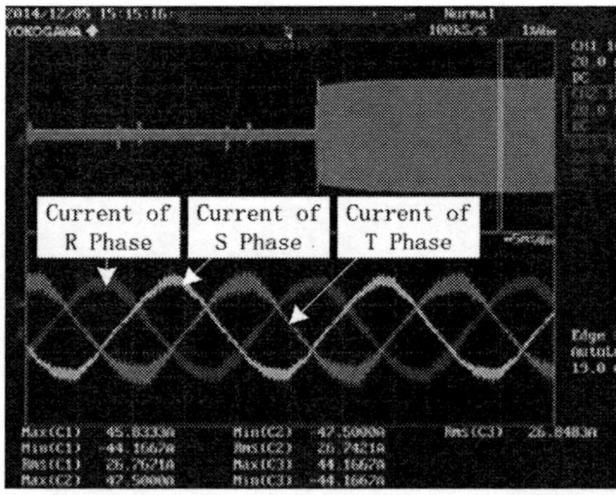

Figure 10. Three-phase input current without inductor compensation

Fig.7 shows the PWM duty-ratio of the converter. When the input voltage is positive, the positive boost works, so as the negative.

Fig.8 shows the high performance of the PFC control. This figure confirms very good current quality, particularly for heavy loads.

Fig.9 shows the transient response of the close-loop control system when a 100% step load change is applied to the converter. And it proves that the compensating coefficient and decoupling strategy is able to accurately maintain the stability of the DC bus.

Fig.10 shows that without the inductor compensation, the current fluctuates at its peak heavily. It is due to the variable inductance value.

5 Conclusion

This paper has presented a novel decoupling strategy for the boost PFC converter. This strategy has decoupled the input voltage and bus voltage, which make the system linear. Due to the proposed strategy, the PFC current is simply determined by the duty-ratio. Thus, the inner current controlled object has been simplified. In addition, a compensation coefficient is proposed to offset the inductance. It performs well not only in the simulation but also in the experiment.

Acknowledgement

This work was supported by National Natural Science Foundation of China under Grant 51475184.

References

1. L. Huber, B. T. Irving, M. M. Jovanović, "Closed-loop control methods for interleaved DCM/CCM boundary boost PFC converters,"Applied Power Electronics Conference and Exposition, 2009. APEC 2009. Twenty-Fourth Annual IEEE. IEEE, 2009 pp.991-997.
2. M. Pahlevaninezhad, S. Pan, S. Eren, et al,"An adaptive nonlinear current observer for boost PFC AC/DC converters".
3. R. B. Ridley, "Average small-signal analysis of the boost power factor correction circuit,"VPEC seminar proceedings. 1989 pp.108-120.
4. H. C. Chen, Z. H. Wu, Y. J. Liao,"Modeling and small-signal analysis of a switch-mode rectifier with single-loop current sensorless control," Power Electronics, IEEE Transactions on, 2010, 25(1) pp.75-84.
5. J. Sun,"On the zero-crossing distortion in single-phase PFC converters," Power Electronics, IEEE Transactions on, 2004, 19(3) pp.685-692.
6. M. Chen, J. Sun, "Feedforward current control of boost single-phase PFC converters,"Power Electronics, IEEE Transactions on, 2006, 21(2) pp. 338-345.
7. D. M. Van de Sype, K. De Gusseme, A. P. M. Van den Bossche, et al,"Duty-ratio feedforward for digitally controlled boost PFC converters," Industrial Electronics, IEEE Transactions on, 2005, 52(1) pp.108-115.
8. J. Calvente, H. Ramirez-Murillo, E. Vidal-Idiarte, et al,"Multisampled average current control of switching power converters,"Industrial Technology (ICIT), 2015 IEEE International Conference on. IEEE, 2015 pp.2120-2124.
9. T. Nussbaumer, K. Raggl, and J. W. Kolar, "Design guidelines for interleaved single-phase boost PFC circuits," IEEE Trans. Ind. Electron.,vol. 56, no. 7, pp. 2559–2573, Jul. 2009.
10. X. Lu, Y. X. Xie, L. Cheng, et al,"Semi-bridgeless boost PFC rectifier for wide voltage input range based on voltage feed-forward control,"Power Electronics and Motion Control Conference (IPEMC), 2012 7th International. IEEE, 2012, 3 pp.1894-1899.
11. F. Haase, A. Kouchaki, M. Nymand. "A low cost rapid prototype platform for a three phase PFC rectifier application,"Industrial Technology (ICIT), 2015 IEEE International Conference on. IEEE, 2015 pp.2024-2029.

MATEC Web of Conferences **55**, 01004 (2016) DOI: 10.1051/matecconf/20165501004
ACPEE 2016

A Fuzzy Predictive PID Control Scheme for the Excitation System of Synchronous Generator

Yang Zheng[1], Jianzhong Zhou[1,a], Chu Zhang[1], Yuncheng Zhang[1] and Wei Zhao[1]

[1]*School of Hydropower and Information Engineering, Huazhong University of Science and Technology, Wuhan, Hubei 430074, P R China*

Abstract. With the rapid development of the process control theories in the electrical engineering, new control strategies which lead to better performances are urgently demanded for the excitation control of synchronous generators. For the sake of improving the convergence rate of the terminal voltage and covering the weakness in the adaptability of operational conditions of conventional controls in disturbances, a fuzzy predictive PID excitation control method is proposed in this paper. This control scheme can be divided into three steps in every sample interval: PID parameter adaptation, rolling state prediction and real-time control movement integration. Numerical simulations have been conducted under different operational conditions with the proposed method as well as the conventional ones, respectively. Experimental comparisons indicate the superiority in voltage regulation performance of the proposed method.

1 Introduction

Although PID control is still being applied in most excitation system of synchronous generators due to its robustness in control and simplicity in design, its deficiencies including dependence on control parameters, weakness in condition adaptation and hardship of parameter tuning make scholars worldwide find ways to improve its performance or design the new control methods with better performance. For the sake of improving the condition adaptation characteristic of PID, parameter tuning tactics such as the swarm intelligence optimization [1], fuzzy logic [2], artificial neural network [3] and their comprehension [4] have been extensively studied over the decades. These efforts do provide us with approaches to make PID parameters more closely match the working conditions in excitation control and to attenuate the deterioration in operating mode transitions. However, these aforementioned methods cannot get rid of the intrinsic defects of PID control, namely, it lacks the ability of prediction. To this end, some scholars have brought the notion of model predictive control into PID control. For instance, Zhang proposed a predictive control optimization based PID control for industrial surfactant reactors described by linear state-space equations [5] and Sato built a GPC-based PID controller for a weigh feeder [6], etc. Most of these predictive integrated PID control methods are either designed for linear systems or possess heavy calculation burden for the inevitably online rolling optimization process [7], thus the application of these control schemes still have many limitations.

---

[a] Corresponding Author: jz.zhou@hust.edu.cn

This paper presents a modified fuzzy predictive PID (FP-PID) control scheme in which the control signal takes account of not only the past and present error of the system output, but also the future predictive state information. In the design of this predictive controller, rolling optimization process in every sampling instance is replaced by a rolling state prediction process based on fuzzy PID control with much lower calculation burden. Confidence probability is introduced to the real-time control signal integration with the predictive control movement vector recording from rolling state prediction, rather than implementing only the first element of the control sequence like ordinary predictive control methods. Numerical simulations have been conducted under voltage step and three-phase short circuit conditions in a power system with a single machine connected to an infinite bus. Simulation results validate the enhancement of the FP-PID control in voltage-tracking performance compared with PID and fuzzy PID controls.

The remainder of the paper is organized as follows: design of the FP-PID control strategy in affine nonlinear systems is explained in detail in Section 2. Then, nonlinear models for the excitation system is stated in Section 3. Section 4 discusses the numerical simulations with the utility of the proposed control scheme and the comparative ones. Eventually, conclusions are summarized in Section 5.

2 Fuzzy Predictive PID Control for Nonlinear Systems

The fuzzy predictive PID (FP-PID) control scheme is designed on the basis of discrete-time affine nonlinear systems that can be described in Equation (1).

$$\begin{cases} \mathbf{X}(\mathrm{k}+1) = \mathrm{f}(\mathbf{X}(\mathrm{k})) + \mathrm{g}(\mathrm{k}) \cdot u(\mathrm{k}) \\ y(\mathrm{k}) = \mathrm{h}(\mathbf{X}(\mathrm{k})) \end{cases} \tag{1}$$

where, k is the sample number, X represents the state vector, u is the input variable, y is the system output.

2.1 Adaptive control Parameters Calculation Using Fuzzy Logic

At every sample time, the real-time PID parameters are updated by fuzzy logic methods so as to make the controller adapt the current working condition of the system. The fuzzy parameter tuning is similar to the way in the extensive studied literature about fuzzy PID control and is briefly stated as follows:

(1) Build fuzzy rule tables of the three PID parameter increments (i.e., $\Delta K_p, \Delta K_i, \Delta K_d$) according to the variation characteristics of the parameters, respectively. Both of the fuzzy sets of error and error deviation are $[NB, NM, NS, ZO, PS, PM, PB]$. The fuzzy membership is conducted with triangular function.

(2) Calculate the fuzzy subset and the fuzzy membership of each subset of $\Delta K_p, \Delta K_i, \Delta K_d$ with the current output deviation and its increment as the fuzzy reasoning machine.

(3) Determine the value of $\Delta K_p, \Delta K_i, \Delta K_d$ by the defuzzination with Center of Gravity Method.

(4) Calculate the current parameters K_p, K_i, K_d by adding the calculated $\Delta K_p, \Delta K_i, \Delta K_d$ to the initial parameters K_{p0}, K_{i0}, K_{d0}.

2.2 Fuzzy PID-based Rolling State Prediction

Considering the heavy burden of calculation in the online optimization in conventional nonlinear model predictive control, we present a rolling state prediction process with lower computations to replace the rolling optimization based on the complex nonlinear quadratic programming. This rolling state prediction makes the controller predict a series of states of system in the future by means of fuzzy PID control and enables the control signal in each sample contains the information about the future dynamics of the system, thus offsetting the deficiency of lacking future prediction capability in conventional controllers. The steps of rolling state prediction at sample time k is explained as follows:

(1) Update current PID parameters K_p(k), K_i(k), K_d(k) with the utility of the principle of fuzzy PID in Section 2.1.

(2) Predict the states at next sample time $\hat{\mathbf{X}}(k+1|k)$ on the basis of the current state $\mathbf{X}(k)$ and the predicted PID control movement $\Delta u(k|k)$ by substituting $\hat{\mathbf{X}}(k+1|k)$ and $u(k|k)$ for $\mathbf{X}(k+1)$ and $u(k)$ in Equation (1), respectively. The above control input $u(k|k)$ satisfies $u(k|k)=u(k-1)+\Delta u(k|k)$.

(3) Assume $i=1$, Updating the PID predictive parameters $K_p(k+i|k)$, $K_i(k+i|k)$, $K_d(k+i|k)$ by repeating step (1) with the last prediction information. Then update $i=i+1$ and repeat the above operations in step (3) until i reaches N_p-1, where N_p is the prediction horizon.

(4) Record all the predictive control movement $\Delta u(k+i)$, $i=0,1,\cdots,N_p$ within the prediction horizon as a control sequence vector $\mathbf{u}(k)$.

The flow chart of the fuzzy PID-based rolling state prediction is illustrated in Figure 1.

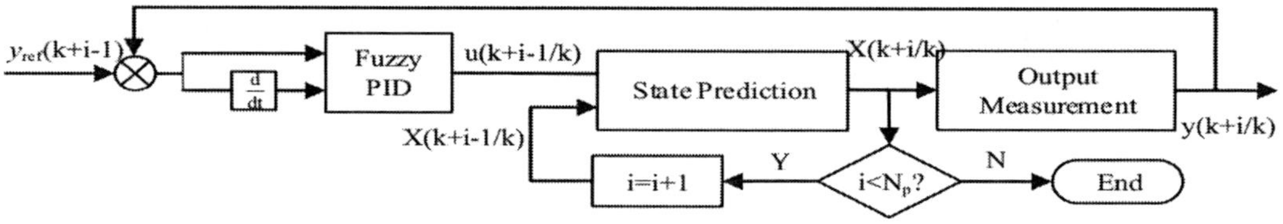

Figure 1. Flow chart of the fuzzy PID-based rolling state prediction

2.3 Real-time Control Movement Integration

The real-time control movement in this paper not only contains the first element of the predictive control sequence but also the rest of them, which differs from ordinary MPC methods. We introduce the concept of the level of confidence to its solution. The confidence probability of each element in the control movement sequence attenuates exponentially with the number of samples away from present. The computational formula of the real-time control movement is stated in Equation (2).

$$\Delta u(k)=\sum_{i=0}^{N_p-1}\alpha^i\cdot\Delta u(k+i|k),\quad i=0,1,\cdots,N_p-1 \tag{2}$$

3 Nonlinear Model for Excitation System

The excitation system is composed of the controller, the power system stabilizer (PSS), the exciter (or the amplifier) and the generator [8], as shown in Figure 2.

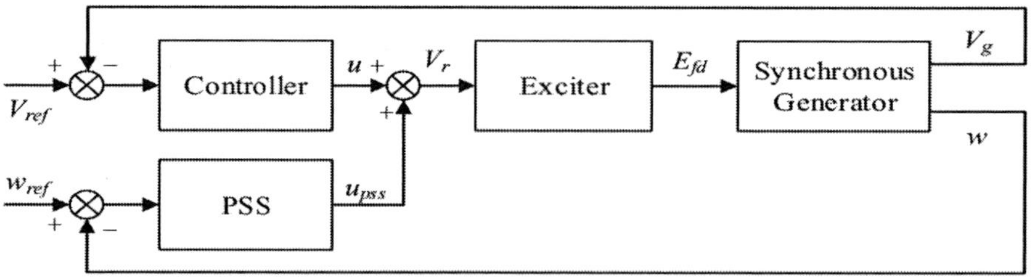

Figure 2. Block diagram of the excitation system

3.1 Power System Stabilizer and Exciter

Power system stabilizer (PSS) is added as an auxiliary control signal with which the damping property of the system is enhanced [9]. In this paper, the model of PSS is stated in Equation (3).

$$\frac{u_{pss}(s)}{\omega_{ref}(s)-\omega(s)}=\frac{T_w s}{1+T_w s}\cdot\frac{1+T_1 s}{1+T_2 s}\cdot\frac{1+T_3 s}{1+T_4 s} \tag{3}$$

where, u_{pss} is the output of PSS, ω_{ref} is the given rotor angular speed, ω is the measured rotor angular speed, T_w is time constant of PSS and $T_{1(2/3/4)}$ are lead-lag time constant.

The exciter in this paper can also be viewed as an amplifier consist of a gain element and a time constant since the excitation voltage u_f and the EMF E_{fd} is simplified to be the same here. Hence, the transfer function model of the exciter is stated in Equation (4).

$$\frac{E_{fd}(s)}{V_r(s)}=\frac{K_a}{T_a s+1} \tag{4}$$

where, V_r is the control signal, K_a is the gain of the exciter and T_a is the constant of the exciter.

3.2 Synchronous Generator and Its Connected Power System

The model of the studied power system consists of a generator linked to an infinite bus through paralleled transmission lines can be expressed in the form of nonlinear state-space equations as shown in Equation (5) [10]. It is apparently in the form of an affine nonlinear system stated in Equation (1).

$$\begin{cases} T'_{d0}\dfrac{dE'_q}{dt}=E_f-E'_q-(X_d-X'_d)\dfrac{U\sin\delta}{X_{q\Sigma}} \\ T_j\dfrac{d\omega}{dt}=P_m-\left(\dfrac{E'_qU}{X'_{d\Sigma}}sin\delta+\dfrac{U^2(X'_{d\Sigma}-X_{q\Sigma})}{2X'_{d\Sigma}X_{q\Sigma}}sin2\delta\right)-D(\omega-1)\omega_0 \\ \dfrac{d\delta}{dt}=(\omega-1)\omega_0 \end{cases} \tag{5}$$

where, U is the constant voltage of the infinite bus, δ is the rotor angle of the generator, E'_q is the transient potential of q axis, T'_{d0} is the transient time constant of d axis, T_j is the inertial coefficient of the generator, X'_d and X_q are transient potential of d axis and synchronous potential of q axis, respectively. $X'_{d\Sigma}$ and $X_{q\Sigma}$ is calculated by $X'_{d\Sigma}=X'_d+X_{TL}$ and $X_{q\Sigma}=X_q+X_{TL}$ in which X_{TL} is the total equivalent reactance on the transmission line. The continuous state-space equations is discretized by fourth-order Runge- Kutta method in the numerical simulation.

4 Numerical Simulations

To verify the effectiveness of the application of the proposed fuzzy predictive PID control method in the excitation system of the synchronous generator, simulations have been conducted in a power system in which a single machine connected to an infinite bus with its initial steady state given: Parameters of the generator are $X_d=0.964,\ X_q=0.578,\ X'_d=0.238,\ D=0.10,\ T'_{d0}=5.43,\ T_j=9.28$.

Parameters of the bus and transmission lines are $X_T=0.125,\ X_L=0.576,\ U=1.00$.

Parameters of the exciter and the PSS are set as $K_a=50, T_a=0.005, K_{pss}=0.1, T_w=5$.

The steady state before the disturbances happen (namely the initial state) in the simulations is $P_0=1.00,\ \delta_0=38.946,\ \omega_0=1.00$. All simulations is running under MATLAB R2014a environment.

4.1 Terminal Voltage Step Case

A 5% step is added on the terminal voltage of the generator at 1 second in the simulation. The FPPID control scheme is compared with the conventional PID and fuzzy PID control scheme. Simulation results viewed in Figure 3 show the significant enhancement in the calm down of the terminal voltage disturbance. Slight extra damping ability is also brought by FPPID in comparison with the other two controls. Note that in this paper, damping property of the excitation system is mainly decided by PSS auxiliary signal and all three controller in the experiment have already been equipped with a conventional PSS of which the transfer function is stated in Equation (3).

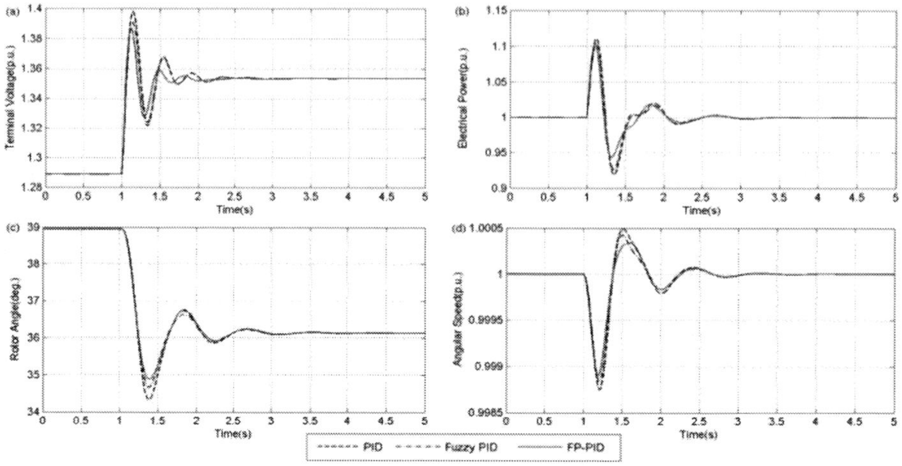

Figure 3. Response of the terminal voltage step of the synchronous generator

4.2 Three-phase Short Circuit Case

In this experiment, a three-phase short circuit fault on the high-voltage side of one of the paralleled transmission lines is taking place at 1 second. The faulted line is cut off within 0.2 second after the fault occurs. As shown in Figure 4, simulation results show the better voltage-tracking capability of the FP-PID over the comparative control methods with lower overshoot and less setting time compared with the conventional PID and fuzzy PID controls. It verifies the voltage recovery ability of the controller in serious disturbance situations, which is the most significant function of the excitation control.

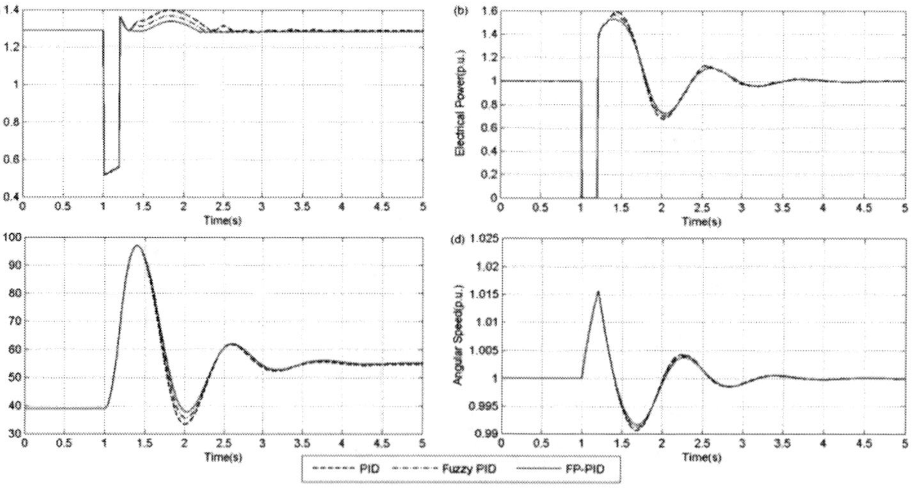

Figure 4. Response of the three-phase short circuit fault on transmission line

5 Conclusions

In this paper, a modified fuzzy predictive PID (FP-PID) control scheme in the synchronous generator excitation system is proposed to enhance the voltage-tracking performance and decrease the computational burden. The fuzzy PID control theory have been utilized in rolling state prediction to replace the conventional rolling optimization process in every sampling instance due to its low calculated amount and control parameter adaptation characteristic. The concept of the confidence level is introduced to the solution of the real-time control movement, which makes it possible for the control signal contains the future dynamic information of the system. Numerical simulations under different disturbance conditions have been conducted using the proposed control method as well as the comparative ones. Experimental results have demonstrated the superiority of the proposed FP-PID control over conventional PID and fuzzy PID controls in control performances.

Acknowledgement

This work was supported by the Key National Natural Science Foundation of China (No. 51239004).

References

1. V Mukherjee, S.P. Ghoshal, Electr. Pow. Syst. Res. **77**, 12 (2007)
2. G. Shabib, Ain Shams Eng. J. **3**,2 (2012)
3. S. Pillutla, A. Keyhani, Int. J. Elec. Power. **19**, 6 (1997)
4. J. Fraile-Ardanuy, P.J. Zufiria, Neurocomputing. **70**, 16 (2007)
5. R. Zhang, S. Wu, R. Lu, et al., Chemometr. Intell. Lab. **135** (2014)
6. T. Sato, Control Eng. Pract. **18**, 2 (2010)
7. K. Aboudolas, M. Papageorgiou, A. Kouvelas, et al., Transport. Res. C-Emer. **18**, 5 (2010)
8. P. Kundur, *Power system stability and control* (McGraw-Hill, NewYork, 1994)
9. E.S. Ali, Int. J. Elec. Power. **61** (2014)
10. Y. Wan, B. Jiang, Automatica. **62** (2015)

MATEC Web of Conferences **55**, 01005 (2016) DOI: 10.1051/matecconf/20165501005
ACPEE 2016

β Style Free-Piston Stirling Engine Control System Research

Jian Xu[1,a], Yongsheng Zhang [2]

[1]*Institute of Mechanical Manufacturing Technology, China Academy of Engineering Physics,621900 Mianyang China*
[2]*Department of Mechanical Engineering Sichuan Engineering Technical Colleg , 618000 Deyang China*

Abstract. For the Free-Piston Stirling Engines (FPSE) control system, a three -phase bridge circuit is reused as the system output about rectifier and start inverter. When FPSE system is in the power stage, the double closed loop control strategy and optimization algorithm of PI control parameters is adopted to ensure the highest system transmission efficiency under the requirements of the system output power and guarantee the stability of the running system. The simulation results prove the effectiveness of the above research content.

1 Introduction

For traditional internal combustion rotary generators, their weaknesses contain low efficiencies, too much noise, big volumes and high emissions. And their loss increases due to the complexity of mechanical transmission structures, which reduces the generating efficiency[1-2]. However, the use of Free-Piston Stirling Engines reduces intermediate mechanical transmission structures, which makes the system more compacted, reliable and small-sized. Meanwhile, the system efficiency rises dramatically. Aiming at Free-Piston Stirling Engines control problems, due to the lag of power electronic technology in early stages, series capacitances are generally used to give reactive power compensation for output voltage of linear generators. Document [3] proposes that the control of stirling engines can be divided into several levels including energy management, control of air-fuel ratio and control of Stirling Engines motion track. Aiming at shortcomings of the direct current voltage non-adjustment of uncontrolled diode rectifier and great current harmonics, Document[4] proposes three-phase bridge PWM rectifier circuit, merely giving the stability control method of DC bus voltage, without doing any power compensation and power factor correction work. Sunpower company proposes two methods of output power control for Stirling Engines system. One controls the system temperature ratio to control output power and the other controls the stroke of the engines[5].Royal Institute of Technology researches the impact on whole stirling generator of forces of different prime motor waves[6-7]. Through verification, the efficiency reaches maximum and power fluctuation reaches minimum if the current waveform is the same as that of velocity.

Based on current FPSE control system researches, the three-phase bridge circuit is reused as the system output rectification and starting inverter multiplex circuit[8-10] .When FPSE is on the generation condition and requirement of reaching system output power, double closed loop control and PI control strategies which can ensure the highest system transmission efficiency are given. It

[a] Corresponding author : tq08psx@qq.com

increases the stability and reliability of the system and promotes the development of FPSE technology as well.

2 FPSE System Structure and Principle

The FPSE control system proposed in this paper has functions including motor starting, electric power transformation with active power factor correction, generation power control of motor, system temperature and working medium pressure control. The control system structure of Stirling motor is shown in Figure 1. The functions above are separately realized by main controllers and secondary controllers. This paper focuses on control progress under the FPSE generation condition, which is controlled by main controllers.

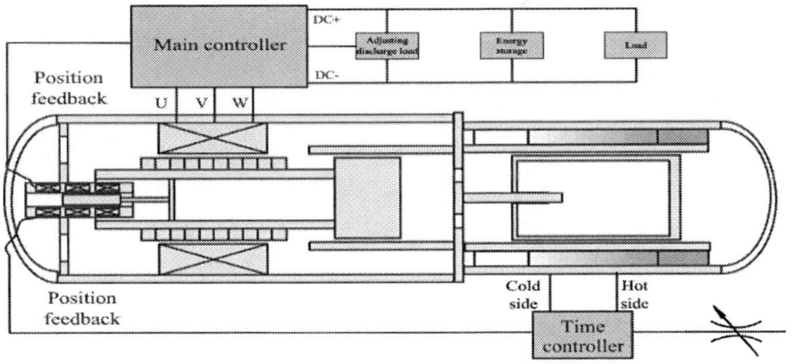

Figure 1. Schematic diagram of control structure

Main controllers are relative to motor starting module, three-phase rectification circuit, power control module, electricity storage module, communication module and so on. Energy storage device controlled by main controllers provides electrical energy for motor staring and starting circuit provides starting current for motor stator coil thus creates magnetic field, providing transient excitation for permanent magnetism and driving the startup of motor. The three-phase rectification module uses rectifier with PFC to convert larger ranging AC to nearly constant DC output voltage(high voltage), and maintain the best power factor, improve efficiency of electricity source. On the premise that the structure parameters, temperature ratio and working medium average pressure of Stirling motor is constant, the stroke of power piston is also relative to system's total power load, which consists of user load and discharge load. The change of total power load will result in the magnetic force change of motor wires, thus affects the stroke of power piston. Therefore, the stability control of power piston is not a direct control. Meanwhile, the system has energy storage part, which can stabilize load pressure at transient time of load variation. Besides satisfying the power consumption of user loads, the system can store surplus power into power storage module, which increases the reserve of emergency power.

3 FPSE System Power Generation State Control and Parameter Optimization

The working states of main controllers can be divided into two stages: starting and power generation stage. The starting stage is not the key attention of this paper. At the power generation stage of Stirling Engine, the AC output of generator needs rectification and filtering in order to be transformed into DC, needed by power consumption equipment. However, due to the introduce of nonlinear part and large filter capacitor, the motor current becomes steep leading and trailing pulsating current, which results in larger current distortion, input and output phase difference, thus causes lower system efficiency and larger system loss. In order to solve this problem and increase Stirling Engine system's power factor, in this paper, the controlled rectifier with power factor correction is used to control the system, which makes the waveform of the system motor current and output voltage consistent and zero-phase difference, improving the system efficiency. In order to satisfy the power demand of different equipment, the voltage part is introduced to adjust output the voltage value in real time, satisfying demand of different users.

The Stirling Engine system power stage control is shown as Figure 2. The "voltage outer ring, current inner ring" double closed loop control strategy is used. The "voltage outer ring" is used to adjust the output value of the system load in real time, which makes it fast and without static error tracking set value. The "current inner ring" is to correct the system power factor. When user loads vary, a discharge load is used to match its variation, making total power load at a constant value and ensuring the power piston stroke nearly unchanged.

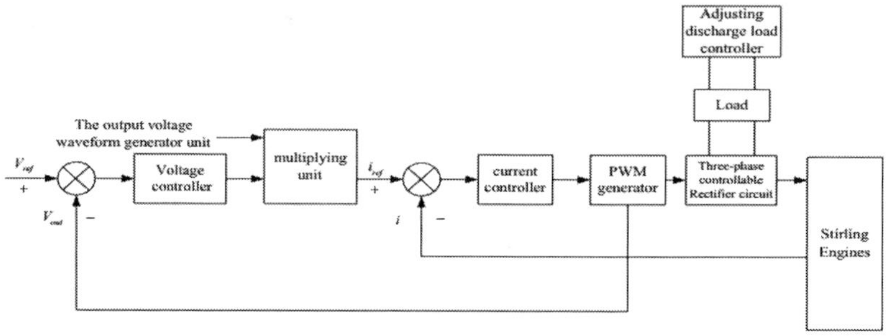

Figure 2. System power stage control chart

Under FPSE system generation condition, the system power density and transmission power are closely related to PI control parameters. The system PI control parameter optimization is an important way to improve the system comprehensive performance. FPSE system is a non-linear, multi-parameter, multi-goal, uncertain and tight coupling system. So, based on non-linear optimization theories, artificial intelligence like neural network and genetic algorithm are mixed to realize the PI control parameter optimization.

Combined with penalty function to deal with restricted problems, dynamic functions are used to improve individual fitness calibration and non-uniform mutation. High-pressure power electronic device with less conduction losses, lower output distortion and lower total load voltage variation are proposed as goals in system's PI control parameter design methods.

a) Optimize objective variables, use penalty functions to deal with restricted problems and decide a suitable weight coefficient to form objective function, which is about to optimize. Less switching loss, lower voltage of high frequency power electronic device and voltage stability of load output are the goal. Assume a_f, a_s, a_c each stands for conduction loss coefficient of f, output current distortion of I_s and total load output voltage variation U_c. System optimizing objective function f_{obj} is formed:

$$f_{obj} = a_f f + a_s I_s + a_c U_c + \sum_{i=1}^{n} b_i \max(g_i, 0), (g_i < 0) \tag{1}$$

where, n stands for numbers of system restricted conditions, g_i is the actual restricted condition. Since a lower f_{obj} is better, when g_i satisfies restricted conditions, on the contrary, $\max(g_i,0)=0$; $,\max(g_i,0)=g_i$. The increase of f_{obj} punishes individuals which violates restrictions.

b) BP neural network mathematical expression mixed with genetic algorithm is shown as follows:

$$\begin{cases} \min \mathrm{E}(w, v, \theta, r) = \dfrac{1}{2} \sum_{k=1}^{M} \sum_{t=1}^{n} \left[y_k(t) - \hat{y}_k(t) \right]^2 \\ s,t: w \in R^{m \times p}, \quad v \in R^{n \times p}, \theta \in R^{p}, r \in R^{n} \end{cases} \tag{2}$$

where, E_1 is the total error of the network, w and θ are weight and threshold value of BP neural network, $y_k(t)$ is the expectancy output, $\hat{y}_k(t)$ is the network actual output.

$$\hat{y}_k(t) = f_{obj}\left\{\sum_{j=1}^{P} v_{\mu} \cdot f_{obj}\left[\sum_{i=1}^{m} w_{ij} \cdot x_i(t) + \theta_j\right]\right\} + r_i \tag{3}$$

Assume $E_2 = \dfrac{1}{N - N_1}\sum_{k=N_1}^{M}\sum_{t=1}^{n}\left[y_k(t) - \hat{y}_k(t)\right]^2$ is the mean square error of test sample, showing the reliability estimate of network output. Lower E_2 means higher network reliability, otherwise it is not that reliable.

c) Combined with conditions a and b, the genetic algorithm is used to do chromosome encoding, selection ,chiasma, disorder and fitness selection repeatedly, thus obtains the optimized variables.

Finally, the satisfied condition of restriction is examined. If not completely satisfied or without solution, weight coefficient, genetic operation parameters or restricted ranges can be adjusted properly. Through multiple optimizations, the optimized variable which fully satisfied restrictions can be obtained.

4 Simulation Result Analysis

4.1 Simulation analysis without control strategy

In order to verify the effectiveness of control strategy analysis and controller design in power stage, Stirling system without control is simulated first, which foreshadows the verifying effectiveness of control strategy, shown as Figure 3. The system model parameters are shown in Table.1.

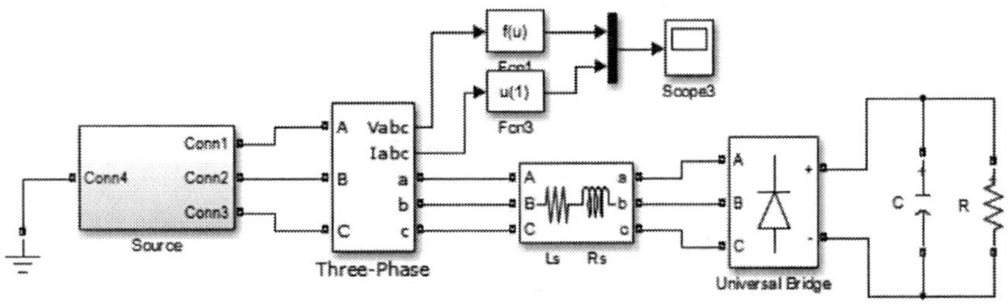

Figure 3. Simulink model without control strategy in power stage

Table 1. model parameters in power stage

name	signs	value
Power frequency	F	80Hz
Supply voltage effective value	U	380V
Internal resistance	Rs	0.02Ω
Inductance	Ls	10mH
Filter capacitor	C	2.5mF
Load resistance	R	600Ω

Simulation waveform in Fig.5 can be obtained by simulation, which is a steady stage waveform. Figure 4 a) is the contrast chart of motor current and back EMF. For better observing the contrast condition, simulation back EMF is 10 times lower and it is observed with oscilloscope. As figures shows, back EMF and motor current differ in waveform and phase. Back EMF is a standard sinewave while the motor current is a pulsate current with a high distortion, which has a large harmonic wave and hardly can be used in engineering application. Figure 4b) shows the output voltage in load ends, with a stable DC output voltage at 473V and output power at 3.7 kW.

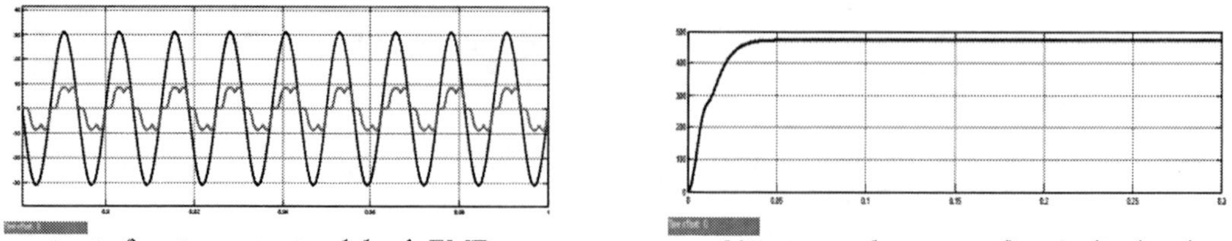

a) contrast of motor current and back EMF **b)** output voltage waveform at load ends

Figure 4. Simulation waveform without control strategy in power stage

4.2 Simulation analysis using double closed loop control strategy

Through the analysis of control strategy in the power stage and PI control parameter optimization, the simulink model in Stirling engine power stage can be established as Figure 5a). Voltage outer ring and current inner ring is used and switch drive signal of controlled rectifier circuit can be produced by PWM algorithm. Control module is a self-defined package module, with the internal simulation module in Figure 5b) . The output voltage value at load ends is compared with set value and produces control variable through PI. This control variable is multiplied by motor current set value and used as motor current given value. The given value is compared with the actual motor current value and three control variables are obtained by P controller. The control model in Figure 5. produces three corresponding control variables according to voltage , current and feeding value. Control variables are amplitude-limited and six switch drive signals are produced by PWM module to control three-phase circuit, where PWM module is a self-defined module. Control variables and STW carrier are compared and square wave with controllable duty ratio can be obtained. Therefore, the switch conduction condition can be controlled. Optimized parameters of double closed loop control network can be obtained according to formula (3), shown as Table. 2.

Table 2. Parameters control in power stage

name	signes	value
Output voltage given	Uref	600V
Voltage ring proportion coefficient	Pu	1
Voltage ring integral coefficient	Iu	30
Current ring proportion coefficient	Pi	25

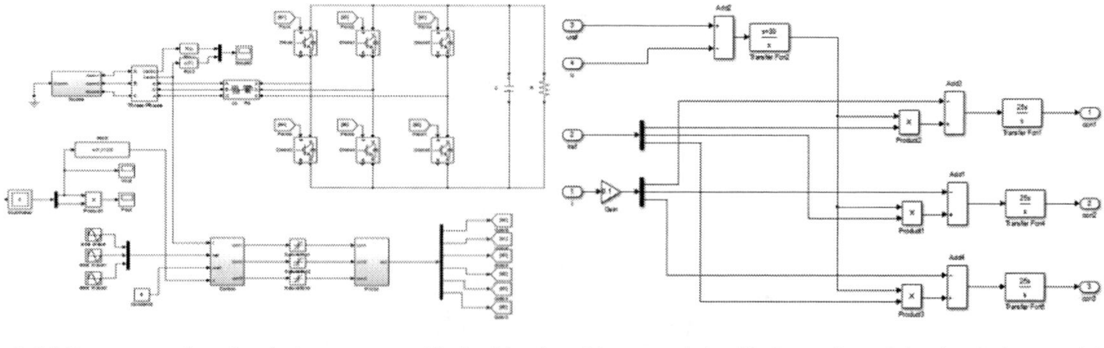

a) Stirling generation simulation system with double closed loop models b) Control module simulation model

Figure 5. Simulink simulation model in power stage

The steady-state simulation waveform in Figure 6. can be obtained by simulation. Figure 6 a) is the contrast figure of motor current and back EMF. The amplitude of back EMF is also lowered by 10 times. By double closed loop control, the motor current is not the previous pulsate form but consistent with back EMF waveform, whose phase difference is zero and the system power factor and efficiency is improved. Figure 6 b) shows the output voltage of load ends. It can be seen the output DC voltage can be stabilized at set value of 600V, output power at 6kW. Under "current inner ring control" , back EMF and motor current can be the same in waveform and phase difference, thus improves the power factor. Under "voltage outer ring control", the output voltage can be effectively adjusted and it can track the set value fast ,static-error free, which satisfies the need of different electrical equipment. Through the above control strategies, the control demand of Stiring engine in power stage is almost realized.

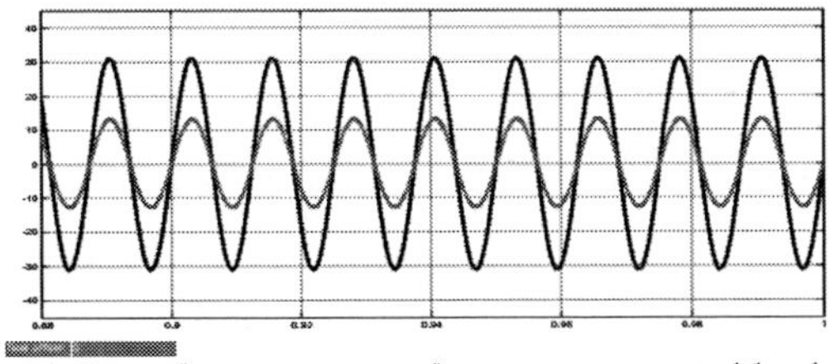

a) Waveform contrast of motor current and back EMF

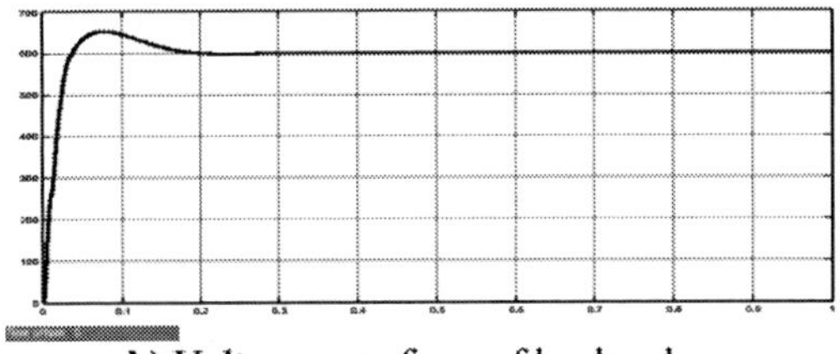

b) Voltage waveform of load ends

Figure 6. Double closed loop control strategy simulation waveform in power stage

5 Conclusion

A three-phase bridge circuit is reused as the system output about rectifier and start inverter. Reaching the system output power demand, double closed loop control strategy which ensures the highest transmission efficiency and its control parameter optimization is given. Aiming at time-varying load resulting in the change of system output voltage, a method which realizes offset control using resistance matrix is proposed, thus maintains the total power load at a constant value. As a result, the piston stroke will hardly change. The simulation result verifies the effectiveness of researches above.

References

1. Li Qingfeng, Xiao Jin, Huang zhen Simulation of a Free Piston Linear Alternator[J] China Mechanical Engineering 2009,20(8):911-915
2. Shoukry E F Numerical Simulation for Parametric Study of a Two-stroke Compression Ignition Direct Injection Linear Engine[D].Morgantown: West Virginia University, 2003
3. Scott S.Gerber, Mike J, Timothy F.Regan, Mary E R. Advanced Controller for the Free-Piston Stirling Convertor[C].13th International Stirling Engineering Conference.2005:1-12
4. Yimin Lu,Huang X F,Zhu Z Q. Direct Power Control of a Permanent Magnet Linear Generator in a Free Piston Energy Converter[C].International Conference on Electrical Machines and Systems.2008:3636-3639
5. Holliday E, Keiter D E. Control Electronics for Palm Power 35W Free-Piston Stirling Engine[C].3rd International Energy Conversion Engineering Conference.2005:1-7
6. Hannson J,Leksell M,Carlsson F. Minimizing Power Pulsations in a Free Piston Energy Converter[C].European Conference on Power Electronics and Applications.2005
7. Liu Y, Leksell M, Arshad W M, Thelin P. Influence of Speed and Current Profiles upon Converter Dimensioning and Electrical Machine Performance in a Free-Piston Generator[C]. Proceedings of the 4th International Symposium on Linear Drives for Industry Application.2003
8. Van Blarigan,P.Free-piston engine[P].United States Patent ,US 6,199,519 B1,Alexandria,VA,2001
9. Scott Goldsborough,Peter Van Blarigan .Optimizing the Scanvenging System for a Two-Stroke Cycle, Free Piston Engine for High Efficiency and Low Emissions:A Computational Approach [C]. International Multidimensional Engine Modeling User Group Meeting at the SAE Congress 2003
10. Jiabin Wang,David Howe.Design Optimization of Radially Magnetized ,Iron-Cored,Tubular Permanent – Magnet Machines and Drive Systems[J].IEEE Transactions on Magnetics Vol.40,NO.5,Sep 2004

MATEC Web of Conferences **55**, 01006 (2016) DOI: 10.1051/matecconf/20165501006
ACPEE 2016

Analysis of the Control Strategies of Parallel Active Power Filter

Limiao Huang [1,a], Guiping Zhu [1]

[1]*Tsinghua University, Department of Electrical Engineering, State Key Laboratory of Power System, 100084 Beijing, China*

Abstract. Depending on detection points of harmonic current, the control strategies of parallel active power filter (PAPF) can be divided into two basic types. Load current detection is the simplest open loop control strategy, but the parallel resonant problem is an inherent defect of it. Source current detection forms an effective closed loop. However, a stability problem would be caused meanwhile. In this paper, firstly, typical system of PAPF was designed and analyzed, and different kinds of current control strategies were deeply discussed. In particular, a long time ignored problem (the amplification of the fundamental current after the compensation) associated with traditional source current detection was pointed out for the first time. Next, we proposed a new double-loop compound control strategy, which detects both the load current and the source current. Through the analysis of each part of the PAPF, the mathematics model of the system was put forward. Furthermore, we used this model to discuss the stability and compensation characteristics of the system, and found out the best working point through large amount of simulation. Finally, simulation results verified the validity of the proposed control strategy.

1 Introduction

The parallel resonance of the grid side LCL output filter of PAPF is a serious problem of the traditional load current detection. Therefore, the compensation effect of PAPF cannot be further strengthened. Some researchers have been devoted to solve this problem from the point of current control strategies. The earliest research could be found in references [1] and [2].Authors put forward the classification criteria of the control strategies of PAPF for the first time and carried on some preliminary theoretical analysis. Then the equivalence between the two basic control strategies was proved [3, 4]. Further, researchers deduced out the unified description of current control strategies of PAPF but without sufficient verification [4]. In reference [5], consideration has been given to both the system stability and compensation characteristics. However, the problem of the amplification of fundamental current after the compensation has been still ignored.

A typical system diagram of PAPF is shown in Fig.1. Harmonic current detection unit extracts harmonic components from the load current or source current as reference current firstly. Then, the main circuit (usually a voltage source inverter, VSI) tracks the reference current by the current tracking control unit and the drive circuit. Thus, PAPF replaces of the power grid in supplying

[a] Corresponding author : 1372742756@qq.com

DOI: 10.1051/matecconf/20165501006

harmonic current to the load to ensure that the power source current stay as undistorted sinusoidal current.

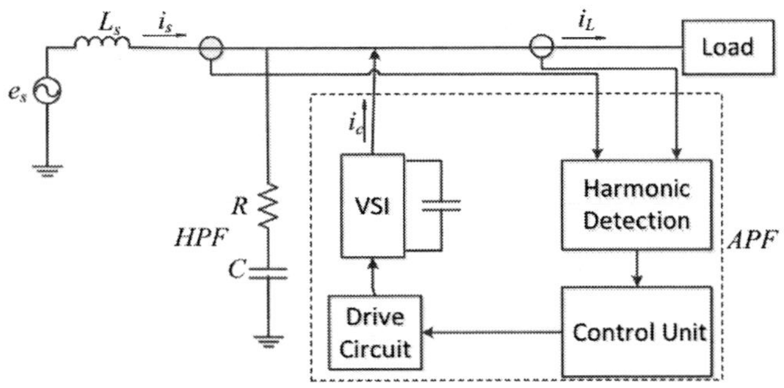

Figure 1. Typical system diagram of PAPF

As in Figure 1, i_S is the source current, i_L is the load current, and i_C is the output compensation current of PAPF. Assuming that i_{Lf} represents the fundamental component of load current, i_{Lh} the corresponding harmonic components. In order to eliminate the harmonic components of source current completely, we need only ensure that the compensation current be equal to the harmonic components of the load current, that is, $i_C = i_{Lh}$.

In this paper, main circuit of the compensation current generator, a voltage type high frequency PWM inverter, was based on three phase four wire system. RC high-pass filter was connected in parallel with the main circuit in AC side to filter out high frequency harmonics of the compensation current near switching frequency. A control strategy was used to track the reference current and realize the DC side capacitor voltage control [3, 6]. It is worth mentioning that we used FBD algorithm [7] to calculate the reference current. FBD algorithm is essentially a projection algorithm and has definite physical meaning. Its accuracy and real-time performance gain an advantage over the traditional instantaneous reactive power algorithm. Then, a three-phase diodes rectifier (with a resistance in the DC side) was served as typical nonlinear load (harmonic sources). Initial total harmonic current distortion of it reached up to 23.71% in the simulation model.

The control strategies of PAPF are usually classified by the detection points of reference current. In the following section, we made an analysis of the load current detection control strategy on the basis of the existing research results of other researchers, and pointed out the inherent defect of it. Next, math model of the source current detection control strategy and the new type of compound control strategy was put forward and discussed. Further, a problem with the source current detection (of course the same as the compound control strategy) which was neglected all the way in previous studies was pointed out, namely, the amplification of the fundamental component of source current after the compensation. Then we proposed a proper solution on the basis of theoretical analysis. Finally, we got a set of optimal control parameters of the new type of compound control strategy, which ensures PAPF to obtain best compensation result within the scope of system stability domain.

2 Grid side output LCL filter and the parallel resonant problem

In order to get the ideal output current, the main circuit of PAPF must be able to track the low frequency reference current accurately, with strong high frequency attenuation characteristic at the same time. This goal is closely related to the grid side output filter. LCL filter is a popular choice in recent studies. The single-phase equivalent circuit of it is shown in Figure 2. L1 is the AC side connection inductance, L2 the grid side inductance, and C the parallel capacitance, which provides

DOI: 10.1051/matecconf/20165501006

pathway for the current harmonics around switching frequency. Some papers have formed a set of algorithm programs for the selection of specific parameters of the LCL filter [8].

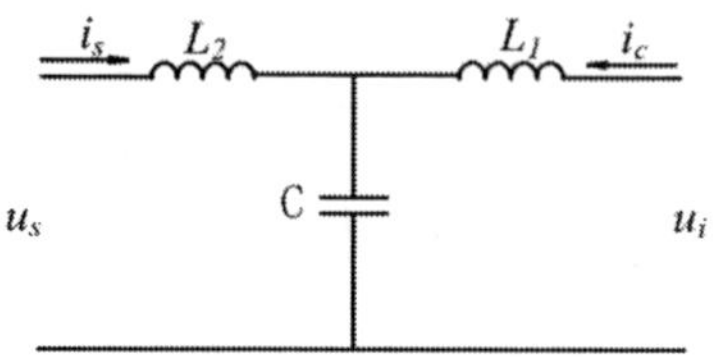

Figure 2. Single-phase equivalent circuit of LCL filter

LCL filter has excellent high frequency attenuation ability, but unfortunately, it results in the impedance resonance at the same time. The single-phase equivalent circuit of the system is shown in Figure 3. Among which, R is the additional damping resistance, i_S is the source current. The sum of compensation current and load current is regarded as a current source, i_{CL} [1].

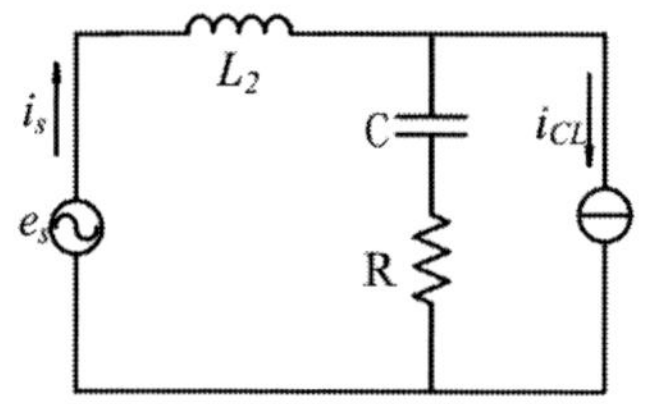

Figure 3. Single-phase equivalent circuit of the system based on LCL

Take i_{CL} as the input, and i_S the output, the transfer function of the intermediate impedance link is:

$$G_I(s)=\dot{i}_S / \dot{i}_{CL} = (R+\frac{1}{sC}) / (sL_2 + R + \frac{1}{sC}) \tag{1}$$

The bode plots of it is shown in Figure 4.

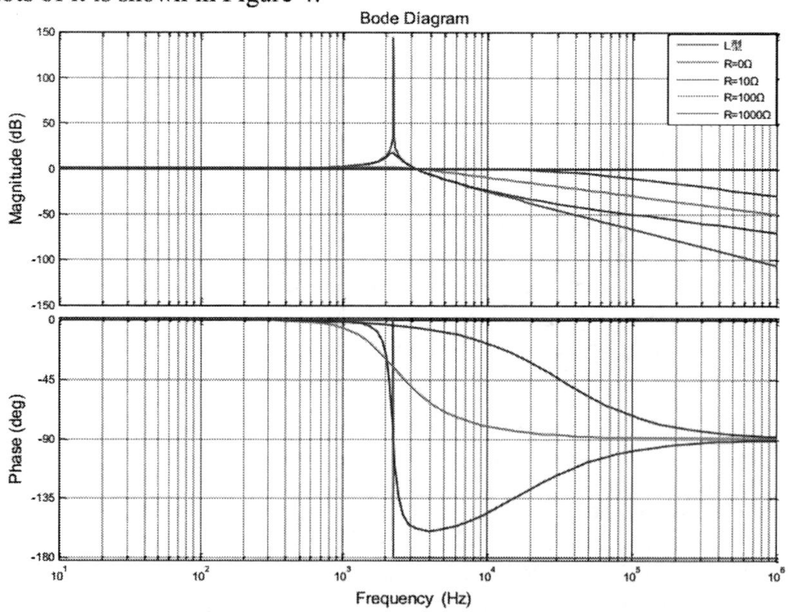

Figure 4. Bode plots of the intermediate link of impedance

DOI: 10.1051/matecconf/20165501006

The blue line represents the traditional L filter, whose high-frequency attenuation characteristic is extremely poor. Other four lines are amplitude-frequency characteristic curves of the LCL filter under different damping resistance. As we can see, the low frequency characteristics of LCL overlap with L filter, and the high frequency attenuation reaches as high as 40dB/decade in the case of non-damping. However, the compensation current components near the resonance frequency (about 42 times power frequency) would be amplified before injection to the source side, and the compensation effect is seriously deteriorated. As the value of the damping resistance increasing from 0 to 1000 Ω, resonant peak becoming more and more small. However, the increase of damping resistance will lead to more and more system loss, and more seriously, weakening of the high frequency attenuation capability. So it is necessary to begin from the Control Strategy to regulate the system response and get good compensation result.

3 Comparison analysis of three control strategies

3.1 Two basic strategies

As has been noted, two basic current control strategies refer to the way of load current detection and the way of source current detection. The unified description of current control strategies of PAPF is deduced in reference [4]. Making use of a set of constraints and graphical method, it is not difficult to prove the equivalence between these two strategies [4]. The load current detection is the most widely used strategy. Fig.5 is the system block diagram of the PAPF. In this diagram, $G_I(s)$ represents the transfer function of the intermediate impedance link as is shown in equation (1), and the main circuit of PAPF is equivalent to proportion link K_{PWM} with time delay T_d. Because the harmonic current detection unit based on FBD algorithm which extracts the harmonic component from the input current(i_{CL}) accurately in ideal situation, so it is omitted directly in Figure 5, being focused on the harmonic component current only.

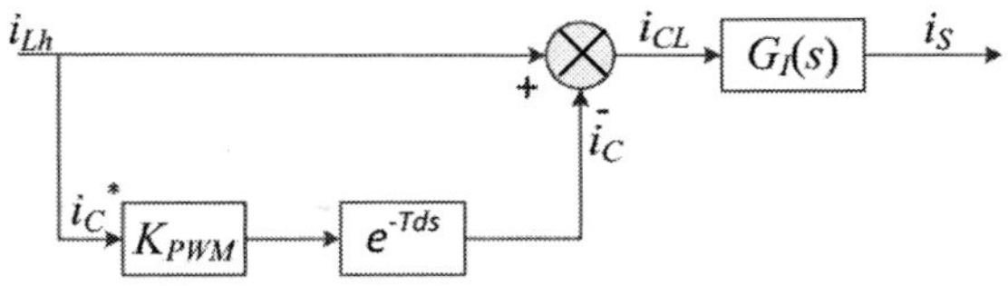

Figure 5. System block diagram of the PAPF with load current detection control strategy

Obviously, $G_I(s)$ is outside of the closed loop, there is no way to form effective feedback to adjust it and eliminate the resonance [1]. So, we can draw the conclusion that under the way of load current detection, even if the harmonic current detection algorithm and the harmonic current tracking control achieve ideal precision and real-time performance, the compensation effect of PAPF cannot be enhanced.

In contrast, as is shown in Figure 6, an effective closed-loop feedback is formed under the way of power source current detection.

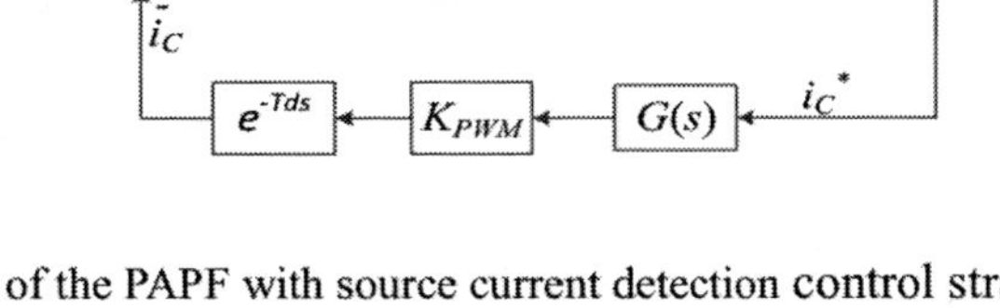

Figure 6. System block diagram of the PAPF with source current detection control strategy

G(s) is the correction unit. In order to improve the system stability, G(s) is usually designed as a differential first-order inertia unit [5]:

$$G(s)=KTs/(Ts+1) \tag{2}$$

K and T are undecided parameters. The impedance resonance of $G_l(s)$ can be eliminated to some extent by choosing the parameters carefully. However, because the value K must be designed as large as 4 in order to acquire the best compensation effect under this way, the stability of the system would be adversely affected. Before making a further analysis on this point in the next part, we will show the compensation results of the two basic control strategies, and points out another problem which was neglected all the way in previous studies based on source current detection control strategy.

Use of three-phase uncontrolled rectifier with resistance as the harmonic source, the fundamental component of the source current (I_f) is equal to 18.50A, and total harmonic current distortion reaches up to 23.71%. The simulation results of two basic strategies are shown in Tab.1. From the table, we can see that source current detection has better harmonic control effect than load current detection. However, at the same time, the fundamental component of source current increases to 21.87A, a data much bigger than the original value 18.50A. Many studies only concentrated on the good compensation performance of the source current detection, but neglected the fact that it can also cause the amplification of the fundamental component of source current.

This paper attempts to explain this problem and suggests a solution for it in the following analysis. In general, when detection point (load side or/and source side) and detection algorithm (namely FBD algorithm in this paper) are determined, the modes of harmonic current detection could be divided into the following two categories [9].

1) Indirect mode. The fundamental component of current is calculated firstly, and its difference from the total current is taken as the output.
2) Direct mode. In this mode, the selected harmonics (of order usually less than 20) are detected directly, and the sum of them is taken as the result.

Obviously, the amount of calculation of mode 1 is much less than mode 2, so mode 1 was adopted in many applications. Inevitably, in mode 1, when the fundamental component is subtracted from total current, a small amount of fundamental component would be left in the output reference current still. Then, as the reference current passing through the correction unit G(s) which includes a amplification factor K(refer to equation(2), K was designed as 4), the remained fundamental component is amplified at the same time. This is exactly the reason that the fundamental component of the source current increases a lot after compensation. To solve this problem, we have carried on simulation research on the basis of mode 2. The results showed that the amplification of the fundamental component of the source current was almost eliminated. Validity of the analysis was proved.

3.2 Double-loop compound control strategy

In practical industrial application, however, it is not the best choice to solve the fundamental current amplification problem at the expense of increasing the amount of calculation. So it is necessary to introduce our new double-loop compound control strategy. We will see in the following that the amplification factor K in feedback loop of source current in this strategy could be designed far smaller than that on using source current detection strategy. Thus, mode 1 can be adopted as usual without the amplification of the fundamental component of source current.

Compound control strategy, detects both the load current and the source current as a combination of the two basic control strategies. The system block diagram of it is shown in Figure 7.

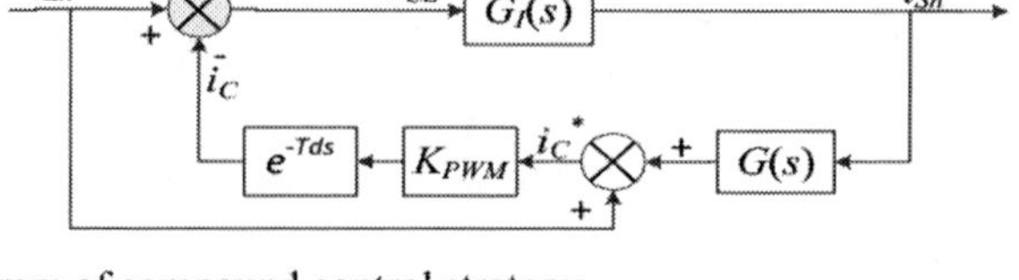

Figure 7. System block diagram of compound control strategy

In this paper, we approximate the proportion link K_{PWM} as 1 and the delay of the main circuit is regarded as one PWM control cycle (T_d=1/f_{sw}, f_{sw}=12.8kHz). Because the switching frequency is high enough than the bandwidth of the PAPF, therefore, the delay unit could be Taylor expanded with second-order approximation.

$$e^{-T_d s} \approx 1/(1+T_d s+(T_d s)^2/2) \tag{3}$$

G_I(s) and G(s) refer to (1) and (2) respectively. Then the closed-loop transfer function of the system is expressed as follows.

$$G_{close} = G_I \bullet (1-e^{-T_d s})/(1+G \bullet G_I \bullet e^{-T_d s}) \tag{4}$$

The characteristic equation of the system would be deduced out according to (4). Then, we can analyse the stability of the system using criterions such as the root locus rule or Routh stability criterion. In Figure 8, the square area is on behalf of the value range of K and T that makes the system stable. The stability margin of the red area (top left corner) is larger than the other area. This shows that the smaller the parameter K, the better is the system stability. In the proposed double-loop compound control strategy, the reference current is mainly extracted from load current, and the source current feedback loop is just aimed to eliminate the impedance resonance. Thus, amplification factor of the source current loop (parameter K) does not need to be designed very large [5]. Therefore, the system could achieve a desired compensation effect while maintaining strong stability at the same time.

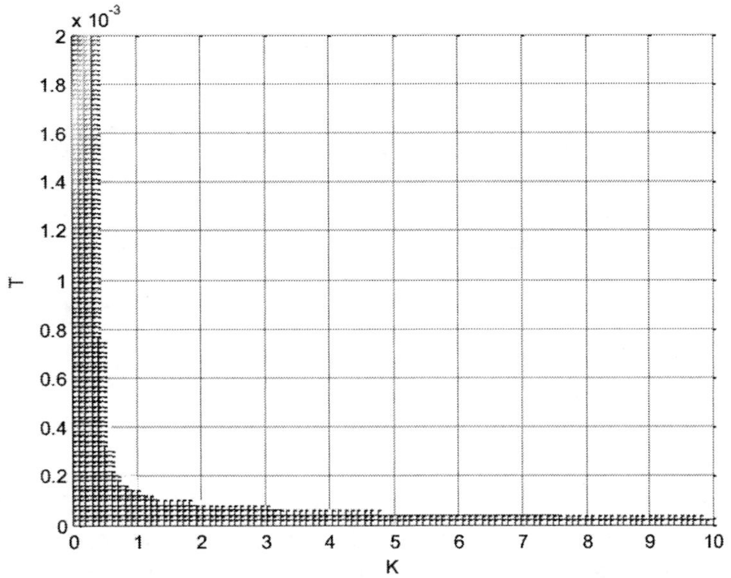

Figure 8. Value range of K and T that makes the system stable

The Comparison of the two basic strategies with the compound control strategy is shown in Table.1.

Table 1. Comparison of two basic strategies with the compound control strategy

Control Strategy	*K*	I_f/A	THD
Load current detection	0	18.97	4.21%
Source current detection	4	21.87	3.73%
Compound control strategy	0.2	19.67	2.89%

Among the three control strategies, the double-loop compound control strategy showed the most ideal compensation effect. Detailed simulation and experimental results are presented in the next section.

4 Simulation results

Through the validation of a large amount of simulations, the parameter set (K, T) was (0.2, 0.00015) which made the system achieve the best compensation performance and met the condition of constraint in Fig.8 at the same time. Parameter K was much smaller than 4, the value designed under source current detection control strategy.

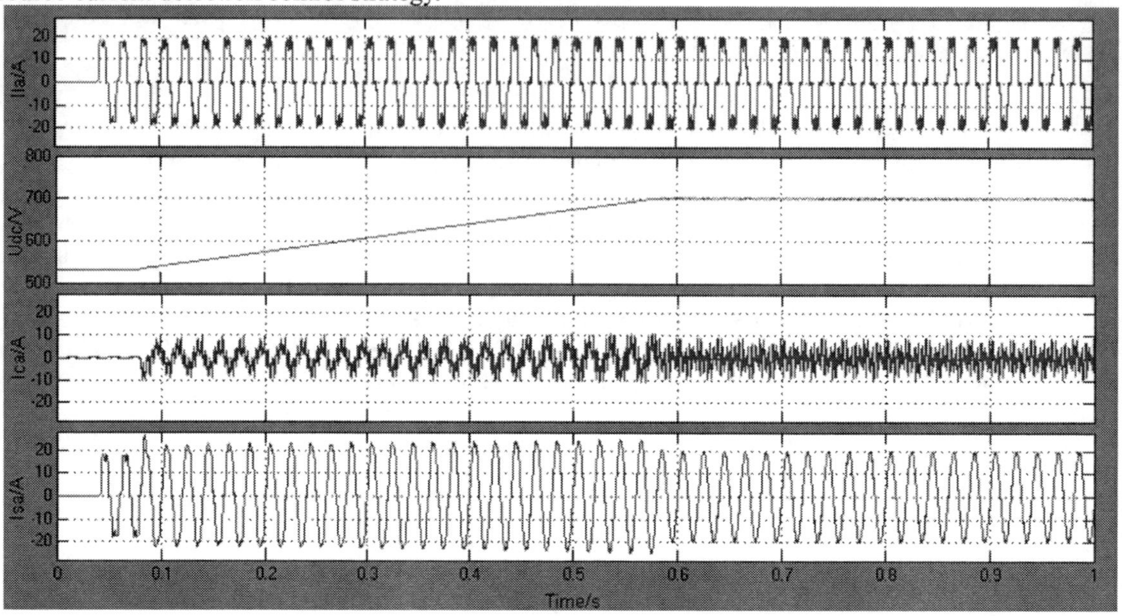

Figure 9. Simulation result under the double-loop compound control strategy
Ila: load current; Udc: DC bus voltage; Ica: compensation current; Isa: source current

Figure 9 shows one simulation result under the double-loop compound control strategy. Four curves are load current, DC bus voltage, compensation current and source current respectively. Source voltage was 380V, and the initial value of the DC bus voltage($U_{dc\text{-}start}$) was designed as 530V. APF soft started at 0.08s, and the DC bus voltage reaches its final value($U_{dc\text{-}end}$=703V, slightly larger than the design value 700V) after 5 seconds. In this process, the reference value of DC bus voltage can be represented by:

$$U_{dc\text{-}ref}=U_{dc\text{-}start}+(U_{dc\text{-}end}-U_{dc\text{-}start})\bullet dU \quad (5)$$

d U is the step size of the soft start process. A voltage control loop was adopted to realize real-time tracking of this reference value. As is shown in Fig.9, the source side impact current reaches it's maximum value 27A at the moment of starting APF. This suggests that the soft start design has restrained the current impacting and protected the IGBT effectively. It is worth noting that the compensation current contained a significant amount of fundamental component during 0.08s to 0.58s. This shows that the dc-link capacitance absorbed part of active power from the source side in the process of soft start. The source current was a nearly perfect sine wave after the system reached steady state at 0.58s.

As is shown in Figure 10, THD of the source current after compensation is reduced from 23.71% to 2.89%. Harmonic components were mainly concentrated on the frequency band below 100 times power frequency. Switching harmonics (near by 256 times power frequency) were effectively suppressed, and the amplitude of them were smaller than 0.04A. This kind of frequency response proved the correctness of the design of the grid side LCL output filter in this study. What's more, the fundamental component of the source current equal to 19.67A, the amplification of the fundamental current under the control strategy of source current detection was well-restrained.

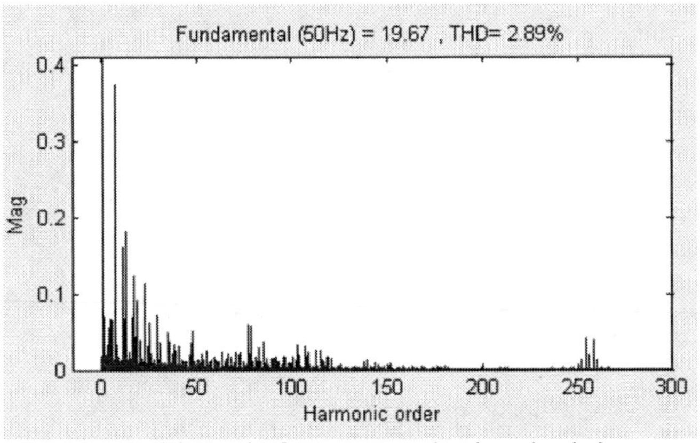

Figure 10. THD of the source current (phase A) after compensation in a simulation

5 Conclusions

The main work of this paper can be summarized as the following respects.

1) Designed the typical system of PAPF. Analyzed the impedance resonance problem of grid side LCL output filter and pointed out the necessity of detecting source current.
2) Built math models of different kinds of control strategies, and studied the unified description of them.
3) Pointed out and proposed a solution for the amplification problem of the fundamental current under the control way of source current detection.
4) Obtained a set of precise optimal control parameters of the proposed double-loop compound control strategy, and proved its superiorities compared with the former two basic strategies via plenty of simulations. Prototype experiments are being conducted as well, the compensation effect has to be further improved to meet the IEEE standards.

References

1. Z.A.Wang. *Harmonics elimination and reactive power compensation.* (China Machine PRESS, Beijing,CHINA, 2006)
2. J.Yang, Z.A.Wang. J.Xi'an Jiaotong Univ, **29(3)**: 97-102 (1995)
3. X.Y.Wang, J.J.Liu, C.Yuan, Z.A.Wang. *IEEE PESC* 2006: 1-7 (2006)
4. X.Y.Wang, J.J.Liu, C.Yuan, and Z.A.Wang. *IEEE APEC* 2007: 675-681 (2007)
5. X.Q.Yan, J.Yang,Z.A.Wang. Transactions of CES, **13(1)**: 41-45 (1998)
6. Z.Y.Yan, G.P.Zhu. Parallel Three-phase Active Power Filter Design and Development[D]. (Tsinghua Univ, Beijing, CHINA, 2012)
7. Q.Chen, J.Y.Zheng. *Proc of 27th CCC*: 232-235 (2008)
8. D.J.Zhang, Z.L.Qiu, Y.L.Li, G.Z.Chen, X.N.He. Transactions of CES, **26(6)**:137-143 (2011)
9. W.W.Liu, H.F.Ding, X.Z.Duan. Proceedings of the CSEE, **31(27)**: 14-20 (2011)

PMSM System Controlled by SVM-DTC

Weiwei Zhang[1,a], Fei Xiao[1] and Mingzhong Qiao[2]

[1] *National Key Laboratory for Vessel Integrated Power System Technology, Navel University of Engineering, China*
[2] *Department of Electrical Engineering, Navel University of Engineering, China*

Abstract. Direct Torque Control (DTC), proposed after Vector Control (VC), has been widely used in speed regulation system due to its good dynamic performance. In order to solving the shortage of poor torque stationary properties of traditional DTC, this paper introduces direct torque control using space vector modulations (SVM-DTC) in Permanent Magnet Synchronous Motor (PMSM) system, which reduces torque ripple by using SVM and maintains fast dynamic response. Simulation model of SVM-DTC under MATLAB condition has been set up and compared with system controlled by traditional DTC. The results testify that the PMSM system based on SVM-DTC has high rate dynamic response, high stationary precision, and good robustness when load has a disturbance.

1 Introduction

Permanent magnet synchronous motor has been widely applied in various AC drive systems for its merits of high efficiency, high power density and high torque/inertia ratio. The research of permanent magnet synchronous motor drive system with high performance produces important economy and society value. While PMSM controlled by traditional DTC has some shortages because of using bang-bang control. Only eight space vector voltage can be chosen lead to the ripple of stator flux and torque in this control. Thus, application of PMSM in high-precision occasion is limited [1].

In order to solve the problem of traditional DTC, many researches has been carried out recently. A number of methods such as space vector modulation, duty cycle control, vectors subdivision have been proposed and obtained some achievements [2]. The main idea of SVM is calculating exact vector voltage needed in every sampling period, so that SVM-DTC has many benefits in reducing ripple of torque except its hysteresis effect. Nevertheless, solutions in paper [3] and [4] have strong dependence on parameters of PMSM and are hard to control. In order to avoid ripple radically, paper [5] utilizes rotating coordinates to obtain definite vector voltage. However, the method depends on complex structure and hard to maintain the fast response of traditional DTC. The SVM-DTC presented in this paper aims to maintain the stationary performance and increase the speed of dynamic response substantially as well as the robustness.

2 Analysis of Torque Ripple in Traditional DTC

[a] Weiwei Zhang : zww_zhang@163.com

Ignoring the core saturation, eddy current loss, hysteresis loss and rotor damper winding, the function of PMSM in α-β coordinate system can be shown as (1) to (4).

$$\psi_\alpha = \int \left(u_\alpha - r_s i_\alpha\right) dt \tag{1}$$

$$\psi_\beta = \int \left(u_\beta - r_s i_\beta\right) dt \tag{2}$$

$$\psi_s = \sqrt{\psi^2{}_\alpha + \psi^2{}_\beta} \tag{3}$$

$$T_e = \frac{3n_p}{2L_s} |\psi_s| \psi_f \sin\delta \tag{4}$$

In the above formulas, u_α, i_α, ψ_α represent the stator voltage, stator current, and stator flux in α axis; u_β, i_β, ψ_β represent the stator voltage, stator current, stator flux in β axis; r_s represents the stator resistance; ψ_s represents the stator flux; ψ_f represents the rotor flux; T_e represents the electromagnetic torque; L_s represents the stator inductance; n_p represents the number of pole pairs; δ represents the angle of stator flux and rotor flux.

According to the relation of space vector with flux in traditional DTC (Figure 1), we conclude that only one vector can be chosen in one control cycle time, so that the ripple of torque can't be eliminated completely.

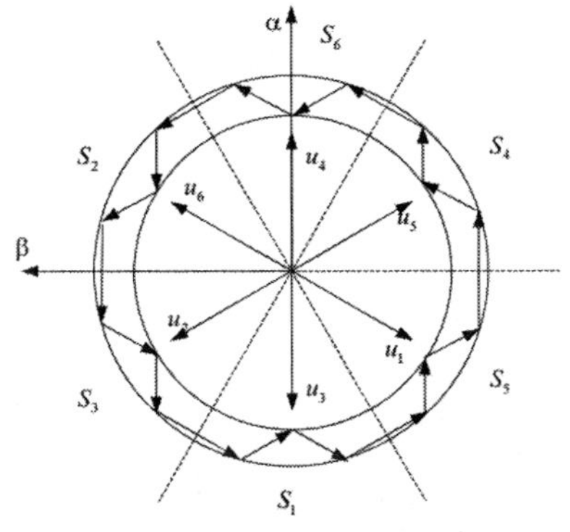

Figure 1. Relation of space vector with flux

3 DTC Using Space Vector Modulations (SVM-DTC)

3.1 Shortage of Existing SVM-DTC

Space vector generating of SVM-DTC proposed in paper [5] is showed in Figure 2. D-axis voltage component in d-q coordinate system can be obtained from PI output of the deviation compared actual flux with given flux; q-axis voltage component in d-q coordinate system can be obtained from PI output of the deviation compared actual torque with given torque. Compared with traditional DTC, SVM-DTC proposed in paper [5] overcoming the larger ripple of torque, while control structure becomes complicated due to introduction of rotating coordinate transformation, and hard to maintain the fast response. What's more, motor parameters and load change affecting PI performance lead to poor robustness.

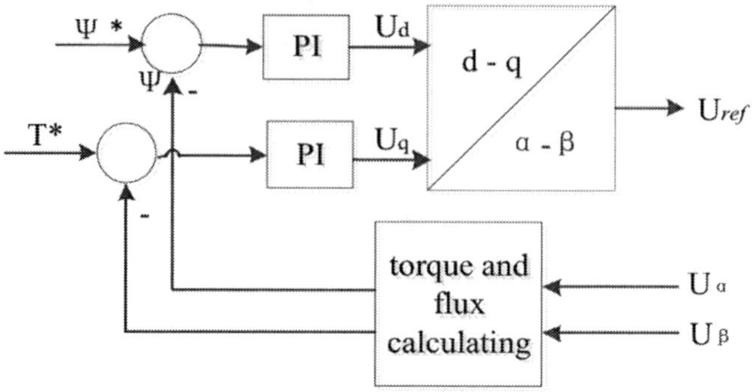

Figure 2. Generating of vector voltage in rotating coordinate

3.2 SVM-DTC Proposed in This Essay

In this paper, the effect of vector voltage on stator and rotor flux in one sampling period is shown in Figure 3. Relation between flux and voltage can be obtained only through mathematical analysis without complex rotating coordinate transformation [6].

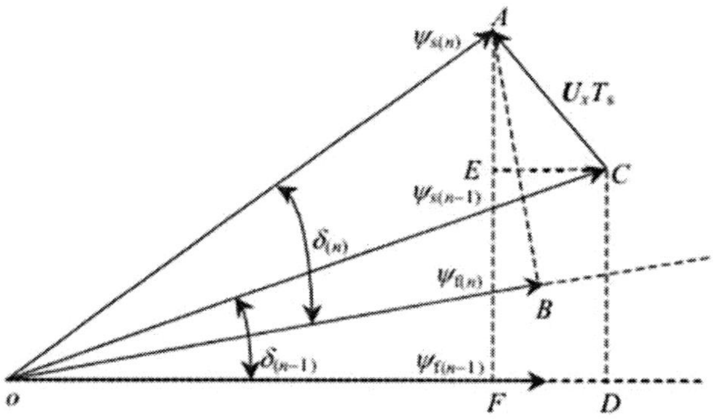

Figure 3. Flux before and after sampling period

By analyzing change of stator and rotor flux, accurate vector voltage can be calculated from functions expressed by the formula (5) to (9).

$$u_{\alpha ref} = \frac{\psi_S^{*}\cos(\theta + \Delta\theta) - \psi_S \cos\theta}{T_S} + r_S i_\alpha \tag{5}$$

$$u_{\beta ref} = \frac{\psi_S^{*}\sin(\theta + \Delta\theta) - \psi_S \sin\theta}{T_S} + r_S i_\beta \tag{6}$$

$$\Delta\theta = \Delta\theta_r + \delta(n) - \delta(n-1) \tag{7}$$

$$\Delta\theta_r = w_r \cdot T_S \tag{8}$$

$$\delta = \sin^{-1}\left(\frac{2T_e L_s}{3n_p \psi_s \psi_f}\right) \tag{9}$$

In the above formulas, $u_{\alpha ref}$, $u_{\beta ref}$ represent the accurate α -axis voltage component and β -axis voltage component in $\alpha-\beta$ coordinate system; ψ^*_s represents the given stator flux; θ represents the angle between stator flux and α -axis; $\Delta\theta$ represents the changed θ after one sampling period; $\delta(n)$, $\delta(n-1)$ represents the angle between stator and rotor flux before and after sampling period; w_r represents the rotor speed; T_s represents the sampling cycle time; the remaining parameters definition is consistent with the previous text.

3.3 Simulation Model

On the basis of the previous analysis, simulation of PMSM system controlled by SVM-DTC is set up and represented in Figure 4.

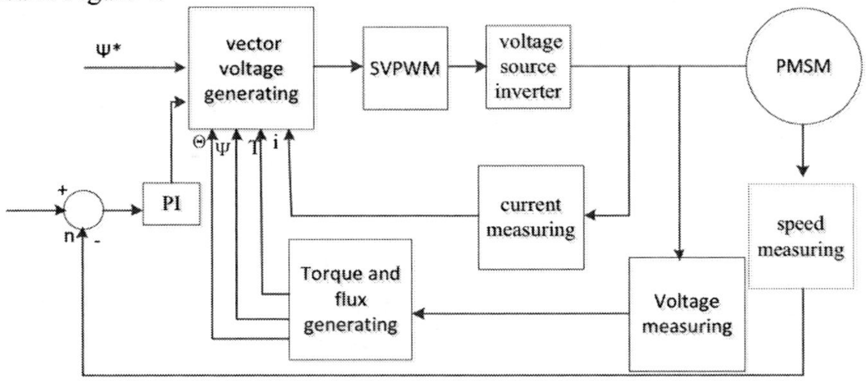

Figure 4. SVM-DTC simulation model

4 Measurements and Simulation

In order to verify the effectiveness of the control system proposed in this essay, simulation of traditional DTC and SVM-DTC in this paper have been done on the same PMSM model. Research of flux waveform and speed wave motion in steady state, torque ripple and load sudden change in dynamic is presented in this essay [7].

Figure 5(a) and 5(b) shows flux waveform controlled by traditional DTC and SVM-DTC in this essay when given stator flux is 0.8 *wb* in steady-state PMSM. The ripple in Figure 5(b) is obviously smaller than Figure 5(a).

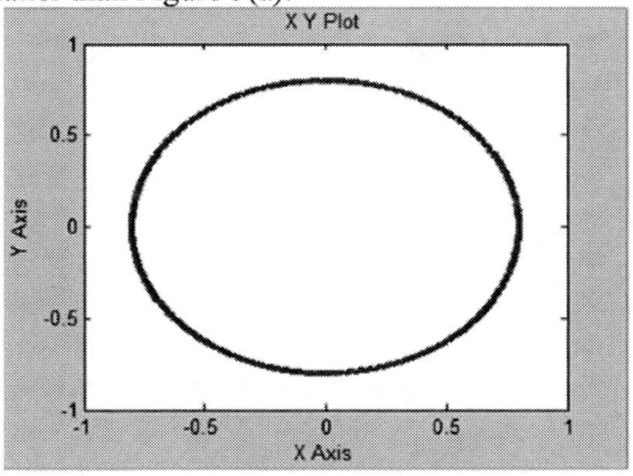

(a)Traditional DTC flux

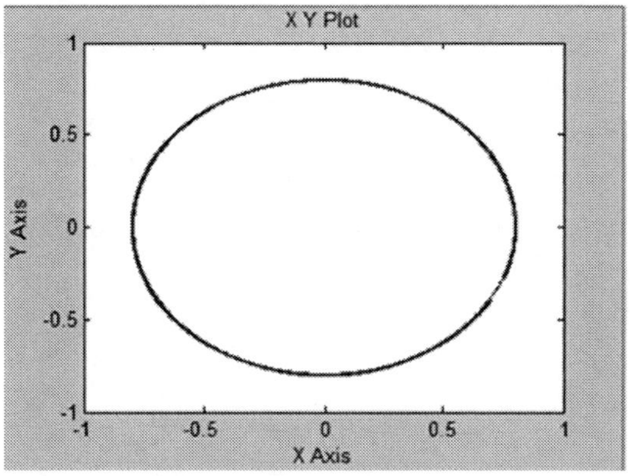

(b)SVM-DTC flux

Figure 5. Flux simulation results

Figure 6(a) and 6(b) shows speed waveform controlled by traditional DTC and SVM-DTC in this essay when given speed is 1000*n*/*rpm* in steady-state PMSM. The ripple in Figure 6(a) is about ± 4 *n*/*rpm*, while it in Figure 6(b) is about ± 2 *n*/*rpm*. Speed is more stable in SVM-DTC than traditional DTC.

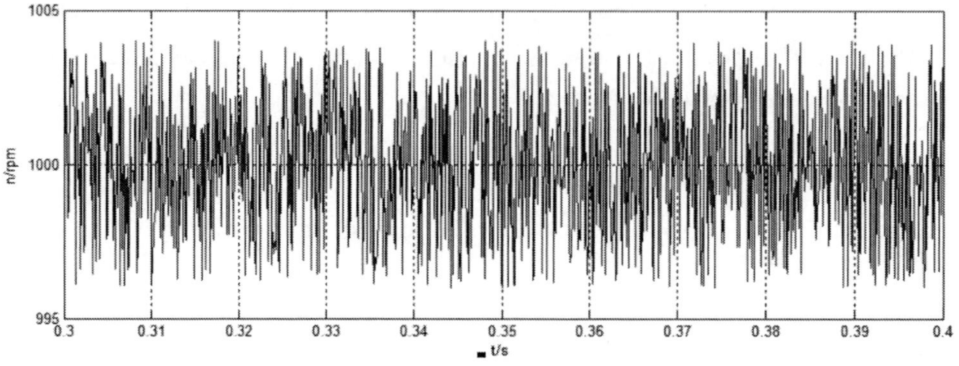

(a)Traditional DTC speed

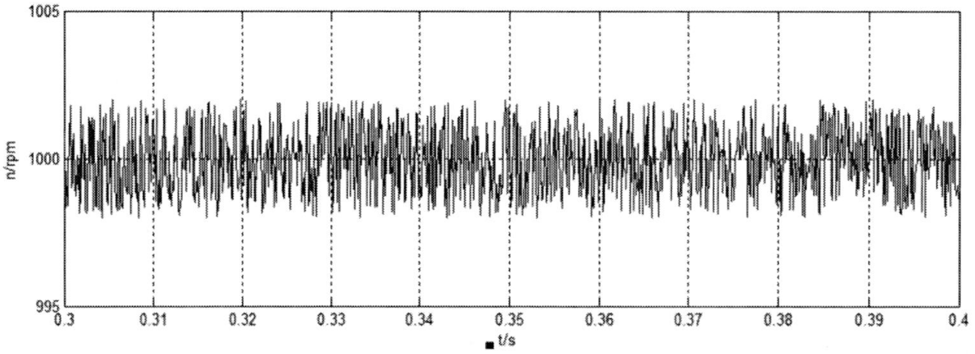

(b)SVM-DTC speed

Figure 6. Speed simulation results

For verifying the ripple of torque and dynamic performance, loads are changed in the same PMSM model controlled by different DTC. Load suddenly increases to15 *N.m* from initial 5 *N.m* at 0.33 seconds and then dips back to 5 *N.m* at 0.37 seconds. The blue waveform represents torque of traditional DTC and the red waveform represents torque of SVM-DTC as displayed in Figure 7.

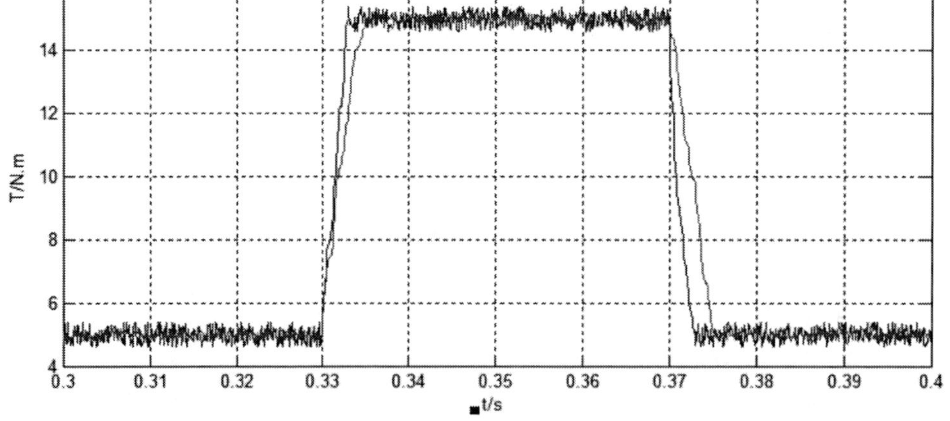

Figure 7. Waveforms of torque dynamic response

In order to analysis the capability to load sudden changes of different DTC, torque ripple coefficient (*TRF*) is delimited as function (10) [8].

$$TRF = \frac{T_{e\max} - T_{e\min}}{T_{eref}} \tag{10}$$

$T_{e\max}$ represents the maximum of torque ripple; $T_{e\min}$ represents the minimum of torque ripple; T_{eref} represents the given torque.

TRF of traditional DTC is about 16%, while 8% of SVM-DTC, as presented in Figure 7, latter is obviously better than the former. On the other hand, response time of load sudden changes of traditional DTC is 3 milliseconds, while 5ms of SVM-DTC, both can response fast.

Conclusions

From the analysis of flux speed and torque in steady-state, results show that SVM-DTC proposed in this essay has smaller change scale flux, more stable speed, and smaller rippled torque, when compare to traditional DTC. By the simulation of load sudden changes, response time almost equal, indicates that SVM-DTC presented in this essay can increase the speed of dynamic response while maintaining the good stationary performance as well as the robustness. SVM-DTC proposed in this essay is briefness, easy to carry out and has a certain practical significance in the field of high accuracy and high efficiency PMSM application controlled by DTC.

References

1. A.Jidin, N.R.N.Idris, A.H.M.Yatim, et al. An optimized switching strategy for quick dynamic torque control in DTC hysteresis based induction machines. J. *IEEE Transactions on Industrial Electronics*, E **58(8)**, 3391-3400 (2011).
2. Yanping Xu, Yanru Zhong. Sectors subdivision and duty ratio control combined DTC for PMSM. J. *Proceedings of the CSEE*, E **29(3)**, 102-108(2009).
3. D.Casadei, F.Profumo, G.Serra, et al. FOC and DTC: Two variable schemes for induction motors torque control. J. *IEEE Transactions on Power Electronics*, E **17(5)**, 779-787(2002).
4. L.Tang, L.Zhong, M.F.Rahman, et al, Novel direct torque controlled interior permanent magnet synchronous machine drive with low ripple in flux and torque and fixed switching frequency. J. *IEEE Transactions on Power Electronics*, E **19(2)**, 346-354(2004).
5. Huaqiang Zhang, Xinsheng Wang, et al. Study on direct torque control algorithm based on space vector modulation. J. *Electric machines and control*, E **06(16)**,13-18(2012).
6. N.R.N.Idris, A.H.M.Yatim. Direct torque control of induction machines with constant switching frequency and reduced torque ripple. J. *IEEE Transactions on Industrial Electronics*, E **51(4)**, 758-767(2004).
7. Dan Sun, Yikang He. Space vector modulated based constant switching frequency direct torque control for permanent magnet synchronous motor. J. *Proceedings of the CSEE*, E **25(12)**, 112-116(2005).
8. A.Jidin, N.R.N.Idris, A.H.M.Yatim, et al. Simple Dynamic Overmodulation Strategy for Fast Torque Control in DTC of Induction Machines With Constant-Switching-Frequency Controller. J. *IEEE Transactions on Industry Applications*, E **47(5)**, 2283-2291(2011).

MATEC Web of Conferences **55**, 01008 (2016) DOI: 10.1051/matecconf/20165501008
ACPEE 2016

Analysis of Non-minimum Phase in Buck-Boost Converter

Huang Jinfeng[1,2,a] and Liu Shulin[1]

[1]*Xi'an University of Science & Technology, 710054 Xi'an, China*
[2]*Shaanxi University of Technology, 723000 Hanzhong, China*

Abstract. In this paper, relations among the parameters of the non-minimum phase for the Buck-Boost converter are studied. It is demonstrated that when the Buck-Boost converter works in Buck mode, the negative regulation voltage is determined by the converter with the pre-designed inductor and load as the duty ratio $D<0.382$,as well as what is determined jointly by the converter with its pre-designed inductor, load and duty cycle as the duty ratio $D>0.382$, and the duty ratio has a limited effect on the negative regulation voltage of the converter ; when the Buck-Boost converter works in Boost mode , the negative regulation voltage is determined jointly by the converter with its pre-designed inductor, load and duty cycle, for which performs that the higher the output voltage is(i.e., the greater the duty ratio is) , the more severe undershoot of the voltage will be. The conclusion of this paper has a guidance meaning for decreasing the negative regulation voltage of the Buck-Boost converter. Finally the correctness of the analysis for the non-minimum phase is verified by simulations and experiments.

1 Introduction

Buck-Boost converter can easily implement a set of applications such as a step-up or step-down of the voltage and a negative voltage output, etc., which is widely used in photovoltaic power generation system, new energy, aviation industry, aerospace technology and other fields. When Buck-Boost converter takes the capacitance voltage as its output for feedback control, it is a non-minimum system which is characterized by a zero point in the right half plane for the small-signal mathematical model. A notable feature of this zero point is that in the case of a sudden increase (or decrease) of the duty cycle, in the early stage, except the case that the output voltage is in overshoot, there will be a change that the voltage takes a rise after a drop(or a drop after a rise), which is namely the non-minimum phase response (negative voltage regulation process).Negative regulation process will deteriorate the dynamic quality of the control system, and extend the time for a transient process of the system. Furthermore, in the negative regulation stage, as the controller receives an error signal and then forms the positive feedback, which seriously affects the transient and steady state performance of the system [1-2].Therefore, to find out the relationship between the converter parameters and the course that leads to a non-minimum phase response will have a significant meaning for

[a]e-mail:jfhuang2000@163.com

optimizing the performance of the converter design, and then further diminishes the influence of non-minimum phase response acting on the system, consequently improves the performance of the converter.

Buck-Boost converter has been extensively studied and has achieved many important results by far. By means of control strategies, deep studies on non-minimum phase response of the Buck-Boost converter are made in [3-6];and the physical meaning of the zero in the right half plane for the Buck-Boost converter is analyzed in[7].The impacts of the parameters including inductance, load and duty cycle on the non-minimum phase of the Buck-Boost converter have not been reported yet.In this paper, such effects of the inductance, load and duty cycle on the non-minimum phase of the Buck-Boost converter are analyzed in detail. The thesis points out that the smaller the inductance value is, the less the load is. For another aspect that the smaller the duty ratio is, the smaller the value of negative regulation load will be.

2 Small-signal mathematical model of a Buck-Boost converter

The circuit topology of a Buck-Boost converter is shown in Figure 1.

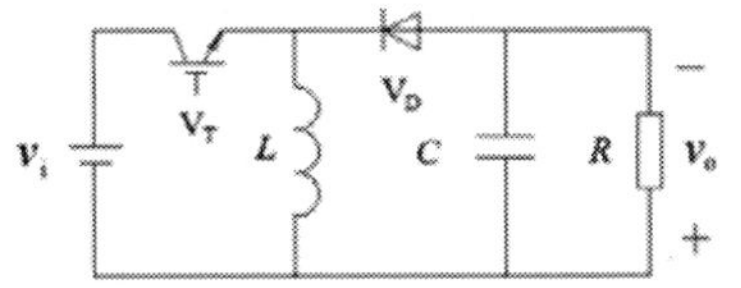

Figure 1. Buck-Boost converter

Buck-Boost converter works in the continuous conduction mode(CCM) for controlling the design-based small signal mathematical model of the output voltage. The description shows in equation (1) can be found in[4].

$$\frac{\hat{v}_o(s)}{\hat{d}(s)} = \frac{V_i(1 - sDL/D'^2R)}{s^2LC + sL/R + D'^2} \tag{1}$$

Through the analysis in equation (1), it can be seen that the mathematical model of a Buck-Boost converter contains a zero in the right half plane, the zero is related to the converter design parameters. The relationship between the non-minimum phase response and the converter design parameters is then analyzed below in detail.

3 Analysis of the relationship between non-minimum phase and the converter parameters

From the analysis of equation(1),it can be seen that in the right half plane of the expression in a Laplace domain, the position of zero changes with the parameters of converter inductance, load and duty ratio. The changes of these parameters can cause a non-minimum phase response. In order to discuss the effect of these parameters on the non-minimum phase, a specific set of parameters of a converter is described. And if one of these parameters is changed , the other parameters would keep constant. Then draw positions for different poles and zeros in the root locus. According to the root locus and the results of circuit simulation, the impact of different component parameters in the same circuit on non-minimum phase of the system is then studied.

3.1 Analysis of inductance response to non-minimum phase of the system

The parameters of the Buck-Boost converter are presented as follows: the input voltage V_i=24V, capacitance C=100μF, load R=8Ω, duty cycle D=0.4 and the switching frequency f_s=50kHz. The following work for the converter which works in CCM with different inductance （L=500μH、

2mH、 5mH) are discussed respectively. Simulation and analysis procedure is accomplished by using the power system simulation software PSIM6.0, simulation results are given to show the case that the non-minimum phase response presents for the output voltage when the system is operated stably with abrupt changes of duty cycle(at 30ms the duty ratio jumps from 0.4 to 0.5 , and at 50ms it jumps from 0.5 to 0.4 again).

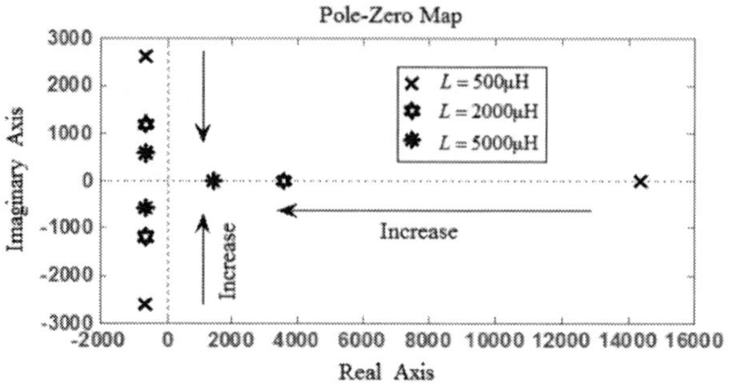

Figure 2. Root locus of zero and pole changes corresponding to different values of inductance

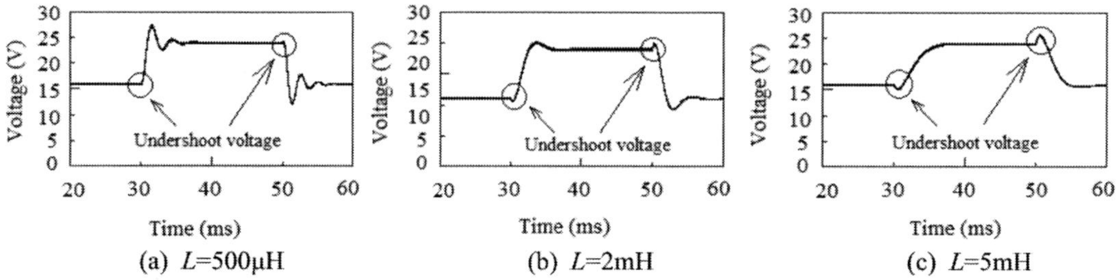

Figure 3. The response for different inductance values with the corresponding non-minimum phase of the system

As can be known from the analysis of Figure 2,with an increase of the inductance from 500μH to 5mH, the pole in the left half plane and the zero in the right half plane of S-domain are getting closer to the origin with the increase of inductance of the system. It can be seen from the simulation results in Figure 3,with the increased approximating of zero and pole to the origin, the non-minimum phase response of the system becomes more severe as the inductance increases, that is to say, when the duty cycle increases (or decreases), the output voltage does not immediately increase (or decrease), instead, it would first decrease (or increase) and then increases (or decreases) again, and with the increase of inductance, the negative regulation course would become more severe, with which is consistent with the features of non-minimum phase in the right half plane.

3.2 Analysis of load response to non-minimum phase of the system

The parameters of the Buck-Boost converter are presented as follows: the input voltage V_i=24V, capacitance C=100μF, inductance L=500μH,duty ratio D=0.4,and the switching frequency f_s=50kHz.The following work for the converter which works in CCM with different resistance (R=8Ω、 5Ω、 3Ω) are discussed respectively. Simulation results are given to show the case that the non-minimum phase response presents for the output voltage when the system is operated stably with abrupt changes of its duty cycle(at 30ms the duty ratio jumps from 0.4 to 0.5 , and at 50ms it jumps from 0.5 to 0.4 again).

As can be obtained from the analysis of Figure 4, with the load resistance decreasing from 8Ω to 3Ω, the zeros of the system in the right half plane are getting close to the origin, as well as the poles of the system in the left half plane are getting close to the origin along the imaginary axis and is getting away from the origin along the real axis in an S plane. As can be obtained from the analysis of the simulation results in Figure 5, with the approximating process from the zeros to the origin, the negative regulation course becomes more significant, i.e., the negative regulation course is getting more severe while the load is getting heavier, this feature is consistent with the mathematical model of non-minimum phase characteristics.

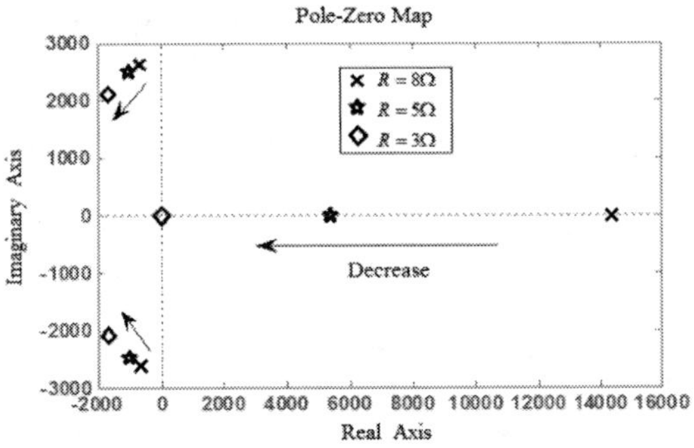

Figure 4. Root locus of zero and pole changes corresponding to different values of resistance

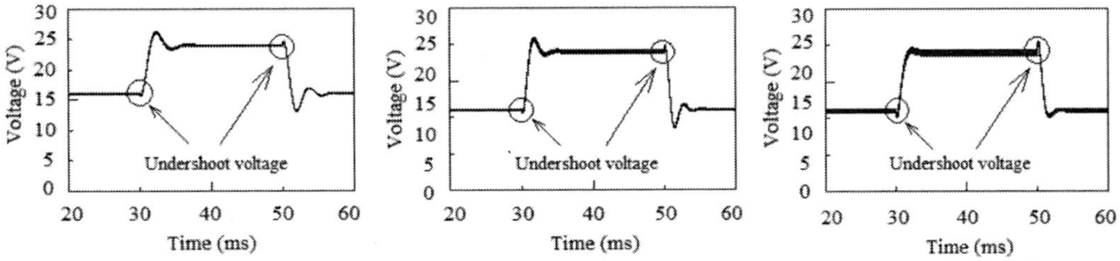

(a) R=8Ω (b) R=5Ω (c) R=3Ω

Figure 5. The response for different resistance values with the corresponding non-minimum phase of the system

3.3 Analysis of duty cycle response to non-minimum phase of the system

Transform equation (1) into equation (2),it can be obtained that

$$\frac{\hat{v}_o(s)}{\hat{d}(s)} = -\frac{V_i(1-\lambda \cdot Ls/R)}{LCs^2 + Ls/R + D'^2} \tag{2}$$

where $\frac{D}{D'^2} = \lambda$, analysis of λ we can get

$$\frac{D}{D'^2} = \frac{D}{(1-D)^2} = 1 \tag{3}$$

As the simplification of equation (3) , equation (4) is expressed as

$$D^2 - 3D + 1 = 0 \tag{4}$$

Solve equation(4), then two solutions of the equation are obtained, where D_1=0.382, D_2=2.618 (abandoned),respectively. Namely when D=0.382, it can be determined that λ=1, therefore, the analysis course of non-minimum phase is divided into two parts by D=0.382 and to be discussed respectively. When D<0.382,it can be can be derived that λ<1, the non -minimum phase response is mainly determined jointly by the inductance and the load; and when D>0.382,it can be can be derived that λ>1,the non-minimum phase response is mainly determined jointly by the duty ratio, inductance and the load. The influence of duty ratio D on λ and V_o is shown in Table 1.

Tab 1. The relationship between D and λ as well as that is between D and V_o

D	0.1	0.2	0.3	0.382	0.4	0.5	0.6	0.7	0.8	0. 9
λ	0.13	0.31	0.61	1.0	1.11	2.5	3.75	7.78	20	90
V_o/V_i	0.11	0.25	0.43	0.618	0.67	1.0	1.5	2.33	4	9

When duty ratio $D<0.5$, the converter works in Buck mode. The analysis can be drawn from table 1,and the analysis results can be divided into two cases, that is, when $D<0.382$,it can be can be derived that $\lambda<1$, the non-minimum phase response is mainly determined jointly by the duty ratio, inductance and the load, the duty cycle would weaken the negative regulation course which is caused by zeros in the right half plane at this point; when $0.382<D<0.5$,the non-minimum phase response is mainly determined jointly by the duty ratio, inductance and the load, the maximum value of the influence of the duty cycle on the negative regulation is 2.5 times than before. When the duty ratio $D>0.5$,the Buck-Boost converter works in Boost mode, and in this mode, the output voltage will increase as the duty ratio increases, however, the change of λ is far more than the ratio of output to the input voltage, that is, the negative regulation course which is caused by the duty cycle becomes more severe, the significance of the negative regulation course will inevitably lead to a slow down for the system output response. The following work is to simulate and analyze the impact of the duty cycle on the non-minimum phase of the system. The parameters of the Buck-Boost converter are presented as follows: the input voltage V_i=24V, capacitance$C$=100μF, inductance L=500μH, load R=8Ω, and the switching frequency f_s=50kHz.Then take a discussion in conditions of different duty ratios(as D=0.5,0.6,0.7, respectively), simulation results are given to show the case that the non-minimum phase response presents for the output voltage when the system is operated stably with abrupt changes of duty cycle(at 30ms the duty ratio abruptly increases 0.1 , and at 50ms it suddenly decreases 0.1 again).

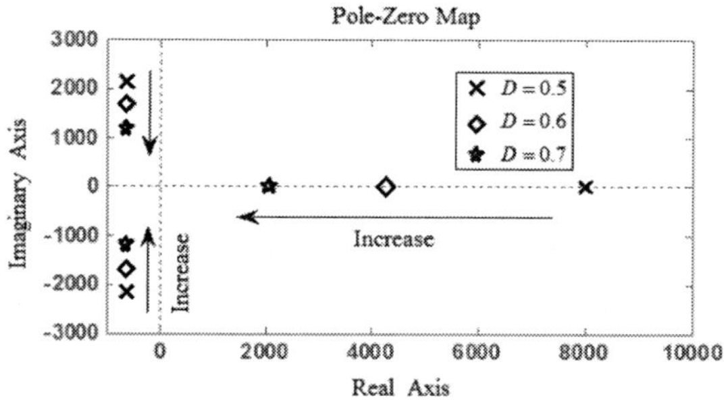

Figure 6. Root locus of zero and pole changes corresponding to different values of duty cycle

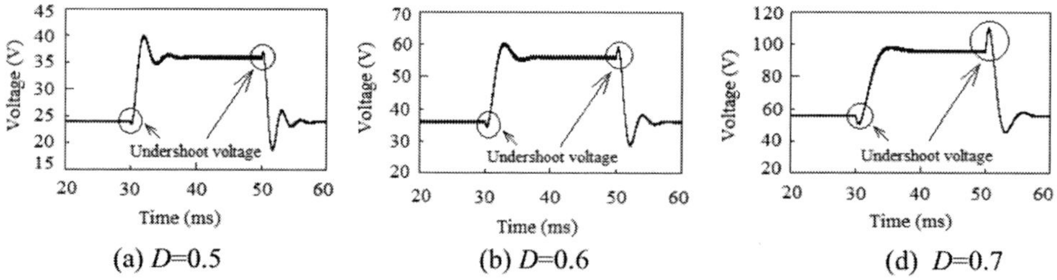

Figure 7. The response for different duty cycle values with the corresponding non-minimum phase of the system

As can be known from the analysis of Figure 6,with an increase of the duty ratio from 0.5 to 0.7, the pole in the left half plane and the zero in the right half plane of S-domain are getting closer to the origin with the increase of duty ratio of the system. It can be seen from the simulation results in Figure 7,with the increased approximating of zero and pole to the origin, the negative regulation course of the system becomes more severe as the duty ratio increases, that is to say, the duty ratio has a great impact on the negative regulation course of the system. And it can be seen that in the simulation results, although the value of the inductance is much less that is barely 500μH,the non-minimum phase response caused by the duty cycle is significant.

4 Experimental tests

In order to verify the correctness of the theoretical analysis for the non-minimum phase response of the Buck-Boost converter, an experimental platform is built for experimental verification. The parameters of the Buck-Boost converter are presented as follows: the input voltage V_i=24V, capacitance C=470μF, inductance L=1mH, the switching frequency f_s=50kHz, and the duty ratios D = 0.4 and 0.6 separately with a variation of the duty ratio $\Delta d = 0.1$ .The experimental results are shown in Figure 8.

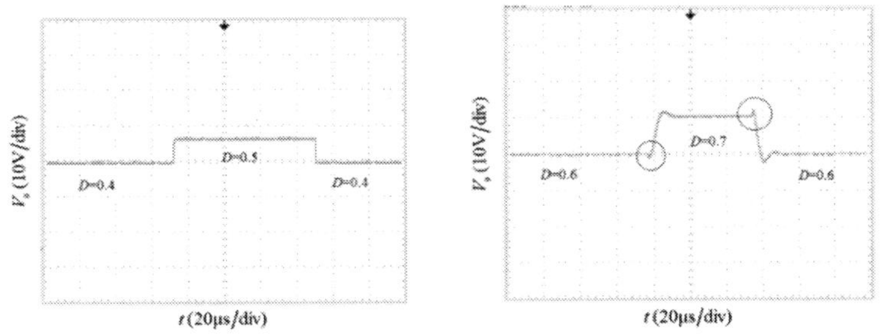

(a) D=0.4 (b) D=0.6

Figure 8. The response for abrupt changes of the duty cycle with the corresponding non-minimum phase

The analysis for the experimental results are presented as follows: in Figure8(a),the duty ratio jumps suddenly from 0.4 to 0.5,and then it abruptly jumps from 0.5 to 0.4 again while the inductance L=1mH. It can be seen from the experimental results that although the inductance value is higher than before, the negative regulation voltage presents a low state; in Figure8(b),the duty ratio jumps suddenly from 0.6 to 0.7,and then it abruptly jumps from 0.7 to 0.6 again while the inductance L=1mH. It can be seen from the experimental results that the negative regulation voltage reaches about 2V,which indicates that the non-minimum phase response of the system will increase as the duty ratio increases.

5 Conclusion

The non-minimum phase response of a Buck-Boost converter which works in CCM is related to its inductance, load and duty ratio. When it works in Buck mode, the duty ratio has a limited effect on the non-minimum phase response of the system, which is mainly determined by the values of its inductance and loads. When it works in Boost mode, the non-minimum phase response of the system is not only to consider the impact of the inductance and the load, but also to consider the impact of the duty ratio on the system. With the increase of the output voltage, the non-minimum phase response becomes even more severe, i.e., the higher the output voltage is, the more severe undershoot of the voltage will be. The correctness of the theoretical analysis above is then verified by simulation and experiments. In this paper, the theoretical approach of analyzing the system for the negative regulation voltage of the non-minimum phase of Buck-Boost converter can be applied to other converter topologies, which has a guidance meaning for suppressing the undershoot voltage in a non-minimum phase system and improving the transient performance for the system.

References

1. C. Rodriguez, J. L. Guzman,Journal of Process Control **24**, 368(2014)
2. H.H. Huang, C.L.Chen, Di.R.Wu,IEEE transactions on power electronics **27**,354 (2012)
3. D.W.Hart,*Introduction to power Electronics*(prentice Hall,inc,1997)
4. Z. Wu, H. Li, P. Zuo, and W. Z. Liu, Proceedings of the CSEE **22**, 110(2001)
5. G. K. Schoneman,D. M. Mitchell, IEEE Trans. Power Electron. 3,31(1988)
6. J. Y. Le, Y. X. Xie, Q. Z. Hong, Z. Zhang, and L. Chen, Proceedings of the CSEE **31**, 16(2011)
7. D. X. Shuai, Y. X. Xie, X. G. Wang, and F. X. Zhou, Proceedings of the CSEE **29**, 15(2000)

MATEC Web of Conferences **55**, 02001 (2016) DOI: 10.1051/matecconf/20165502001
ACPEE 2016

Dynamic Control Based Photovoltaic Illuminating System

Chengkai Zhang[1,a] , Ziyu Li[2] ,Wenhai Zhou[1] and Qili Zhang[1]

[1] *Hohai University, College of Energy and Electrical Engineering, Nanjing, Jiangsu, China.*
[2] *Hohai University, College of Environment, Nanjing, Jiangsu, China.*

Abstract. Smart LED illumination system can use the power from whether the photovoltaic cell or the power grid automatically based on the SOC (State Of Charge) of the photovoltaic cell. This paper proposes a feedback control of the photovoltaic cells and a dynamic control strategy for the Energy system. The dynamic control strategy is used to determine the switching state of the photovoltaic cell based on the illumination load in the past one hour and the battery capacity. These controls are manifested by experimental prototype that the control scheme is correct and effective.

1 Introduction

With the development of the economy, China's environmental problems have become increasingly serious. At present, the use of new energy is regarded as one of the major developing projects in many countries. And, it is studied that the lighting load will occupy 20% of total electric load approximately. The fast growth of electricity for lighting not only spends much funds, but will also produce environmental pollution. So, it's considerable to combine the photovoltaic cell and illuminating system.

Nowadays the combination of the photovoltaic cell [1] and illuminating system has been developed a lot. The traditional combined system determines a fixed switching SOC [2] of the photovoltaic cell. When the control system finds that the SOC of the cell is under a preset value, the system will cut out the photovoltaic cell and use the power from the power grid. When the control system finds that the SOC of the cell is above a preset value, the system will cut out the power grid and use the power from the photovoltaic cell[3].

However, this mode of the combination of the photovoltaic cell and illuminating system is rigescent. A fixed switching state of the photovoltaic cell can't reflect the conditions of the system. In the traditional combined system [4], although the cell can still power the lower-load illuminating system for several hours, It's common that the cell has to be cut out because its SOC is under the preset value.

This paper proposes a feedback control of the photovoltaic cells and a dynamic control strategy for the Energy system [5]. When the system is under the dynamic control strategy, it can be avoided that the cell has to be cut out for its SOC is under the preset value when the cell can still power the lower-load illuminating system for several hours.

[a] Corresponding author : zhangcklovelizyforever@gmail.com

 DOI: 10.1051/matecconf/20165502001

2 Traditional Mode of The Combination of The Photovoltaic Cell And Illuminating System

The voltage of the storage cell can reflect the SOC of the storage cell [6]. The traditional combined system sets the rated voltage of the illuminating system as the switching voltage of the system. When the system finds that the voltage of the discharging cell is under the switching voltage [7], it'll cut out the photovoltaic cell and input the power grid. Also the traditional system sets the working voltage of the system. The working voltage of the system [8] means that, after being input to power the system, the voltage of the no-load cell is at the state that the cell can achieve the switching voltage.

Although this method of the combination of the photovoltaic cell and illuminating system can make the energy system stable and practical, the system with this combination can't make the most of the solar energy and lack the control of the photovoltaic cells' charging . As for these questions, this paper proposes a feedback control of the photovoltaic cells and a dynamic control strategy for the energy system to ease them.

3 Dynamic Mode of the Combination Of The Photovoltaic Cell And Illuminating System

3.1 Feedback Control of the Photovoltaic Cells

To make the most of the solar energy, this dynamic control based photovoltaic illuminating system uses two cells in the energy system [9]. The charging current of the photovoltaic is monitored by AD module whose measurement data is transferred to the SCM. Once the SCM finds that the charging current is greater than the rated charging current of the cell, it'll output a signal to control the relay switch to put one more cell in the charging circuit. Also, when two cells have been put into the charging circuit, if the SCM finds that the charging circuit is half of the rated charging current of a single cell, then, it'll output a signal to control the relay switch cutting one cell out of the charging circuit. Through this feedback control of the photovoltaic cells, the dynamic control based photovoltaic illuminating system can make the most of the solar energy and increase battery lifespan. Furthermore, this feedback control is easy to be brought into life and costs little. **Figure 1** Shows the Feedback control based Energy system. **Figure 2** Shows the control flow chart of the feedback control [10].(I=the charging current of the photovoltaic; I_N=rated short circuit current)

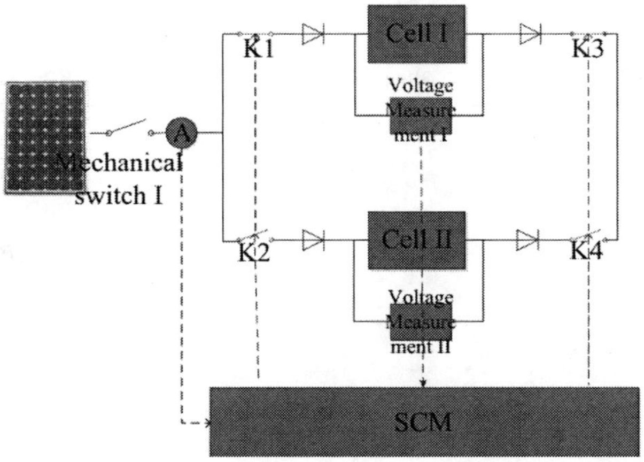

Figure 1. Feedback control based Energy system

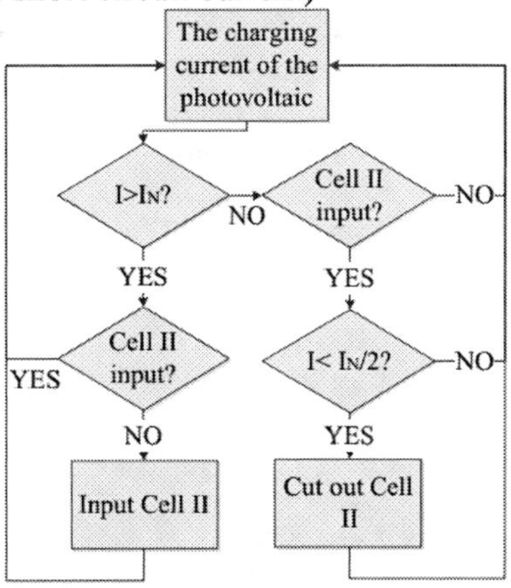

Figure 2. The control flow chart of the feedback control

3.2 Dynamic Control Strategy of The Combination

It is shown that the relationship between the SOC of the cell and the voltage of the cell is about linear by actual experiment. According to the charging and discharging curve of the cell and Battery parameters, the cell's discharging function can be calculated out approximately(This function is nearly a liner function).

When the cell is under the rated voltage of the illuminating system, we set the cell's discharging function as

$$U(t) = At + B \tag{1}$$

U(t)= the voltage of the cell

t= the discharging time

A,B=the cell's discharging parameters

Furthermore, the voltage of the cell is used to reflect the SOC of the cell in this paper.

$$\mathrm{SOC} = \frac{U' - U_{\mathrm{MIN}}}{U_{MAX} - U_{\mathrm{MIN}}} \tag{2}$$

U'=the real-time voltage of the cell

U_{MAX}=the maximum voltage of the cell during the discharging phase

U_{MIN}=the minimum voltage of the cell during the discharging phase

The system uses a dynamic control strategy to make the combined system more flexible. The SCM keeps receiving the metrical data from the Voltage Measurement. Meanwhile the SCM calculate out the average load of the illuminating system in the coming one hour —Wav.

$$A' = \frac{U' - U''}{1} \tag{3}$$

$$t^* = \frac{U' - A' - B}{A'} \tag{4}$$

$$W_{av} = \int_{t^*}^{1+t^*} \frac{(A't + B)^2}{R} tdt \tag{5}$$

A'=the modified gradient of the discharging curve based on the past one –hour change of the voltage

U'=the real-time voltage of the cell

U''=the voltage of the cell one hour ago

t*=the starting prediction moment of the one-hour discharging phase

R= the rated resistance of the illuminating system

The dynamic control strategy is based on the prediction model of the coming one-hour. An auxiliary parameter is set to predict the consumption of the illuminating system in the coming one hour.

$$G_{\mathrm{load}} = \frac{W_{\mathrm{av}}}{Q} \tag{6}$$

Q=the whole battery power

$$V_X = \left(\frac{U' - U_{\mathrm{MIN}}}{U_{MAX} - U_{\mathrm{MIN}}} - G_{load}\right)(U_{MAX} - U_{\mathrm{MIN}}) + U_{\mathrm{MIN}} \tag{7}$$

Once the dynamic control is taken by the combined system, the SCM will keep calculating the Vx of the system[11]. After the calculation of the Vx , the SCM will judge whether the Cell I is in the power system or not. If the SCM finds that the voltage of the working cell is smaller than the Vx or the voltage of the non-working cell is smaller than 11.4V(which is the working voltage of the lowest level), the SCM will then check whether the voltage of the non-working cell II is greater than 11.4V. If it does be greater than 11.4V, the system will output a signal to control the relay switch to bring the

 DOI: 10.1051/matecconf/20165502001

cell II into the power system and cut the cell I out of the power system. Else it won't bring Cell I & Cell II in until the voltage of cell I can be greater than 11.4V. **Figure 3** shows the control flow chart of the dynamic model.

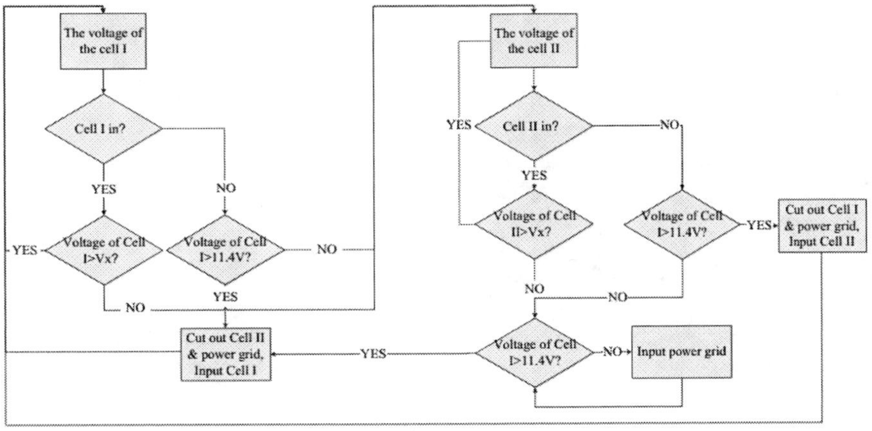

Figure 3. The control flow chart of the dynamic model

4 Experimental Check

To validate the effectiveness of the proposed algorithm, a simulation environment is established, as is shown in **Figure 4**. This model simulated the characteristics of the traditional combined systems including the inefficient use of the solar energy. Under the inefficient use of the solar energy working condition, the proposed dynamic control is used to make the good use of the solar energy.

4.1 Feedback Control of The Photovoltaic Cells

The simulation environment adopts 12v lead-acid batteries. The rated storage of this battery is 5A.h, the rated charging current of which is 0.5A. The feedback control is taken into use.

The SCM controls the amount of the charging cell automatically. Once the charging current is greater than 0.5A, two cells will be put into the charging system. Once the charging current is lower than 0.25A, only one cell will stay in the charging circuit. It's manifested by experimental prototype that the control scheme is correct and effective and the experimental data is shown in the **Table 1**, the Experimental Environment is shown in the **Figure 4**.

Table 1. The Experimental Data of the Feedback Control Tests.

Charging Current (A)	State of the Charging Circuit	Action	Switching Action Time (Second)	Charging Current after Actions(A)
0.60	One cell is in the charging circuit	One more cell is put into the charging circuit	10.16s	0.29
0.19	Two cells are in the charging circuit	One cell is cut out of the charging circuit	9.56s	0.41
0.41	One cell is in the charging circuit	No action		0.41

Figure 4. Experimental environment

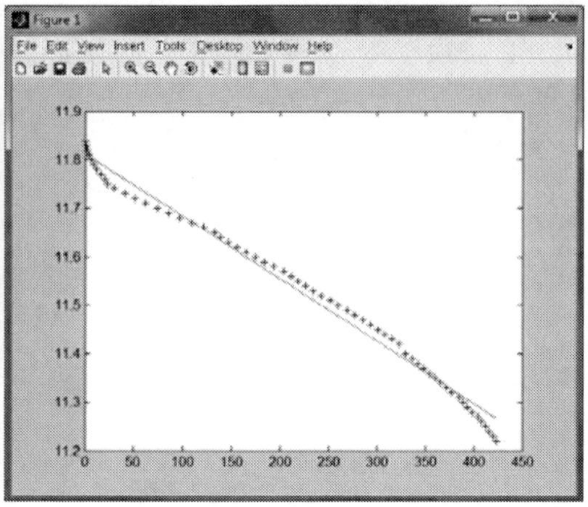

Figure 5. The discharging curve of the lead-acid battery under 1W load

4.2 Dynamic Control Strategy of the Combination

When the lead-acid battery is under 1W load, the discharging time and the voltage changes are recorded in the table. Figure 5 shows the discharging curve of the lead-acid battery under 1W load based on the experimental data.

It is shown in the Figure 5 that the relationship between the voltage of the lead-acid battery and discharging time is about linear by actual experiment .The relationship between those two is

$$y = 11.8123 - 0.0013t \quad (8)$$

y=the voltage of the lead-acid battery

t=the discharging time

So, through this experimental prototype, the A in formula (1) is -0.0013; The B in formula (1) is 11.8123; The SCM is set to achieve the goal of the dynamic control of the combined system. The experimental environment is shown in the Figure 4. First, the traditional control is adopted by the system. When its voltage comes to the rated voltage of the system, the cells are cut out of the power circuit, and the power grid begins to power the system. At that time, the voltage is the rated voltage of the state when the load is the maximum. Then, under the same condition, the dynamic control is adopted. When its voltage comes to the rated voltage of the system, the cells still power the system and the illuminating system can keep the same illuminating condition of the experimental environment for another 19.9min. The experimental data is shown in the Table 2.

Table 2. The Experimental Data of the Comparison Between the Traditional Control And the Dynamic Control.

Control strategy Adopted	Voltage of the system(V)	The cut-off voltage (V)	State of the illuminating system	Kind of the power energy	Time of the solar energy use(min)
Traditional control	11.56	11.5	One lamp on	Solar power	48.8
	11.5*	11.5	One lamp on	Power grid	0
Dynamic control	11.5	11.44	One lamp on	Solar energy	19.9

*11.5V is the switching voltage of the illuminating system

5 Conclusions

Based on the experimental data of the feedback control tests, the feedback control of the photovoltaic cells is capable to improve the charging efficiency of the photovoltaic cells and improve the battery lifespan. Meanwhile, based on the experimental data of the feedback control tests, the dynamic control strategy is capable to make the most of the solar energy. When the feedback control is not applied, the charging current can sometimes get greater than the rated charging current which will shorten the battery's lifespan. When the dynamic control is not applied, although the solar cell can still power the illuminating system, it is common that the cell has to be cut out of the power system because of the traditional fixed control.

After the feedback control of the photovoltaic cells is applied to the system, the charging circuit can be changed automatically based on the value of the charging current. Also, after the dynamic control strategy is applied to the system, the switching value is changed automatically based on the real-time voltage of the photovoltaic cell, the illumination load and the calculation of the dynamic model.

Unlike the traditional fixed control, the feedback control and dynamic control are more intelligent and efficient. In dynamic control based photovoltaic illuminating system, the control center controls the system based on the system load and the system power. In addition, by taking the new dynamic control, the system gets more stable and makes the collections and the use of the solar energy more efficient.

References

1. Sochor, P. Theoretical Comparison in Energy-Balancing Capability Between Star- and Delta-Configured Modular Multilevel Cascade Inverters for Utility-Scale Photovoltaic Systems, IEEE TRANSACTIONS ON POWER ELECTRONICS, 31: 1980-1982. (2016)
2. Meng, Jinhao. Lithium Polymer Battery State-of-Charge Estimation Based on Adaptive Unscented Kalman Filter and Support Vector Machine, IEEE TRANSACTIONS ON POWER ELECTRONICS,31:2226-2228. (2016)
3. Chengkai Zhang. 2016. Fuzzy Control based Self-Adaptive Illumination System, Advances in Power and Energy Engineering. (to be published)
4. Yao, H. The control strategy for improving the stability of a powertrain for a compound hybrid power excavator, PROCEEDINGS OF THE INSTITUTION OF MECHANICAL ENGINEERS PART D-JOURNAL OF AUTOMOBILE ENGINEERING, 229:1944. (2015)
5. Liu, XS. A Highly Efficient Ultralow Photovoltaic Power Harvesting System With MPPT for Internet of Things Smart Nodes, IEEE TRANSACTIONS ON VERY LARGE SCALE INTEGRATION (VLSI) SYSTEMS , 23:3015. (2015)
6. Abraham, KM. Prospects and Limits of Energy Storage in Batteries, JOURNAL OF PHYSICAL CHEMISTRY LETTERS,06:830. (2015)
7. Hammond, GP. Indicative energy technology assessment of advanced rechargeable batteries, APPLIED ENERGY, 138:559. (2015)
8. Wu, Y. Theoretical analysis of the effect of static charges in silicon-based dielectric thin films on micro-to nanoscale electrostatic actuation, JOURNAL OF MICROMECHANICS AND MICROENGINEERING,14:989. (2004)
9. Adriano, S. Experimental analysis of the auxiliaries consumption in the energy balance of a pre-series plug-in hybrid-electric vehicle, ATI 2013 - 68TH CONFERENCE OF THE ITALIAN THERMAL MACHINES ENGINEERING ASSOCIATION,45:779. (2014)
10. Jinkang Huang. The Design and Realization of Indoor Illumination Control Based on MCU, Heilongjiang Science,03:37. (2014)
11. Yuejun Zhang. Research and development of intelligent lighting system [D]. Zhejiang University(2006)

MATEC Web of Conferences **55**, 02002 (2016) DOI: 10.1051/matecconf/20165502002
ACPEE 2016

Decentralized Control Strategy for PV Based DC Microgrid with Hybrid Storage

Ishita Biswas[1, a], Prabodh Bajpai[1]

[1]*Department of Electrical Engineering, Indian Institute of Technology Kharagpur, 721302, West Bengal, India*

Abstract: Due to the rapid development of power electronics technology and modernization of DC based loads, DC microgrids (DCMG) are becoming most attractive solution in terms of their increased efficiency and high power quality. However they face certain challenges like intermittency of renewable energy sources, limitations in dynamic behaviour of storage systems and various stability issues during islanding. Thus it is mandatory to select the DCMG components properly with a robust control strategy for power converters in order to maintain power balance despite of variation in source and load power in order maintain the system power quality. In this work, a DC bus signalling based decentralized control strategy is proposed for a PV based DCMG, which eliminates the risk associated with DC bus voltage fluctuation. Fuel cell is as an auxiliary power source along with solar PV and storage devices, reduces the possibility of black out and helps to increase system reliability. The proposed control strategy has been validated through MATLAB/Simulink based simulation studies for variable weather condition and load demand on DCMG system.

1 Introduction

Now a day's DC based power systems are becoming one of the most emerging solutions. However lots of challenges are associated with designing of DCMG system like stability of the system, effective control of DC bus voltage variation, optimal load sharing mechanism, paralleling of various distributed generation (DG) sources. A wide range of investigation, for configuration and control of DCMG with various combinations of renewable sources has already been carried out for both AC and DC based loads [1-4]. But the DGs with AC output increases power conversion steps. Also the use of PV, wind both may cause problem in terms of higher system size, cost and unavailability of proper site. PV with battery as storage device is also becoming one of the most emerging structure, as PV has the natural DC coupling which makes it easier to connect to the common DC bus [5, 6]. In addition to PV and Wind, Fuel cell (FC) is also becoming popular carbon free source due to high efficiency [7, 8]. Along with the DCMG system configuration, control of DG sources is another important aspect which is an emerging area of research in recent years. Hierarchical control approaches associated with ACMG can be applied to DCMG system also, which is mainly divided into three categories as primary, secondary and tertiary control [9, 10]. Issues of poor voltage regulation and circulating current among DG sources [11] can be avoided in case of secondary control which is mainly based on

[a] Corresponding author : i.biswas4@gmail.com

energy management strategy. Secondary control is of two types, centralized and decentralized. Centralized controller mainly mitigates the voltage imbalance caused by primary control, but it needs a real time communication system for operation. Decentralized control can be modelled with or without communication system. Thus the communication problem can be eliminated with DC bus signalling technique [12].

In this paper, a PV based DCMG system with hybrid energy storage system (HESS), comprising of battery and SC both is considered for study. In addition, a FC is considered for the emergency condition to minimize load shedding. A DC bus signalling based decentralized control algorithm is formulated for this DCMG system where the DC bus voltage is used as the key parameter to decide the system operating modes and the control signals for local controllers are computed.

2 Structure of DCMG and operating modes

The proposed DCMG system considering PV-FC-Battery-SC is shown in Fig. 1(a). The system component sizing is based on optimal results of PSO-EO heuristic method [13]. The sizing details are not covered here due to space limitations. It consists of main source, solar PV (520kWp) connected to the DC bus with DC-DC buck converter-I. The controlled source of energy, utility grid, is connected to the DC bus by grid connected voltage source converter (G-VSC-I) (500kW) and a static transfer switch (STS), an intelligent bypass switch (Normally closed) to continuously monitor the mode of operation of DCMG [14]. SC bank (63F×5) and Battery (5kAh, 400V) are used as HESS. Two bidirectional DC-DC converters (converter II and III) connect HESS elements to DC bus. FC (450kW) is interfaced with DC bus by means of DC-DC boost converter-IV. A combination of one constant AC load-1 (150kW) and three variable DC load 2, 3, 4 (125kW×3) is considered. DCMG operation is broadly classified in two modes, depending on the grid connection condition with DCMG [15] as follows:

2.1 Mode-1/Grid connected (GC) mode

This mode corresponds to the DCMG operation in connection with utility grid. In this mode, Solar PV always delivers maximum power (P_{pv}) and storage devices (SC and Battery) can take charging/ standby power (P_{sc}, P_{bat}) to attain their maximum SoC level. Any surplus or deficit power within the DCMG is balanced by grid power (P_g) by power control mechanism of G-VSC-I as presented in equ.1, where P_{load} is the load power [16,17].

$$P_{pv}+P_g=(P_{load}+P_{sc}+P_{bat}) \quad (1)$$

2.2 Mode-2/Standalone (SA) mode

In this mode, PV operates at MPPT, SC power (P_{sc}) maintains the DC bus voltage, battery maintains a certain SC voltage level by supplying required charging/discharging power to the system according to equ.2. In this work, FC supplies the load and charges the HESS in case of lower battery SoC level. FC power (P_{fc}) maintains system power balance as expressed in equ.3, even if PV power and HESS are not available enough to support.

$$P_{pv}+P_{bat}=(P_{load}+P_{sc}) \text{ , when FC is OFF} \quad (2)$$

$$P_{pv}+P_{fc}=(P_{load}+P_{sc}+P_{bat}), \text{ when FC is ON} \quad (3)$$

3 Power management strategy for DCMG system

Two possible scenarios are considered as the cause of grid outage. One is a short period AC grid fault, which results in grid side AC voltage dip and reduction of power transfer limit of G-VSC-I. The second possibility can be collapse of grid or grid isolation due to maintenance purpose. In both situations, G-VSC-I becomes incapable of maintaining system power balance. Thus HESS must control the system and system switches to SA mode. The flowchart presented in Fig.1 (b) describes the changeover logic with power management strategy.

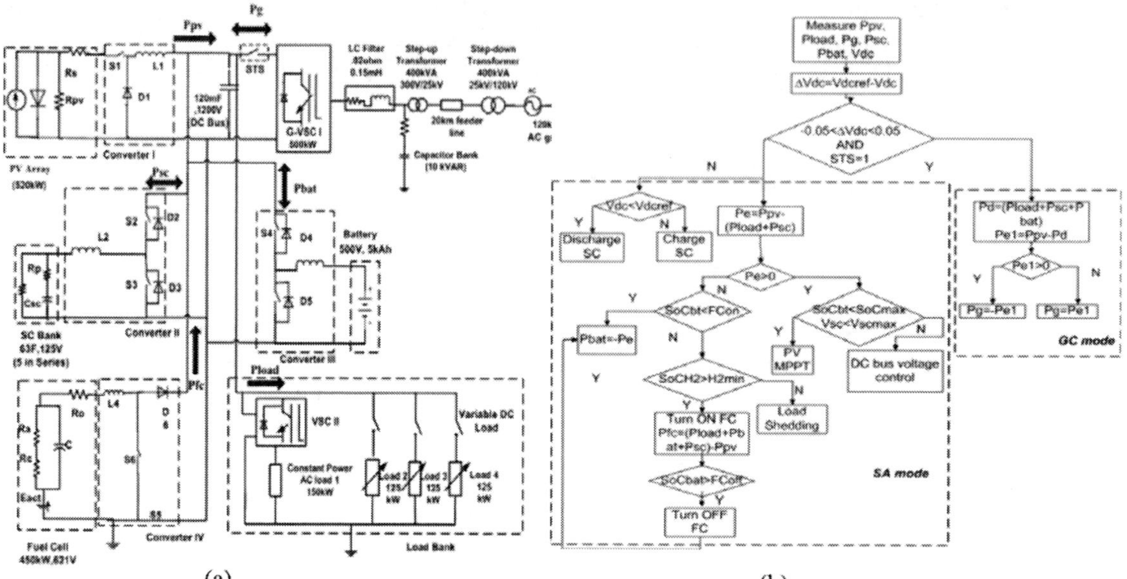

(a) (b)

Figure 1. (a)Structure of proposed DCMG with system rating, (b) Changeover logic for DCMG system

The algorithm initiates with the measurement of PV power (P_{pv}), load power (P_{load}), battery power (P_{bat}), SC power (P_{sc}), grid power (P_g), DC bus voltage (V_{dc}) and the state of STS (0/1). The DC bus voltage (V_{dc}) and STS state determine the mode of operation. In GC mode, when bus voltage lies within $\pm$ 5% and STS is closed, the excess/deficit PV power is balanced by grid, in order to supply the load and to maintain maximum storage level. But, if the bus voltage violets its regulation range and STS also gets opened due to the detection of AC fault/grid failure, the system operation shifts to SA mode. In SA mode, HESS plays key role to maintain power balance, where SC maintains DC bus voltage by charging/discharging during transient load condition and battery takes care of steady state load. In SA mode, the excess PV power (P_e) charges the battery and SC, with MPPT operation, however, if HESS is fully charged, MPPT operation is suspended. If P_e <0, battery first takes the charge of load, until it discharged up to a certain SoC level, where FC operation is scheduled. At this level FC turns ON and supplies the load, and charge HESS up to maximum SoC level. As soon as HESS gets charged, FC turns OFF and system operation continues with PV, SC and battery only.

4 Simulation results and discussions

The proposed control strategy has been validated for GC and SA modes of operation of hybrid storage DCMG system through simulation studies.

4.1 Normal GC operation

The proposed hybrid storage DCMG control strategy has been validated for GC mode with desired power sharing as shown in Fig.2. PV power is varying with solar irradiance, but load end DC bus voltage is maintained constant.

Fig.2 (a) illustrates power sharing among the PV, grid, HESS and load grid across the DC bus. In Part-1 and 2, grid exports power to DCMG to maintain power balance, since PV power is not enough to supply the load as well as fully charge the HESS. In part III, P_g approaches to zero and start importing power from DCMG as $P_{pv} > P_{load}$ and HESS is almost fully charged. The PV power decreases, but load demand has increased in Part 4 (1.57-2.5sec), so grid again starts exporting power to DCMG. Fig. 2 (b), (c), (d) and (e) are showing the variation in solar irradiance, SC voltage, DC bus voltage and battery SoC level respectively.

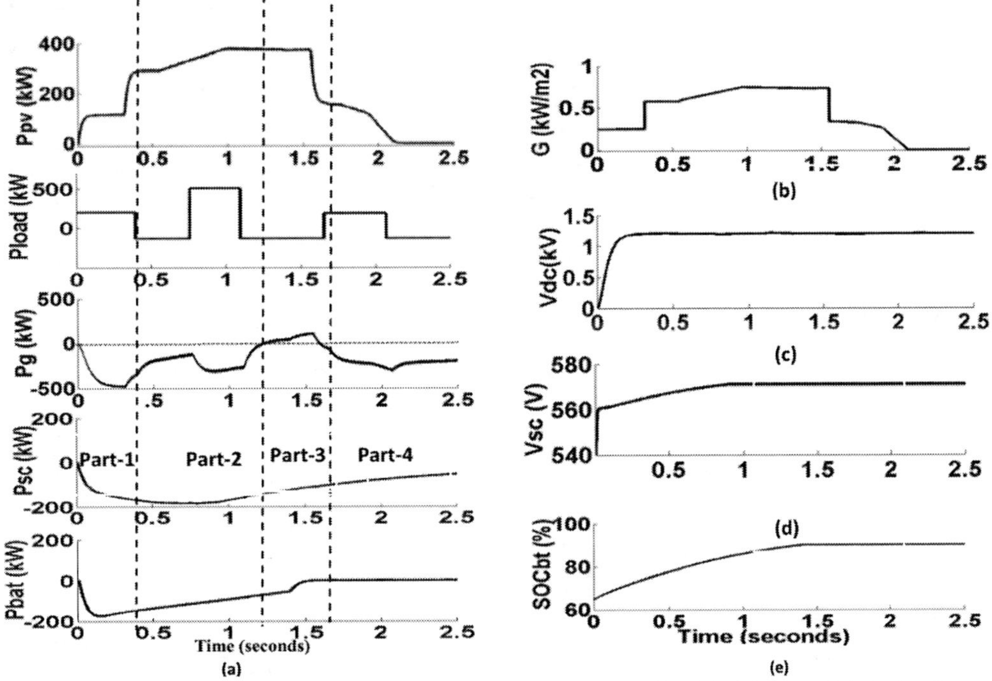

Figure 2 . (a) Power sharing strategy across DC bus among sources, HESS and load in GC mode(b) Solar irradiance profile (c) DC bus voltage (d) SC voltage (e) Battery SoC (%)

4.2 Impact of AC fault / grid collapse

Changeover operation during AC fault followed by grid collapse is described in this section through various plots in Fig.3. The power plots presented in Fig.3 (a) is divided into four parts. In Part-1, grid exports power to DCMG to maintain power balance in the system. Part-2 (0.5-0.7sec) simulates a short duration single line to ground fault at the end of transmission line on grid side, at 0.5 sec. Pg approaches to zero as STS detects the fault. So SC discharges to maintain V_{dc} constant (at 1.2kV) and battery charge SC for voltage control of SC. In Part-3 (0.7-1.5sec), as soon as the fault gets cleared at 0.7sec, the system moves back to GC mode, and due to low PV power and insufficient storage level, grid again starts exporting power to DCMG (Pg is negative). Grid gets disconnected again at 1.5sec in Part-4 due to grid collapse and Pg approaches to zero value. STS detects grid failure and HESS supply the load with partial support from PV. Fig.3(c) shows that V_{dc} is maintained constant during changeover from GC to SA mode and back.

4.3 Standalone operation

DCMG operation in SA mode based on proposed power management strategy is validated in this section considering FC as auxiliary power source. For this simulation, FC ON and FC OFF levels are chosen as 50% and 90% of battery SoC level respectively. Fig.4 (a) is illustrating power sharing among sources and load in three different parts. Now P_g plot is replaced with P_{fc} plot from previous section. In the Part- 1, P_{pv} is less that P_{load}, so battery and SC support to meet the load and maintain

DC bus voltage constant. In Part-2, P_{pv} is zero and SoCbt discharges to 50% as shown in Fig.4 (e). Thus FC gets ON at 0.85 sec, and starts supplying a constant power of 450kW. In Part-3, FC turns OFF at 11 sec when battery SoC reaches 90% SoC. So battery and SC meet the load and 1.86 sec. PV power is available so it starts sharing the load power with battery. But Ppv and Pbat are together enough for load demand so SC starts charging with P_{sc} in negative region.

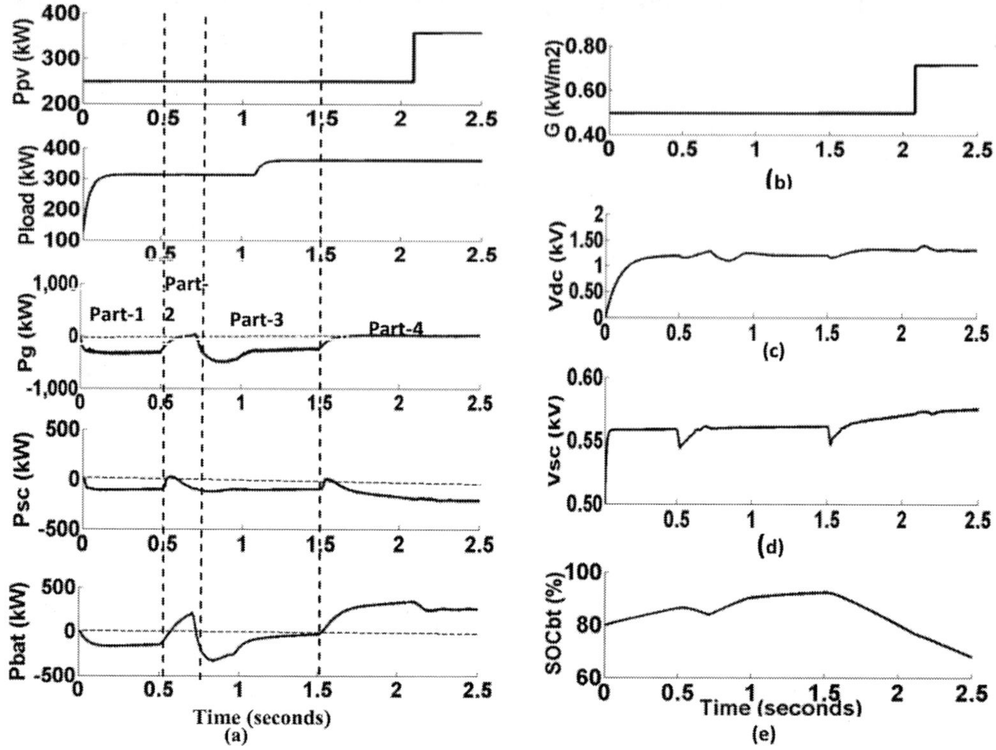

Figure 3. (a) Power sharing strategy across DC bus among sources, HESS and load in case of AC fault/grid collapse (b) Solar irradiance profile (c) DC bus voltage (d) SC voltage (e) Battery SoC (%)

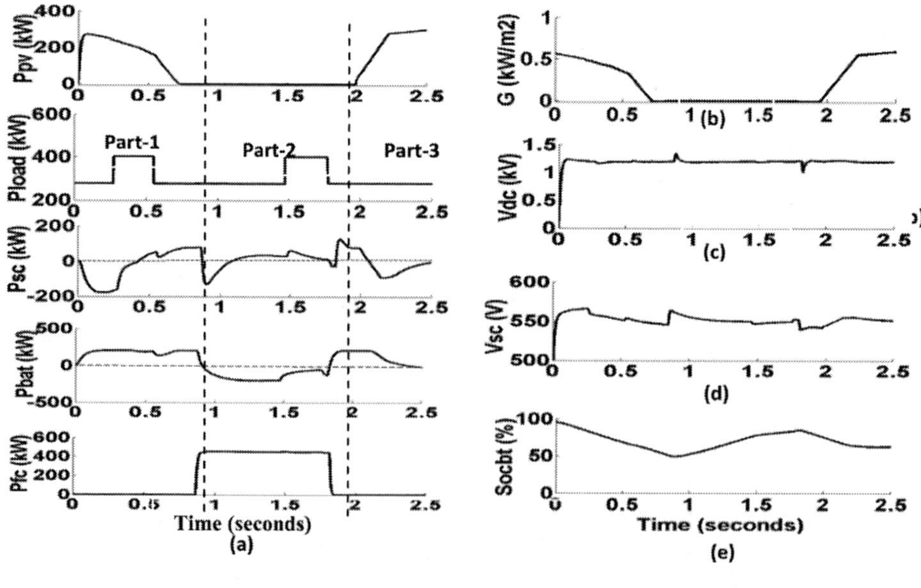

Figure 4. (a) Power sharing strategy across DC bus among sources, HESS and load in case of SA mode (b) Solar irradiance profile (c) DC bus voltage (d) SC voltage (e) Battery SoC (%)

5. Conclusion

In this paper, a DCMG system is designed to get high reliability in terms of uninterruptable power supply and the quality of power is maintained by always limiting load end voltage variation. A decentralized control strategy to maintain constant DC bus voltage of the DCMG is validated with various simulation studies under GC mode, SA mode and changeover between them. The desired performance is achieved with proposed HESS in DCMG and FC as auxiliary power source.

Acknowledgement

The authors are thankful to the authority of Electrical Engineering department, IIT, Kharagpur and VICET, for their encouragement and support for this work.

References

1. A. Khorsandi, M. Ashourloo, H. Mokhtari, A Decentralized Control Method for a Low-Voltage DC Microgrid, IEEE Trans. Energy conversion, **29**(4), 793-801(2014)
2. D. Chen, L. Xu, Autonomous DC voltage control of a DC microgrid with multiple slack terminals,IEEE Trans. Power System. **27**(4), 1897–1905 (2012)
3. L. Xu, D. Chen, Control and design of a DC microgrid with variable generation and energy storage, IEEE Trans. on Power Delivery, **26**(4), 2513-2512 (2011)
4. J. Guerrero, J. Vasquez, J. Matas, M. Castilla, L. de Vicuna, Control strategy for flexible microgrid based on parallel line-interactive ups systems, IEEE Trans. Ind. Electronics, **56**(3), 726–736 (2009)
5. X. Wang, M. Yue, E. Muljadi, Modelling and control system design for an integrated solar generation and energy storage system with a ride-through capability, ECCE, 3727-3734 (2012)
6. S.K. Kim, J.-H. Jeon, C.-H. Cho, J.-B. Ahn, Modelling and simulation of a grid connected PV generation system for electromagnetic transient analysis, Solar Energy, **83**, 664-678 (2009)
7. M. Mohammadi, M. Nafar, Fuzzy sliding-mode based control (FSMC) approach of hybrid micro-grid in power distribution systems, J. Elect Power and Energy System, **51**, 232–242 (2013)
8. A. Ghazanfari, M. Hamzeh, H. Mokhtari, H. Karimi,Active power management of multihybrid fuel cell/supercapacitor power conversion system in a medium voltage microgrid, IEEE Trans. on Smart Grid, **3**(4), 1903–1910 (2012)
9. J.M. Guerrero, J.C. Vasquez, J. Matas, L.G. de Vicu˜na, M. Castilla, Hierarchicalcontrol of droop-controlled AC and DC microgrids—a general approach toward standardization, IEEE Trans. Ind. Electron. **58** (1), 158–172 (2011)
10. S. Anand, B.G. Fernandes, M. Guerrero, Distributed control to ensure proportional load sharing and improve voltage regulation in low voltage DC microgrids, IEEE Trans. Power Electron. **28** (4), 1900–1913(2013)
11. C. N. Papadimitriou, E. I. Zountouridou, N. D. Hatziargyriou, Review of hierarchical control in DC microgrids, Electric Power Systems Research, 122 (0), 159–167 (2015)
12. J. Bryan, R. Duke, S. Round, Decentralized generator scheduling in a nanogrid using DC bus signalling, IEEE PES GEM. 977–982(2004)
13. Biswas, I, Dash. V, Bajpai, P, Sizing optimization of PV-FC-Battery system with hybrid PSO-EO algorithm," in India Conference (INDICON), 2012 Annual IEEE, 869-874, 2012.
14. X. Wang, M. Yue, E. Muljadi, Modelling and control system design for an integrated solar generation and energy storage system with a ride-through capability, ECCE, 3727-3734 (2012)
15. N. Eghtedarpour E. Farjah, Power control and management in a hybrid AC/DC microgrid, IEEE Trans. Smart Grid, **5**(3), 1494–1505 (2014)
16. E. Serban, H. Serban, A control strategy for a distributed microgrid application with voltage & current-controlled source converter, IEEE Trans. on Power Elect., 25(12), 2981–2992 (2010)
17. S. R. Nandurkar, M. Rajeev, Design and Simulation of three phase Inverter for grid connected Photovoltaic systems, IEEE APECE, 80–83 (2012)

MATEC Web of Conferences 55, 02003 (2016) DOI: 10.1051/matecconf/20165502003
ACPEE 2016

Performance Analysis of Photovoltaic Panels with Earth Water Heat Exchanger Cooling

Sanjeev Jakhar[1,a], M.S. Soni[1] and Nikhil Gakkhar[1]

[1]*Center for Renewable Energy and Environment Development (CREED), Department of Mechanical Engineering, Birla Institute of Technology and Science, Pilani, Rajasthan, India 333031*

Abstract. The operating temperature is an important factor affecting the performance and life span of the Photovoltaic (PV) panels. The rising temperature can be maintained within certain limit using proper cooling techniques. In the present research a novel system for cooling of PV panels named as Earth Water Heat Exchanger (EWHE) is proposed and modelled in transient analysis simulation tool (TRNSYS v17.0) for the conditions of Pilani, Rajasthan (India).The various parameters which include cell temperature, PV power output and cell efficiency are observed with respect to variation in mass flow rate of fluid. Simulation results of the system without cooling show that the maximum PV panel temperature reached up to 79.31 °C with electrical efficiency dropped to 9% during peak sunshine hour. On the other hand, when PV panels are coupled with EWHE system, the panel temperature drops to 46.29 °C with an efficiency improving to 11% for a mass flow rate of 0.022 kg/s. In the end the cooling potential of EWHE is found to be in direct correlation with mass flow rate. The proposed system is very useful for the arid regions of western India which are blessed with high solar insolation throughout the year.

1 Introduction

The solar energy is an abundant energy source which can be harnessed by PV system to produce electricity and is a commercially proven technology [1-2]. Out of total solar insolation, only 10-20% is converted into electrical energy while the remaining solar radiation is either converted into heat or is reflected back [3]. Thus results in an increment in the PV cells operating temperature. The increase in the operating temperature of the cell beyond certain limit not only decreases the efficiency of the cell but also decreases the life span due to the thermo-mechanical stress involved [4-5]. It is reported in the literature [6] that drop in efficiency by 0.45% may occur with every degree rise in the cell operating temperature above a certain limit. Therefore, the temperature control by cooling of PV becomes important and crucial factor for its better performance [7]. From the past few years a significant amount of work has been done on different PV cooling techniques by many researchers. One of them is Photovoltaic/ Thermal (PV/T) cooling technology in which thermal collectors are attached with PV panels to produce both thermal and electrical energy.

Hegazy [8] presented four types of PV/T air system and measured their thermal and electrical performance. He concluded that PV between single pass channels consumed the least fan power.

[a] Corresponding author :sanjeev.jakhar@pilani.bits-pilani.ac.in, sanjeevj450@gmail.com

Solanki et al. [9] developed a cost effective setup of PV/T air collector and tested for indoor conditions. They found the thermal and electrical efficiency as 42% and 8.4%, respectively. Cuce and Cuce [10] have done a numerical and experimental analysis of silicon PV cells with air cooling in order to improve their thermodynamic parameters. Dubey and Tiwari [11] presented an integrated combined system of a PV/T (glass to glass type) solar water heater of capacity 200 lts. Their experimental results showed that the instantaneous efficiency of system increased from 33% to 64% due to increase in glazing area. Bilbao and Sproul [12] conducted an experimental study on PV/T water cooling system with an improvement the 5% efficiency of the system. Ji et al.[13] presented a sensitivity analysis of a hybrid PV/T water heating system with natural circulation. Their results showed saving of around 65% of primary energy on the daily basis with a PV cell covering factor of 0.63.

Literature also cites various other cooling techniques which are used for different thermal systems other than PV. One of them is geothermal cooling which works on the principle that at a depth of about 3.5 m or more, the soil temperature remains fairly constant throughout the year and is approximately equal to the average annual ambient air temperature [14]. Some researchers have used earth air tunnel heat exchanger and EWHE for air conditioning using air and water as a cooling medium respectively [15-18]. In the EWHE hot water is sent through the pipes which are buried at a certain depth below the ground, where the heat is transferred from the hot water to the earth resulting in decrease in the outlet temperature[19-20]. This system of cooling water using buried pipes may be used for various cooling applications of which PV may be one.

In this current research, EWHE is coupled with PV panels as a cooling technique. The modelling and simulation of such coupled system for the conditions of Pilani, Rajasthan, has been done on TRNSYS (v17.0) using its metrological data files. The main aim of this research is to estimate the cooling performance of EWHE when couples with PV panels. For the same, various performance parameters of system are identified which include PV panel temperature, electrical efficiency etc.

2 Description of Proposed System

The schematic diagram of proposed coupled system is shown in Fig. 1. It consists of a PV/T, EWHE, valves and pump connected in series. In the PV/T system copper tubes are bonded to an absorber plate which is fixed underneath the PV panels as shown in Fig.1 (b). The whole PV/T is enclosed in a casing and insulated from the sides and back. The back side of PV panel consists of copper tubes which carries water to take away the excess heat from the panel thus resulting in increase in its own temperature. The outlet of PV/T is connected to EWHE through pipes and valves. It comprises of 30 m long horizontal High Density Polyethylene (HDPE) pipe, buried in flat land with dry soil at a 3.5 m depth. In the EWHE, heat is transferred from hot water (PV/T outlet) to the soil. As a result, the water temperature at the outlet of EWHE is much lower than the ambient during peak summer period. The outlet from the EWHE is then sent back to PV/T inlet through copper tubes for cooling.

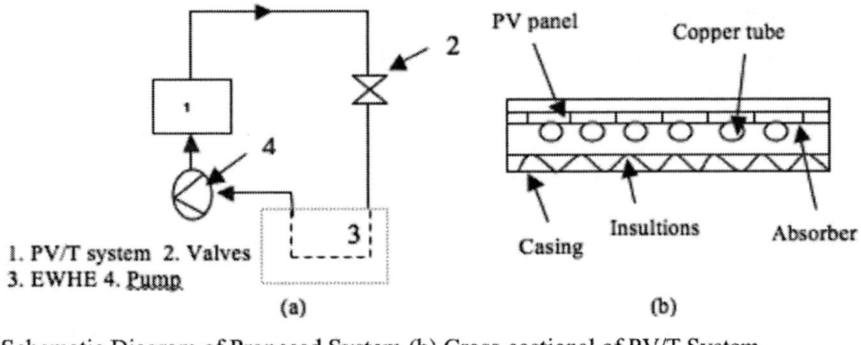

Figure 1. (a) Schematic Diagram of Proposed System (b) Cross-sectional of PV/T System.

3 Modelling and Simulation

The proposed system is modelled and simulated using TRNSYS software which is widely used for renewable energy system analysis [21]. Although its primary use is to model solar thermal applications, a number of attempts have been made to couple solar energy systems with other appliances. Here, a numerical simulation of the PV/T coupled with EWHE was carried out for a local weather condition using its inbuilt Meteonorm files. The model includes inbuilt components or Types, which can be interconnected through a flow chart, and takes parameters and time dependent inputs to give transient condition output. In the current system, the transient model for PV/T coupled with EWHE was used to estimate its performance over a particular time.

The following standard TRNSYS component models (Types) were used in the simulation:

- Type 560- PV/T collector
- Type 3- Variable speed pump
- Type 952- Earth water heat exchanger
- Type 15 - Weather data processor
- Type 65- Online plotter
- Type 28- Simulation summary
- Type 25- Printer
- Type 2 Controller

Type 560 is important model in electrical library of TRNSYS which supplied both electrical and thermal energy. Type 952 which is horizontal heat exchanger, interacts thermally with soil by considering conductive heat transfer and convective heat transfer within the pipes. The physical and thermal parameters used for the simulation, are shown in Table 1.

Table 1. Physical and Thermal Parameters Used in Simulation.

Parameters	Properties
PV/T collector length	1.20 m
PV/T collector width	1.59 m
Thermal conductivity of absorber plate	385 W/m K
Copper tube diameter (OD)	0.012 m
PV efficiency at reference condition	12 %
PV panel reference temperature	25 °C
HDPE pipe thermal conductivity	0.40 W/m K
Fluid density	1000 kg/m^3
Fluid thermal conductivity	0.55 W/m K

4 Results and Discussion

The performance of PV panel coupled with EWHE was analyzed by using TRNSYS (v17.0) for Pilani, Rajasthan (India). The simulation was conducted for 10 hours of system operation, which is average sunshine duration during peak (June 21) summer period. For the EWHE the performance of system was evaluated at the depth of 3.5 m where soil temperature remains almost constant throughout the year, while the length of pipe as considered 30 m.

Fig. 2 shows the PV panel temperature variation without cooling or with cooling for different mass flow rates. PV temperature varies between 35.43 °C to 79.31 °C without cooling for 10 hours of system operation. It can be seen from the Fig. 2. that the PV panel temperature drops with the EWHE cooling and it ranges between 30.33 -55.01°C, 29.99-53.82°C, 28.40-44.71 °C and 28.33-46.29 °C for 0.006 kg/s, 0.01 kg/s, 0.022 kg/s and 0.026 kg/s flow rates respectively. It is also observed that the temperature remains almost constant for 0.018 kg/s, 0.022 kg/s and 0.026 kg/s flow rates.

 DOI: 10.1051/matecconf/20165502003

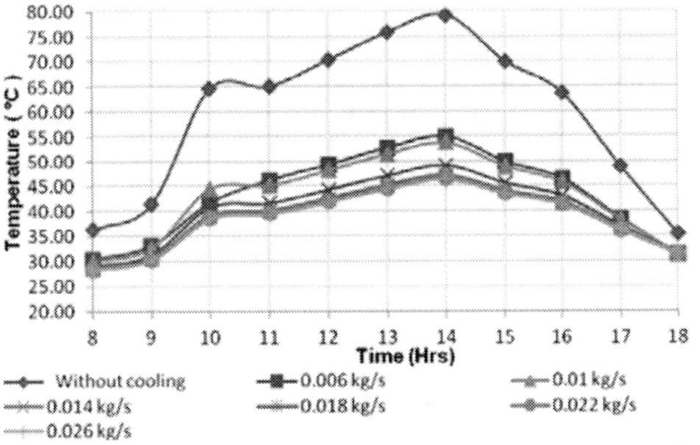

Figure 2. PV Panel Temperature Without Cooling and with Cooling for Different Mass Flow Rates.

Fig. 3 reveals the PV power output for different mass flow rates. The PV power over 10 hours remains within the range of 39.4 W to 140.3 W without cooling. While it varies from 38.3 W to 164.23 W with cooling for 0.026 kg/s flow rate. It can be observed that the PV power increased by 23.93 W with EWHE cooling for same mass flow rate as compared to without cooling.

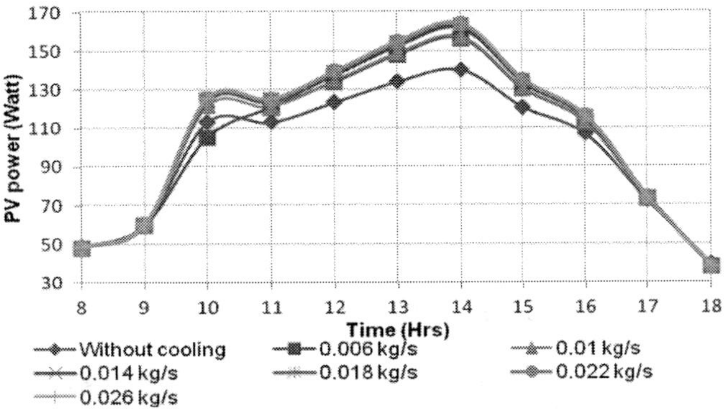

Figure 3. PV Power Output Without Cooling and with Cooling for Different Mass Flow Rates.

The simulated results of PV electrical efficiency for both configurations (without cooling and with cooling) at various flow rates have been shown in Fig. 4. In case of without cooling, the electrical efficiency was 9% for most of the time due to high cell temperature. Further with EWHE the efficiency is maintained at 11% for 0.022 kg/s and 0.026 kg/s flow rates. It can be concluded that the electrical efficiency of system is increased by 2% for 0.022 kg/s and 0.026 kg/s, while it increases by 1% for 0.006 kg/s and 0.01 kg/s flow rates as compared to without cooling.

The cooling potential (kWh) of the EWHE for different flow rates is shown in Fig. 5 and was calculated by using following equation.

$$Q_c = \dot{m}\, C_w\, \Delta T \tag{1}$$

where $\dot{m}$ is the mass flow rate of water in kg/s, C_w is the specific heat of water in kJ/kg K and ΔT is the temperature difference between inlet and outlet of EWHE. The cooling potential calculated after 6 hour operations are 2038.9 kWh for 0.006 kg/s and 3503.3 kWh for 0.022 kg/s. It reveals that with increase in mass flow rate the cooling potential of EWHE increases as expected. This proves that higher mass flow rates (up to 0.022 kg/s) are beneficial for EWHE when it is coupled to PV/T system.

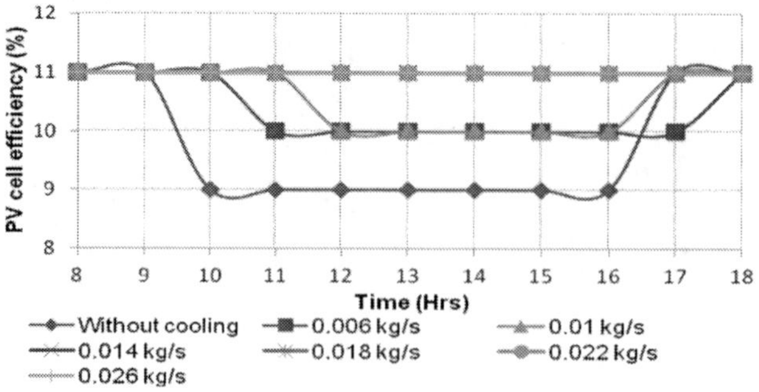

Figure 4. PV Electrical Efficiency Without Cooling and with Cooling for Different Mass Flow Rates.

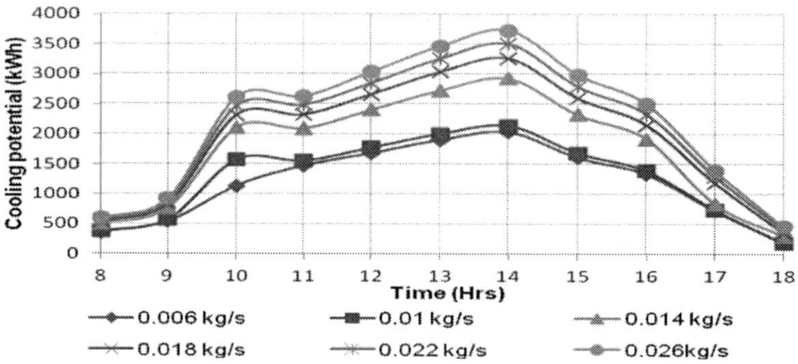

Figure 5. Cooling Potential of EWHE for Different Mass Flow Rates.

The EWHE system along with PV/T may be used for the arid regions of western Gujarat and Rajasthan (India) where solar irradiation is plenty and can be utilized by installing PV power plants. Such type of plants would provide electrical as well as thermal energy. This thermal energy may be dissipated with the help of an effective cooling system. Since, the ambient temperature of these regions goes up to 47 °C to 50 °C, leaving very small scope for utilization of thermal energy. While rejecting heat to the ambient will also be great challenge. Thus, EWHE will be better solution under such climatic conditions to achieve better performance of PV systems.

5 Conclusion

In the present paper an integrated system (EWHE and PV/T) has been simulated using transient simulation tool (TRNSYS v17.0) for local weather conditions of Pilani, Rajasthan. In the analysis the burial depth for EWHE as taken 3.5 m while the length of pipe as considered 30 m. Results show that the maximum panel temperature dropped significantly from 79.31 °C to 46.29 °C for mass flow rate of 0.022 kg/s when it is operated with EWHE. The electrical conversion efficiency of PV/T panel was increased by 2% when it is coupled with EWHE. The maximum power output from PV was found to be 140.3 W without cooling, while with cooling it is increased to 163.94 W for 0.026 kg/s flow rate. It is observed that for 6 hours of operation of system, the cooling capacity of EWHE increases from 2038.9 kWh to 3732.8 kWh as mass flow rate increased from 0.006 kg/s to 0.022 kg/s respectively. Simulation results of the study confirmed that the EWHE system with PV/T is quite effective for PV cooling during summer season as it increases the electrical efficiency up to 11%.

References

1. M.S Soni, N. Gakkhar, Renew Sustain Energy Rev., **40,** 326-334 (2014)
2. A. Ummadisingu, M.S Soni, Renew Sustain Energy Rev., **15**, 5169-5175 (2011)
3. M.Ozgoren, M.H. Aksoy, C. Bakir, S. EPJ Web Conferences, **45**, 01106 (2013)

4. A. Royne, C. J. Dey, D. R. Mills, Sol Energy Mater Sol Cells, **86**, 451-483 (2005)
5. S. Jakhar, M.S Soni, N. Gakkhar, Renew Sustain Energy Rev., **60**, 41-59 (2016)
6. D. Du, J. Darkwa, G. Kokogiannakis, Sol Energy, **97**, 238-254 (2013)
7. P. S. Raj, M.S Soni, S. Jakhar, *Proceedings of the International Congress on Renewable Energy, "Powering National Growth Through Solar", (ICORE)*, 282-289 (2014)
8. A. A. Hegazy, Energy Convers Manag, **41**, 861-881 (2000)
9. S.C. Solanki, S. Dubey, A. Tiwari, Appl Energy, **86**, 2421-248 (2009)
10. E. Cuce, P.M. Cuce, Int J Ambient Energy, 1-7 (2013)
11. S. Dubey, G.N. Tiwari, Sol Energy, **82**, 602-612. (2008)
12. J.I Bilbao, A.B. Sproul, *Proceeding of the 50th Annual conference, Austrailian Solar Energy Society*, 1-11(2012)
13. J. Ji, J.P. Lu. T.T. Chow, W. He, G. Pei, Appl Energy, **84**, 222-237 (2007)
14. *Handbook, A.S.H.R.A.E., Fundamentals. American Society of Heating, Refrigerating and Air Conditioning Engineers Inc.*, NE Atlanta, GA 30329 (2009)
15. M.S. Sodha, A.K. Sharma, S.P. Singh, N.K. Bansal, A Kumar, Build Environ, **20**, 115-122 (1985)
16. F.A. Ajmi, D.L. Loveday, V.I. Hanby, Build Environ, **41**, 235-244 (2006)
17. V. Bansal, R. Misra, G. D. Agrawal, J. Mathur, Energy Build, **42**, 645-648 (2010)
18. S. Jakhar, R. Misra, V. Bansal, M.S. Soni, Energy Build, **87**, 360-369 (2015)
19. S. Jakhar, R. Misra, M.S. Soni, N. Gakkhar, Parametric simulation and experimental analysis of earth air heat exchanger with solar air heating duct, Engineering Science and Technology, an International Journal (2016), doi: 10.1016/j.jestch.2016.01.009
20. C.T. Joen, L. Liu, M.D. Paepe, *Proceeding of International Refrigeration. Air Conditioning. Conf. Purdue Univ.*, 2194-2204 (2012).
21. TRNSYS 17: Transient System Simulation Program, University of Wisconsin, Madison (USA): Solar Energy Laboratory.

MATEC Web of Conferences **55**, 02004 (2016) DOI: 10.1051/matecconf/20165502004
ACPEE 2016

Wavelet Study of Meteorological Data Collected by Arduino-Weather Station: Impact on Solar Energy Collection Technology

Maria Teresa Caccamo[1,a], Emanuele Calabrò[1], Antonio Cannuli[1] and Salvatore Magazù[1]

[1]*Department of Mathematics and Computer Science, Physics and Earth Science, Messina University, Viale F. Stagno D'Alcontres, S. Agata, Messina, Italy*

Abstract. Meteorological data collected by an automated LSI Lastem weather station connected with an Arduino device for remote acquisition are reported and discussed. Weather station, located at 38° 15' 35.10'' N latitude and 15° 35' 58.86'' E longitude, registered data which were analysed by wavelet transform to obtain time-frequency characterization of the signals. Such an approach allowed to highlight the correlation existing among the registered meteorological data. The results show a positive correlation between the minimum temperature and the maximum temperature values whereas a negative correlation emerges between daily rainfall and minimum temperature values as well as for daily rainfall and maximum temperature values. These results suggest the possibility to estimate the global and diffuse solar radiation using more reliable climatologic parameters for optimizing solar energy collected by solar panels.

1 Introduction

The photovoltaic technology is considered well-suited technology, particularly for distributed power generation. The literature provides that solar power supplied by the modules depends on many extrinsic factors, such as insolation levels, temperature, load conditions and orientation of the panel. In particular, meteorological parameters used as predictors include the amount and distribution of clouds or other observations such as the fractional sunshine and water content [1,2]. Some variables can be changed to maximize the solar energy acquired by the panel. In fact, the design of a solar energy module involves complex tradeoffs due to the interaction of several factors such as the characteristics of the solar cells, power supply requirements and power management features of the embedded system, application behavior, inclination and orientation of the panel, geographical latitude. In particular, various optimum tilt angle values as a function of geographical latitude were provided in literature for fixed solar panels [3-9]. The significant disagreement among these values may be due to the different algorithm used for the determination of the optimum slope of a solar panel and the different empirical models that were used for the determination of diffuse solar radiance. Indeed, the diffuse solar component is the most difficult quantity to determine, because the distribution of the sky-diffuse radiance strictly depends on the local condition of the sky and the most considerable cause of

[a] Corresponding author : mcaccamo@unime.it

 DOI: 10.1051/matecconf/20165502004

error in the computation of the optimum tilt and orientation of a solar panel depends on the model of diffuse solar radiation which is used [10].

Meteorological data from all parts of the world are needed to know the horizontal global solar radiation, and for some regions, measured data may only be applied within a radius of about 50 km from weather stations. This circumstance lead to interpolate parameters between stations.
Some empirical models for estimating the diffuse solar radiation have to be used to obtain the solar radiation impinging on a tilted surface so that the relationship between the global solar irradiation on horizontal planes and that on tilted planes can be evaluated [11-15]. Nevertheless, the large uncertainties regarding climatological parameters appearing in these empirical models suggest us to search more reliable parameters.

Meteorological data study is generally used for the environmental monitoring, in particular for weather forecasts, to analyze air pollution and to give useful information for farmers, for alternative energies, such as photovoltaic and wind systems, to establish their correct position and for all actions and processes required to monitor and characterize the environment quality [16]. It follows several standards for climatic observations, for the selection of the sites, accuracy and specifications, for the weather station installation and calibration and data interpretation developed by the World Meteorological Organization (WMO) [17]. In general, automated weather stations, such as LSI-LASTEM station, are meteorological stations, which perform climatic observations and data transmission automatically, with a specifically microcontroller, such as Arduino. Measurements, read out by a central data acquisition unit, are more reliable and can be performed much more frequently compared to human weather observations.

In general, it is possible to classify meteorological data acquisition systems in real time and no real time weather stations. The first record data at a precise time and on particular request, established by the user, that are very important in emergency conditions and help to monitor all critical warning states (tide and river levels, storms). The second are offline stations that provide data on data storage devices.

In order to analyze the meteorological parameters collected by an Arduino-weather station system, such as maximum and minimum temperatures and the total daily rainfall, for the year 2014, a Discrete Wavelet Transform (DWT) has been employed. From this analysis it is possible to correlate the meteorological parameters, what it emerges is a positive correlation between the maximum temperature and the minimum temperature values, and a negative correlation between daily rainfall and maximum temperature values, as for daily rainfall and minimum temperature values [18].
This methodology can be used in the near future to obtain more reliable climatological parameters to estimate better the diffuse solar radiation.

2 Data Engineering Acquisition System: Arduino-Weather Station

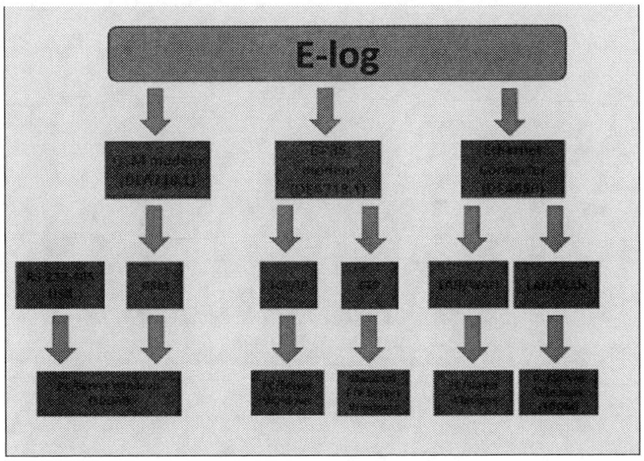

Figure 1. Environmental Data Logger (E-log) Communications Protocols.

The data were carried out by a LSI LASTEM weather station, integrated by a microcontroller for the remote connection, that allows to register typical meteorological parameters such as temperature, wind speed and direction relative humidity, barometric pressure, solar radiation and rainfall [19]. More precisely, it has been employed the MODBUS protocol to establish a communication between the weather station and the microcontroller to read and write the data [20,21]. In Fig. 1 a sketch of an environmental data logger (E-log) communications protocol is reported.

3 Wavelet Results Data Analysis

In the last years wavelet transforms find a lot of various applications in numerous field such as physics, meteorology, engineering, mathematics, and spectroscopic data analyses [22-27]. Differently from Fourier transform, the wavelet analysis allow to highlight which frequencies are present and where, or at what scale, they are. Furthermore, the wavelet transform allows to decompose the registered signal into its wavelet components by means of a set of translated and scaled of mother functions [28]. The meteorological data has been collected by means of a meteorological station situated at 38° 15' 35.10'' N latitude and 15° 35' 58.86'' E longitude, i.e. Messina University; and then, the registered meteorological quantities, i.e. maximum temperature, minimum temperature and total daily rainfall for the year 2014 has been analyzed through the Discrete Wavelet Transform (DWT). It's well known that Messina's temperature is moderated by the warm Mediterranean sea, with dry summers and wet winters, and its location in proximity to the equators. In Fig. 2, the average of minimum temperature and the average of maximum temperature for the year 2014 are shown. It clearly emerges that the warmest month is August, with a value 28,90° and the coolest months is March, with a value 11,22 °C.

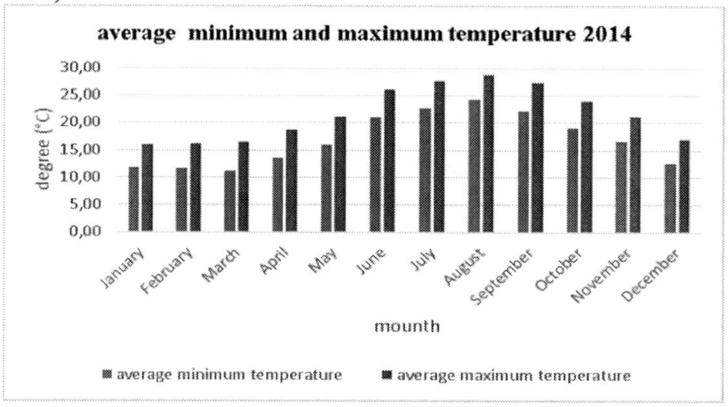

Figure 2. Average minimum temperature and average maxima temperature for the year 2014 (see the text for details).

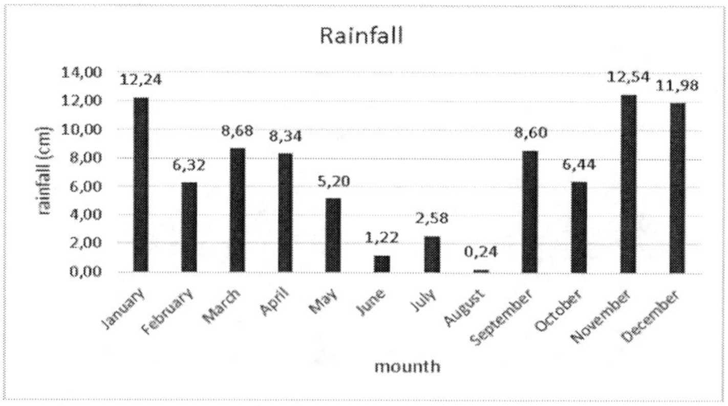

Figure 3. Total annual rainfall averages for the year 2014.

DOI: 10.1051/matecconf/20165502004

The annual precipitation in Messina are due to the alternating jet streams in the Mediterranean, Fig. 3 reports the total annual precipitation averages for Messina in the year 2014. As it can be seen, November is the wettest month (12,54 cm) and August is the driest month (0,24 cm).

In order to investigate meteorological parameters, the Symlet Wavelet was taken into account. More specifically the Symlet wavelet allows to decompose the analyzed signal into approximation coefficient, that represent the low frequency of the signal and into detail coefficients which correspond to the bands of higher frequencies, as shown in Fig. 4.

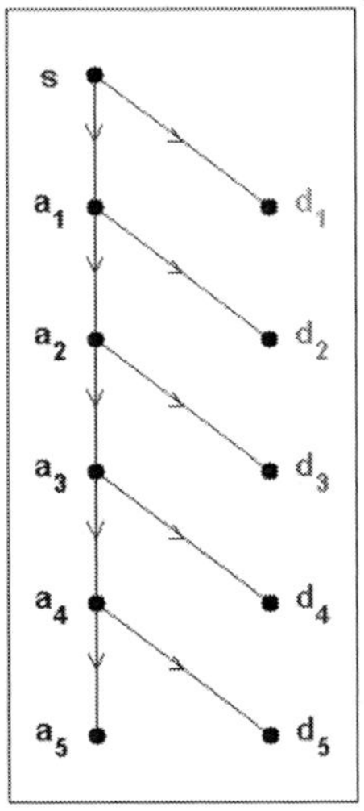

Figure 4. Decomposition of the signal into approximation (a1, a2, a3, a4, and a5) and detail (d1, d2, d3, d4, and d5) coefficients.

Using Matlab Wavelets Toolbox, it has been possible to make, by means of Symlet wavelet transform at level 5 , the decomposition of minimum temperature, maximum temperature and total daily rainfall [29,30]. Figg. 5-7 report the analysis of investigated signals: at the top the input investigated meteorological parameter, the second raw represents the approximation coefficient: a5 which is connected to the general parameter trend, and finally the details coefficients: d1, d2, d3, d4 and d5 take account of the hifher frequencies [31].

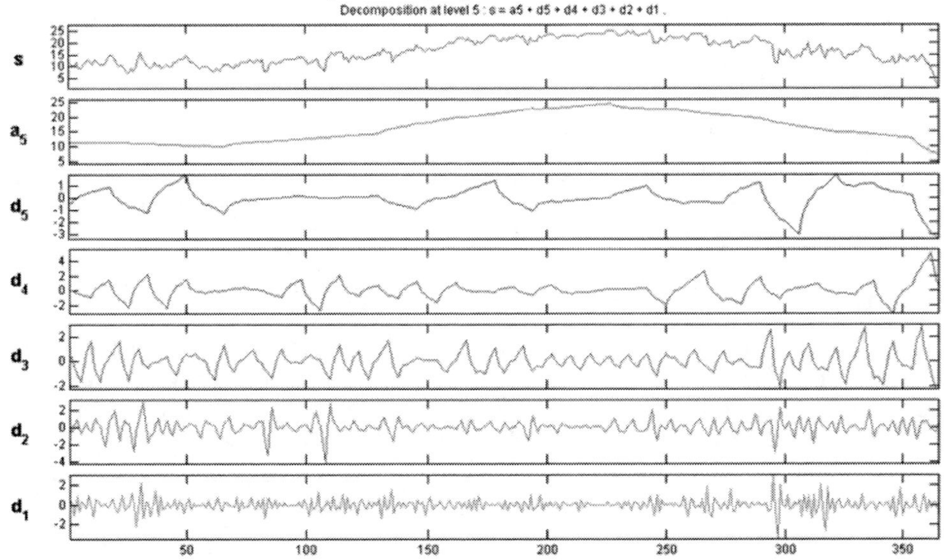

Figure 5. Decomposition of minimum temperature data by means of the mother wavelet Symlet at level 5.

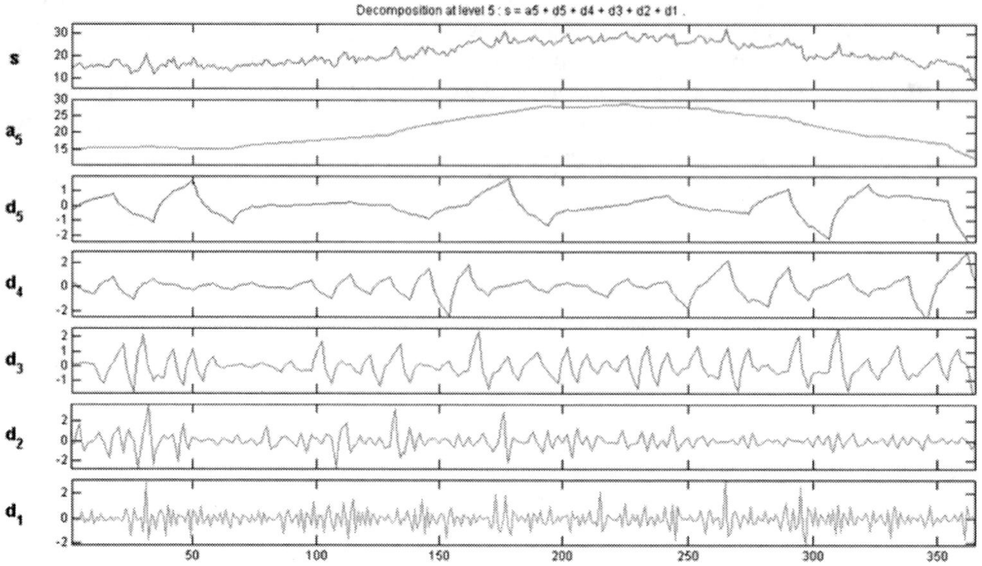

Figure 6. Decomposition of maximum temperature data by means of the mother wavelet Symlet at level 5.

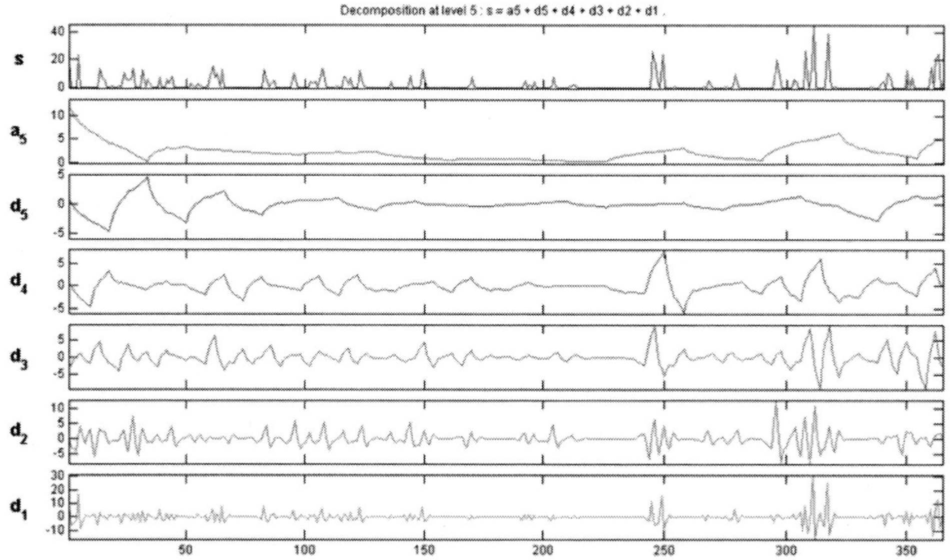

Figure 7. Decomposition of rainfall data by means of the mother wavelet Symlet at level 5.

From such an analysis, it emerges a positive correlation between maximum temperature and minimum temperature for the year 2014, due to a similar behavior for the two quantities: the approximation and the detail coefficients as a function of time are in most the cases each other in phase. On the other hand, since the approximation and the detail coefficients as a function of time are in most the cases each other out-of-phase, between daily rainfall and minimum temperature as well as between daily rainfall and maximum temperature negative correlations emerge [32,33].

Conclusions

In this paper DWT was applied to data registered by means of the LSI-LASTEM interconnected to a microcontroller for remote data aquisition. The analysis, performed through the mother wavelet Symlet shows a positive correlation, for the year 2014, between the minimum and maximum temperature parameters. On the contrary, a negative correlation between rainfall and maximum temperature as well as between rainfall and minimum temperature is registered. The employed methodology can be used to collect climatological parameters with the aim of studying more reliable empirical models to better estimate the diffuse solar radiation and hence to optimize solar radiation collection technology.

References

1. ASHRAE *handbook: HVAC applications* (Atlanta (GA): ASHRAE, 1999).
2. M. Iqbal, Sol. Energy **20**, 101 (1978)
3. J.I. Yellott, ASHRAE J. **15**, 31 (1973)
4. P. J. Lunde, *Solar Thermal Engineering: space heating and hot water systems* (J. Wiley and Sons, New York, 1980)
5. G. Lewis, Sol. Wind Technol 4, 407 (1987)
6. H. P. Garg, *Advances in Solar Energy* Technology (Reidel, Boston, 1987)
7. G. Qiu, S. B. Riffat, Int. J. Ambient Energy 24 13 (2003)
8. E. Calabrò, J. Ren. Sust. Energy **1**, 033104 (2009)
9. E. Calabrò, Smart Grid Ren. Energy **3**, 253 (2012)
10. E. Calabrò, J. Ren. Energy, 2013, Article ID 307547, 12 pages, (2013)
11. B.Y.H. Liu, R.C. Jordan, Sol. Energy **4**, 1 (1960)
12. R. Perez, P. Ineichen, R. Seals, Sol. Energy **44**, 271 (1990)
13. L. Robledo, A. Soler, Energy **23**, 193 (1998)
14. F.J. Olmo, J. Vida, I. Foyo, Y. Castro Diez, L. Alados-Arboledas, Energy **24**, 689 (1999)
15. L. Wenxian, G. Wenfeng, P. Shaoxuan, L. Enrong, Energy, **20**, 723 (1995)
16. G. Booch, C. Robert, M. J. Newkirk, *Case Study: Weather Station in Object Oriented Analysis and Design with Applications* (2d. ed. 1998)
17. R. Artz, G. Ball, K. Behrens, et al. "*Geneva: World Meteorological Organization" Measurements at Automatic Weather Stations* (WMO-No. 8, 2nd ed., 2010)
18. K. M Lau, H. Weng, Bull. Am. Meteo. Soc. **76**, 2391 (1995)
19. G. Hoogenboom, D.D Gresham, Auto weather station network , Proc. of the Georgia water resource Conference. Athens, **1**, 483 (1997)
20. K. Montgomery, K. Chiang, A New Paradigm for Integrated Environmental Monitoring, ACM International Conference Proceeding Series (2010).
21. Y. Zhengtong, Z. Wenfeng, The research of environmental pollution examination system based on the Cloud Computing, International Conference on Communication Software and Networks, 514 (2011)
22. S. Magazù, F. Migliardo, M.T. Caccamo, J.Phys. Chem. B, **116**, 9417 (2012)
23. S. Magazù, F. Migliardo, B.G. Vertessy, M.T. Caccamo, Chem. Phys, **424**, 56 (2013)
24. F. Migliardo, M.T. Caccamo, S. Magazù, J. Non Crys. Solids, **378**, 144 (2013)
25. F. Migliardo, M.T. Caccamo, S. Magazù, Food Biophys. **9**, 99 (2014)
26. S. Magazù, F. Migliardo, M.T. Caccamo, Adv. Mat. Science Eng. **2013**, 1(2013)
27. F. Migliardo, M.T. Caccamo, S. Magazù, Int. J. Adv. Comp. Science & Appl. **5**, 50 (2014)
28. S. G. Mallat, IEEE Trans. Pattern Analysis, **2**, 674 (1989)
29. R. Singh, R. E. Vasquez, R.Singh, Visual Inf. Proc. **6**, 151 (1997)
30. S. Kumari, R. Vijay, ACIJ, **3**, 137 (2012)
31. E.A. Robinson, S. Treitel, *Geophysical Signal Analysis* (Prentice-Hall, New Jersey, 1980)
32. P. Kumar, E. Foufoula-Georgiou, Rev. Geoph. **35**, 385 (1997)
33. R. Bhardwaj, Inter. J. Curr. Eng. Tech. **2**, 178 (2012)

MATEC Web of Conferences 55, 03001 (2016) DOI: 10.1051/matecconf/20165503001
ACPEE 2016

Lightning Protection Performance Assessment of Transmission Line Based on ATP model Automatic Generation

Luo Hanwu[1], Li Mengke[1], Xu Xinyao[1], Cui Shigang[1], Han Yin[1], Yan Kai[1], Wang Jing[1], Le Jian[2 a]

[1]State Grid East Inner Mongolia Electric Power Maintenance Company, Tongliao 028000, Neimenggu Province, China
[2]School of Electrical Engineering, Wuhan University, Wuhan 430072, China

Abstract. This paper presents a novel method to solve the initial lightning breakdown current by combing ATP and MATLAB simulation software effectively, with the aims to evaluate the lightning protection performance of transmission line. Firstly, the executable ATP simulation model is generated automatically according to the required information such as power source parameters, tower parameters, overhead line parameters, grounding resistance and lightning current parameters, etc. through an interface program coded by MATLAB. Then, the data are extracted from the generated LIS files which can be obtained by executing the ATP simulation model, the occurrence of transmission lie breakdown can be determined by the relative data in LIS file. The lightning current amplitude should be reduced when the breakdown occurs, and vice the verse. Thus the initial lightning breakdown current of a transmission line with given parameters can be determined accurately by continuously changing the lightning current amplitude, which is realized by a loop computing algorithm that is coded by MATLAB software. The method proposed in this paper can generate the ATP simulation program automatically, and facilitates the lightning protection performance assessment of transmission line.

1 Introduction

Transmission line plays a key role in the transmission of electrical power from power plant to end-users, its reliability has an immensely impact on the safe and reliable operation of the whole power system[1]. The transmission line corridor in China is characterized by the diversity and multiplicity in environment due to the vast land area and complex topography and geomorphology distribution[2,3]. Many transmission lines, especially those with rated voltage level is 110kV and above, are subjected to the test of complex terrain and climate[4]. Lightning accident is a kind of severe natural disaster, its frequent occurrence is more likely to cause the flashover accident of transmission line, thus seriously threatens the safe and reliable operation of power system[5,6]. Therefore, the lightning protection performance of transmission line and the performance evaluation is the foundation for the design of lightning protection measures[7]. In this sense, to evaluate the lightning protection level using computer and automation technologies to obtain the lightning withstand level at different places is the

[a] Corresponding author : lej01@ mails.tsinghua.edu.cn

key aspect for the comprehensive assessment of the lightning protection performance of transmission line[8,9].At present, the methods to evaluate the lighting performance of transmission line include regulation calculate method and simulation analysis method[10,11]. Regulation calculate method that is based on operation experience has been widely used in engineering due to fact that it is simple and easy to use, However it applies inductance model to handle the lightning current pathway on the tower and take no account on the wave process of the tower, thus lead to large error between the calculation and actual result[12]. The simulation analysis method need to establish simulation model at first, and then run the model on computer to obtain the lighting performance. The lighting back-strike model involves the choose of the method to calculate the lightning counter-attack withstand level, while the shielding failure model involves the choose of analysis model to calculate the shielding failure rate and the shielding failure trip rate[13,14]. It need to establish the simulation model for a transmission line manually, and then change the lightning current manually many times to obtain the initial breakdown current of the transmission line, which is a complicated and time consuming work.

In this paper, a method to solve the initial breakdown current of transmission line by integrating the ATP and MATLAB simulation software effectively is presented. It can improve the automatic level and accuracy in evaluating the lightning protection performance of transmission line.

The remainder of this paper is structured as follow: In Section II, the characteristics and simulation process of ATP and MATLAB is introduced briefly for reference purpose. The analysis on the structure of ATP/EMTDC simulation program and the principle of the automatic simulation program creation is given in Section III. The framework and process of the method that automatically solves the initial breakdown current of transmission line, and an example to illustrate the application and validate the effectiveness of the proposed method are present in Section IV and V, respectively. Finally in Section VI, the main findings of the paper are summarized.

2 Simulation Software Introduction

ATP(Alternative Transients Program) /EMTDC simulation software is a kind of electro-magnetic transient analysis program that has perfect compatibility to run in multi-types of platforms and has been widely applied in power system simulation.

ATPDraw, which is a preliminary data processing part of ATP, is used to generate the input date file for ATP/EMTDC program with correct format. Essentially, it is a kind of visualized programming interface through which many simulation models can be established intuitively and graphically. The established simulation mode in graphical form will be translated by compiler into ATP file(*.atp) that is in text form, and this ATP file is the actual simulation model that can be executed by ATP/EMTDC program.

MATLAB software, which is developed by Math Works corporation in the United States, is one of the most widely used mathematical simulation software for its powerful mathematical analysis functions. MATLAB can be used for the purpose of image fitting and rendering, numerical analysis and calculation, also as an interface to interact with many other software.

In this paper, the program used to automatically create ATP simulation model and solve the initial lightening breakdown voltage of transmission line is coded by MATLAB for its convenience and multiple interface in interacting with other program.

3 ATP-EMTP Model Auto-creation

3.1 ATP file parsing

The code in ATP-EMTP file typically is in the type of comment line and code line, and the role of the comment line is to explain the procedures and instructions, thus do not affect the function of simulation model.

The code line is distributed in different data cards, which are connected in sequence and forms the executable ATP simulation model. According to the order of sequence in ATP file, these data cards are: starting data card, switch data card, MODEL data card, TACS data card branch data card, power data card, output data card and blank data card. Fig.1 shows the content of a typical ATP file containing above data cards.

```
BEGIN NEW DATA CASE
C -------------------------------------------------------
C Generated by ATPDRAW  25-Jun-2015   --> the comments begin with C
----  --> the omitted content
$DUMMY, XYZ000
  1.E-9   1.E-4
     10      1      1      1      1      0      0      1      0
C 345678901234567890123456789012345678901234567890123456789012345678901234567890
/TACS
TACS HYBRID
11CurrI          5.                                              1.4e-5
98X1L014   =CurrI
----
C 345678901234567890123456789012345678901234567890123456789012345678901234567890
/MODELS
MODELS
INPUT
----
OUTPUT
----
ENDMODELS
/BRANCH
C < n1 >< n2 ><ref1><ref2>< R  >< L  >< C  >
  XXRS11                  350.                                        0
----
/SWITCH
C < n 1>< n 2>< Tclose ><Top/Tde ><   Ie   ><Vf/CLOP >< type  >
  X1L030X1L012                                   MEASURING            0
----
/SOURCE
C < n 1><>< Ampl.  >< Freq.  ><Phase/T0><   A1   ><   T1   >< TSTART >< TSTOP  >
14X0051A 0179629.248      50.      300.                          -1.      100.
----
/OUTPUT
BLANK TACS
BLANK MODELS
BLANK BRANCH
BLANK SWITCH
BLANK SOURCE
BLANK OUTPUT
BLANK PLOT
BEGIN NEW DATA CASE
BLANK
```

Figure 1. Content of a typical ATP file

Starting data card. This card contains the initial statement and some simulation parameters, including simulation step size, simulation time, plot step size and so on.

TACS(Transient analysis of Control Systems) data card. This cad is used in control system transient simulation.

MODEL data card. The card is a custom component that can be implemented with different functions.

Branch data card. The role of this card is to number the elements in simulation model in sequence to determine the connection relationship between them. The code to call for Lib file is at the end of this card. The Lib file is the library file of the overhead line model. The LIB file of overhead line needs to be created before the creation of ATP file.

Switch data card. The function of this card is to set the open/close status of individual switch in simulation model.

Power data card. The function of the card is to set the parameters of the power source in simulation model.

Output data card. The function of the card is to output the values of the variables measured by voltage or current meter.

Blank data card. This card is used to represent the end of the simulation model. Any complete ATP file should be ended with this card, the ATP program will terminate when it finds an ATP model without blank data card.

3.2 Auto-creation implementation

By the above analysis on the structure of ATP file, the auto-creation of ATP file can be achieved by the following steps as shown in Fig.2: 1) Reading required data for the establishing of simulation model from data files, and processing these data to create *.mat file; 2) Reading date related to overhead line to create *.lib file; 3) Generating each data card used in ATP file; 4) Connecting the data cards in right sequence to create ATP file.

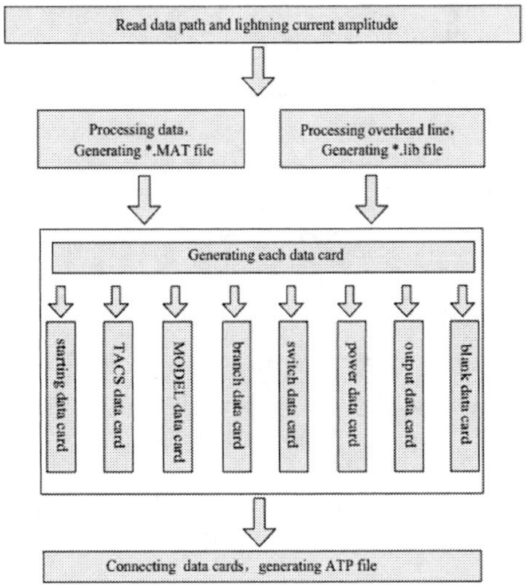

Figure 2. ATP file auto-generation flow

ATP-EMTP program has very strict requirement on the format of ATP file, for example, different types of component may has different character length, and the requirement on the arrangement of characters is very strict. The input data should be translated in to string with certain length. When the length of the input data is too long, it should be converted into the form of scientific notation and saved in the form of a string. While when the data length is not long enough, blanks should be added at the beginning of the data and then saved as a string .

Overhead line model is handled specially in ATP file. It should be proceed by LCC module provided by ATP program to create LIB file that can be called by ATP file. Fig.3 shows the processing of overhead line model.

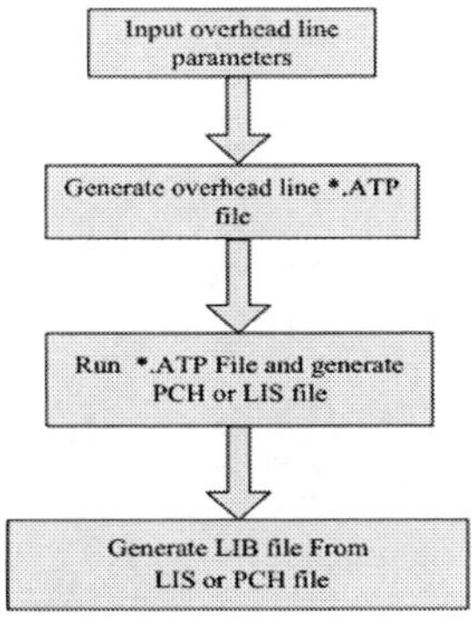

Figure 3. Generation of the overhead line LIB file

4 Solution of the Initial Breakdown Voltage

Once the ATP simulation model is established automatically, the initial breakdown current of transmission line can be solved by using the steps shown in Fig.4.

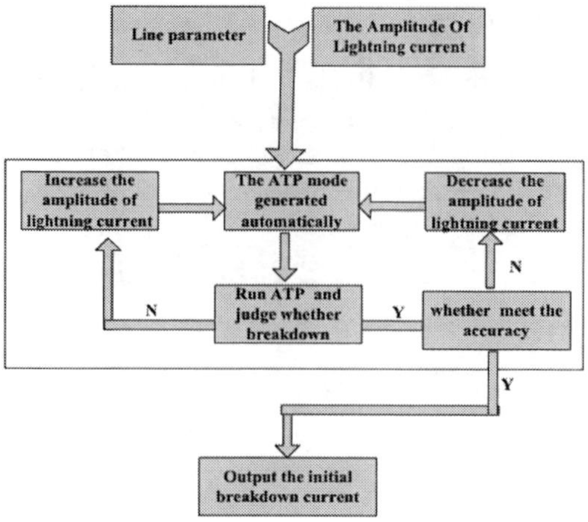

Figure 4. Process for solving initial breakdown current

The principle of this algorithm is to set a initial amplitude for lightning current, and create the ATP simulation model automatically according to the line parameters, then run the simulation model and obtain the result. If no breakdown occurs, then increase the lightning current amplitude, otherwise decrease, then create the simulation model and run again it automatically to find whether the breakdown occurs or not. Repeat above process until the amplitude difference between the two adjacent lightning current is less than the preset threshold, then the initial breakdown current amplitude is solved.

5 Application of the Proposed Method

Fig.5 shows the parameters of the components in simulation model of overhead line lighting strike, and Fig.6 shows the part code of the created simulation model according to these parameters. Executing the created model automatically and automatically change the lightning current amplitude every time until the initial breakdown current of transmission line is solved.

SOURCE PARAMETES	179629.2478	50	30	0					
	179629.2478	50	300	0					
TOWER PARAMETERS	1	113.3	225000000	0.7	113.3	225000000	0.7		
	1	334.1	225000000	4.07	334.1	225000000	4.07		
	1	107.4	225000000	4	107.4	225000000	4		
	1	251.2	225000000	2.75	251.2	225000000	2.75		
	1	102.5	225000000	5	102.5	225000000	5		
	1	102.8	225000000	15.7	925.8	225000000	23.55		
	2	113.3	225000000	0.7	113.3	225000000	0.7		
	2	334.1	225000000	4.07	334.1	225000000	4.07		
	2	107.4	225000000	4	107.4	225000000	4		
	2	251.2	225000000	2.75	251.2	225000000	2.75		
	2	102.5	225000000	5	102.5	225000000	5		
	2	102.8	225000000	15.7	925.8	225000000	23.55		
	3	113.3	225000000	0.7	113.3	225000000	0.7		
	3	334.1	225000000	4.07	334.1	225000000	4.07		
	3	107.4	225000000	4	107.4	225000000	4		
	3	251.2	225000000	2.75	251.2	225000000	2.75		
	3	102.5	225000000	5	102.5	225000000	5		
	3	102.8	225000000	15.7	925.8	225000000	23.55		
'OWER LINE PARAMETERS	5	2328.6	0.005	0.17	0	10	10	500000	50
	5	2328.6	0.005	0.18	0	10	10	500000	50
	5	2328.6	0.005	0.18	0	10	10	500000	50
	5	2328.6	0.005	0.18	0	10	10	500000	50
LIGHTNING PARAMETERS	-50000	2.6E-06	-1000	0.00005	0.00001	1	-1	2	2
GROUND PARAMETERS	24	500	2500						
	24	500	2500						
	24	500	2500						

Figure 5. Parameters of the component in model

```
C       1         2         3         4         5         6         7         8
C 345678901234567890123456789012345678901234567890123456789012345678901234567890
/BRANCH
C < n1 >< n2 ><ref1><ref2>< R  >< L  >< C  >
C < n1 >< n2 ><ref1><ref2>< R  >< A  >< B  ><Leng><><>0
  XXRS11                  350.                                                 0
  XXRS12                  350.                                                 0
        XXRS21            350.                                                 0
        XXRS22            350.                                                 0
  X0001B                  4.E4                                                 3
91XX0002      TACS  XX0026                                                     0
-1XX0023XX0004                  113.32.25E8    .7 1 0                          0
-1XX0024XX0004                  113.32.25E8    .7 1 0                          0
-1XX0028XX0004                  334.12.25E8  4.07 1 0                          0
-1XX0004XX0025                  334.12.25E8  4.07 1 0                          0
-1XX0028XX0032                  107.42.25E8    4. 1 0                          0
-1XX0025XX0030                  107.42.25E8    4. 1 0                          0
-1XX0032XX0027                  102.52.25E8    5. 1 0                          0
-1XX0030XX0027                  102.52.25E8    5. 1 0                          0
-1XX0027XX0029                  102.82.25E8  15.7 1 0                          0
-1XX0027XX0029                  925.82.25E8 23.55 1 0                          0
-1XX0031XX0032                  251.22.25E8  2.75 1 0                          0
-1XX0030XX0033                  251.22.25E8  2.75 1 0                          0
  XX0031XX0008               .1                                                0
  XX0031X0001A                       1.E-6                                     0
  XX0004XX0012               .1                                                0
  XX0004X0001C                       1.E-6                                     0
  XX0033XX0016               .1                                                0
  XX0033X0001B                       1.E-6                                     0
91XX0017      TACS  XX0038                                                     0
-1XX0035XX0019                  113.32.25E8    .7 1 0                          0
-1XX0036XX0019                  113.32.25E8    .7 1 0                          0
-1XX0040XX0019                  334.12.25E8  4.07 1 0                          0
-1XX0019XX0037                  334.12.25E8  4.07 1 0                          0
-1XX0040XX0044                  107.42.25E8    4. 1 0                          0
-1XX0037XX0042                  107.42.25E8    4. 1 0                          0
-1XX0044XX0039                  102.52.25E8    5. 1 0                          0
-1XX0042XX0039                  102.52.25E8    5. 1 0                          0
-1XX0039XX0041                  102.82.25E8  15.7 1 0                          0
-1XX0039XX0041                  925.82.25E8 23.55 1 0                          0
-1XX0043XX0044                  251.22.25E8  2.75 1 0                          0
-1XX0042XX0045                  251.22.25E8  2.75 1 0                          0
91XX0020      TACS  XX0047                                                     0
-1X01002XX0022                  113.32.25E8    .7 1 0                          0
-1X01003XX0022                  113.32.25E8    .7 1 0                          0
-1XX0049XX0022                  334.12.25E8  4.07 1 0                          0
-1XX0022XX0046                  334.12.25E8  4.07 1 0                          0
-1XX0049XX0053                  107.42.25E8    4. 1 0                          0
-1XX0046XX0051                  107.42.25E8    4. 1 0                          0
-1XX0053XX0048                  102.52.25E8    5. 1 0                          0
-1XX0051XX0048                  102.52.25E8    5. 1 0                          0
-1XX0048XX0050                  102.82.25E8  15.7 1 0                          0
-1XX0048XX0050                  925.82.25E8 23.55 1 0                          0
-1XX0052XX0053                  251.22.25E8  2.75 1 0                          0
-1XX0051XX0054                  251.22.25E8  2.75 1 0                          0
$INCLUDE. C:\Users\Yinlong\DOCUME~1\ATPdata\work\130 7.lib. X00S1A. X00S1B $$
```

Figure 6. Part of code created automatically

6 Conclusions

In this paper, a method of automatic generation of ATP simulation model is put forward, which can solve the problems such as heavy workload and easy to make mistake in establishing the ATP simulation model. On this basis, the method of the iterative solution of the initial breakdown current of overhead transmission line is proposed based on the presented simulation model automatic generation method, which can improve the efficiency and accuracy. The proposed method can be applied to other occasions based on ATP simulation, which has high practical value and application prospect.

References

1. Haiyang Wang. Journal of jinling institute of technology. **27(4)**, 17-22,(2011),:.
2. Deji Shang. The public science and technology.**5**,120-122,(2010).
3. Gang Chen, Zhang Peichai.. Relay. **34(22)**,5-8,(2006).
4. Yan Wang, Guo Huaide.. Shandong electric power college journals. **13(5)**,8-10,(2010).
5. Xi Li, Lin Tang.. The software of time and space.**22(3)**,212-214,(2006).
6. Moran Wang.. BeiJing: Electronic industry press. 2003.
7. Wang Yu,Fu Li, Shikun Zhang.. The computer technology and development. **(2)**, 26-28,(2004).
8. Yinan Geng, Rong Zeng, Yu Li.. High voltage engineering. **36(6)**, 1501-1505, (2010).
9. Jiahong Chen, Jun Lv, Zhiyin Qian.. High voltage engineering. **35(12)**, 2891-2901,(2009).
10. Zhao Chun, Lei Mengfei, Su Jie. Electrotechnical Application. **33**(21), 85-88(2014)
11. Dong Chen, Jianhua Liu, Chenxi Jia. Insulators and surge arresters. **(5)** ,8-12,(2011).
12. Haiyan Yuan, Zhengcai Fu, Bengang Wei.. Proseedings of the CSEE. **(5)**: 111-117,(2009).
13. Yu Wang, Xishan Wen, Lei Lan.. High voltage engineering. **37(10)**, 2471-2476 (2011).
14. Richeng Luo, Wen Li, Zhiqian Li.. High voltage engineering. **41(8)**, 2780-2787 (2015).

MATEC Web of Conferences **55**, 03002 (2016) DOI: 10.1051/matecconf/20165503002
ACPEE 2016

Electric Power Saving Awareness System at School Using ICT

Koichi Kuzume[a], Tomonori Tabusa and Haruka Sawa

Information Science and Technology Department, National Institute of Technology, Yuge College, 1000 Kamishima-cho, Ehime-ken, Japan

Abstract. Nowadays, education on energy saving at schools attracts attention to reduce energy consumption. In this paper we proposed a novel system to promote energy conservation activity at schools using a sensor network. The system consisted of a wall-type clock embedded with several kinds of sensors for temperature and light in the classroom, and wireless devices to connect to a cloud computer network. The system had the following superiorities: offered various services for energy conservation activity, needless of CT (Current Transformer) sensor, low price, ease of installation and extension, and visualization of power consumption in conjunction with school timetable in real time. We presented the system and a user study conducted to evaluate its usefulness.

1 Introduction

In order to reduce electric power consumption, it is necessary to examine our lifestyles and develop awareness of electric power conservation. It has been pointed out that activities to enlighten people on power saving or energy conservation are very important. Educational institutions, such as primary schools and junior and senior high schools, appeal as organizations particularly suited to teaching energy conservation. Many researchers have already reported on energy conservation at schools[1-2]. Some schools set up a smart meter, displaying electric power consumption, allowing school children to check power usage in real time. They could identify potential for saving power in their schools, by becoming aware of energy consumed through their actions. Communication between people plays an important role in deepening understanding of energy saving. Miura reported on the usefulness of the power meter for electric power saving activity [2]. In elementary schools, power saving activity by children mainly involved turning on or off lights and air-conditioners in classrooms and toilets. The amount of energy consumed in school was not so large. However, school children spoke at home of their experiences in school, including the energy conservation campaign. As a result of their action, the energy conservation campaign could be expected to have had a wider and wider impact in the school zone or community in which the children lived. If energy saving activity or education on global warming were more advanced in schools, we could realize huge savings in energy consumption beyond our forecasts.

Visualization technology for power consumption is very effective to promote energy conservation. There have been developed a lot of visualization systems. Florian and Kevin developed a system

[a] Corresponding author : kuzume@info.yuge.ac.jp

called Ubiquitous Smart Energy Management (USEM), which combined home automation and energy usage monitoring technology [3]. The system made use of Current Transformer (CT) sensors to measure the current of electric appliances. It was necessary to attach it to all electric appliances intended to be monitored for power consumption and to connect to a sensor network to send the data. However, neither smart meter nor USEM are suitable for promoting power conservation activity in schools because USEM needs a lot of sensors to measure the electric current of each appliance and they are expensive in terms of construction work.

In a previous paper, we proposed a novel system for promoting energy saving activity in schools, without CT devices, with ease of construction and low price, and confirmed its basic operation [1].In this paper we review our modified system and present experimental results from a user study to confirm its usefulness.

This paper is organized as follows: In section 2, we present the concept of the system proposed and a novel visualization technology for electric power consumption using a sensor network called Eco Assist Clock (EAC). Then the architecture of the EAC and its networking are described. In section 3, we detail the support service for energy saving activity in conjunction with the school lecture timetable. Section 4 presents the results from a user study to evaluate the performance of the system. Finally, section 5 outlines our conclusions and future work.

2 Electric Power Saving Awareness System

2.1 Architecture of Electric Power Saving Awareness System

Figure1 shows the architecture of the electric power-saving awareness system we proposed. A clock was installed, containing several kinds of sensors, such as for temperature, light and wind, to monitor the amount of energy consumed in a classroom. The energy saving support system at school consisted of the sensor network constructed from several EACs, the database for storing data and the management system offering several services to promote energy saving activity. The EACs, which could sense the on/off switching state of the lights and air-conditioners, sent the data to the management PC via Zigbee wireless devices [4]. The data collected could be stored on the management PC and also immediately sent to the Cloud Web Server. The accumulated data could be visualized on the web site. School children and teachers could access the web and check the power usage situation from anywhere in real time. School children could brow seanytime and anywhere, including with their families, and communicate with their families or their communities about events at schools using the internet. This could help promote energy conservation activity. The software for developing our system was as follows: MySQL was used for the database system, Apache for constructing the web server, and PHP for web programming. Visualization of power consumption on the web was realized using the graphic library provided by PHP.

2.2 Eco Assist Clock

Smart meters have usually been utilized for visualization of power consumption. They have many functions, not only the display or reporting of power consumption, but also the control of electric appliances. Suchun necessary functions and high installation cost are not needed for promoting power conservation in schools. We proposed a novel device, ECA, for visualization of power consumption. Figure2 and 3 show a circuit diagram for the EAC and its prototype model. The clock used in our system was a commercial clock of 26cm in diameter, costing 5,500 yen. It was installed in the most conspicuous place in the classroom, on the wall above the blackboard. The EAC contained light, temperature and wind sensors, LEDs, a wireless module, and a microcomputer (Arduino) [5] embedded in the clock. The reason for adopting a wall-type clock was that school children looked at it frequently during lectures. The clock embedded in the EAC ran independently from that of the clock itself. Moreover, it warned students of energy waste by flashing the LED lights.

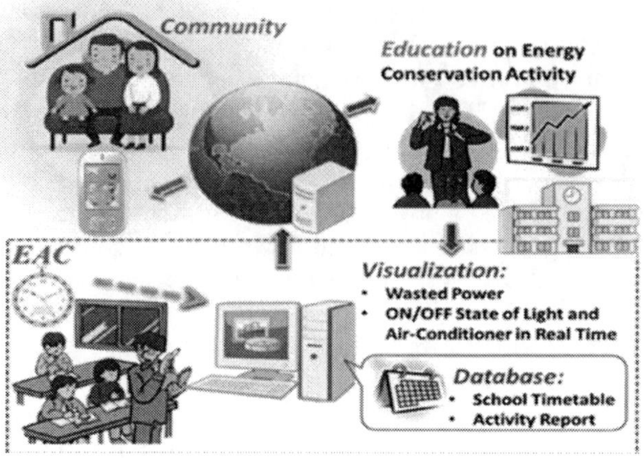

Figure 1. Architecture of Electric Power Saving Awareness System.

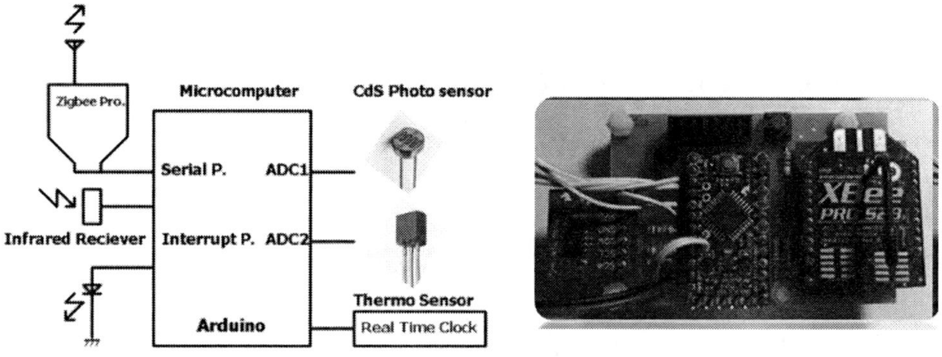

Figure 2. Architecture of EAC and Fabricated Print Circuit Board.

2.2.1 Detection of ON/OFF Switching of Lights and Air-Conditioner in Classroom

A CdS-photo sensor was used to detect the on/off switching state of the lights. It was fabricated at the top of the clock. It was not necessary to accurately measure power consumption, approximate power consumption being estimated from the on/off switching data. There are 24 fluorescent lights in the classroom and each light has 30W power consumption. Therefore, the power consumption is about 0.72kW (24×30W). The wind sensor can detect the air movement generated by an air-conditioner. The on/off signal of the air-conditioner is sent to the EAC by an infrared transmitter. The wind sensor could detect air movement generated by an air-conditioner. The on/off signal of the air-conditioner was sent to the EAC via infrared.

2.2.2 Data Transmission by Wireless Communication

Useful information, such as the temperature, the on/off switching state of the lights and air-conditioner etc. were sent to the personal computer via a wireless communication device, Zigbee Pro., which could communicate with the base station 1.5km distance from the EAC outside. The reason for using the wireless network was that it could extend the area of installation toother places in the campus area with lower construction cost.

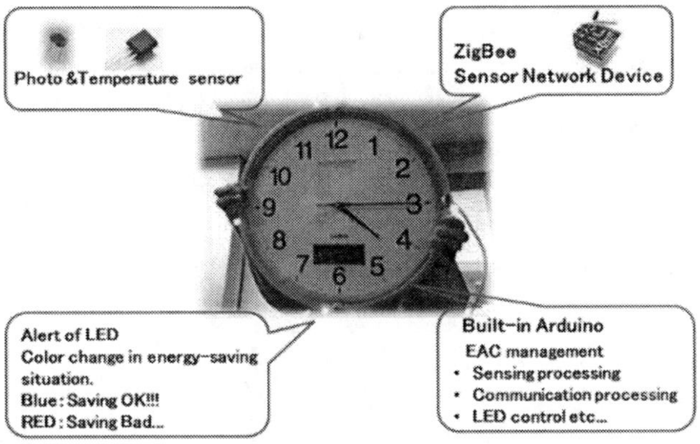

Figure 3. Prototype of Eco Assist Clock.

3 Support Service for Energy Saving Activity

3.1 Evaluation of Power Saving Activity in Conjunction with School Timetable

Figure 4 shows the top webpage. This page contained the following information. (i) Power consumption of both classrooms and whole school. (ii) On/off switching state of lights in classrooms installed with the EAC. If the lights were turned on, the picture of the light on the web changed to yellow. (iii) A report on power consumption and wasted power consumption over a day, a week or a month, could be displayed. Wasted power consumption could be calculated and the time and duration when the lights in the classroom were turned on applied in red to the school lecture timetable. See Figure 5.

3.2 Definition of Performance Factor of Power-saving Activity

At primary schools and junior and senior high schools in Japan, school children are assigned to learn in a set classroom. They usually study in their classroom, except for special subjects, like experiments and gymnastics, etc. We introduced the performance factor for evaluation of power-saving activity below,

$$PF_{save} = \frac{\Sigma T_p}{\Sigma T_{out}} \times 100\ [\%],$$

where T_p is energy saving interval time and T_{out} is time period when the classroom was unoccupied. T_p could be measured by our system and T_{out} could be ascertained from the school lecture timetable. If all lights in the classroom were turned off while school children were outside the classroom, the power-saving factor, PF_{save}, was equal to 100[%]. This factor could be utilized for evaluating energy saving activity. The web page could be browsed by a mobile phone, as shown in Figure 6.The result of measured power consumption and PF_{save} over 6 days are shown in Figure 7.

4 User Study for Evaluation of System Performance

4.1 Methodology and Study Participants

The 1st grade, 46 students of our college's Information Science and Technology Department, participated in the study. First we set an EAC on the wall over the blackboard in the classroom without giving notice and monitored power consumption over a week. Next we explained about our study and showed our webpage. We asked them to check the web site using their mobile phones more than once a day during the 2nd week. We then compared PF_{save} for the two periods. Finally, after the completion of the test, the participants were asked to complete a questionnaire.

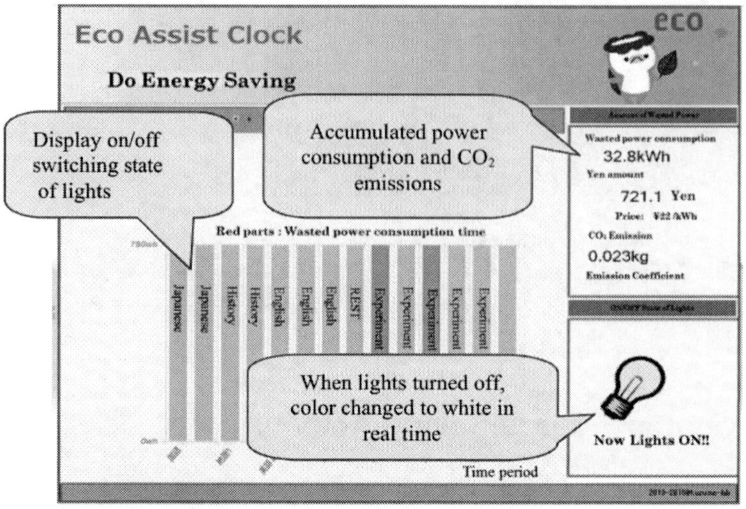

Figure 4 Top Page of the Power Saving Activity Support System

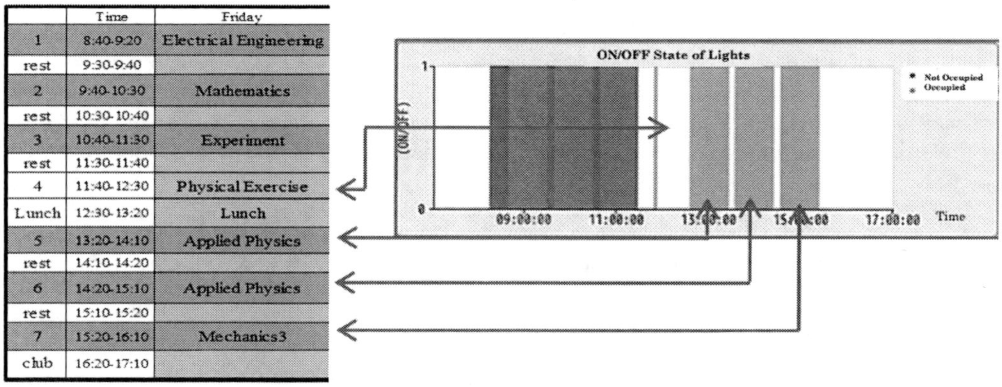

Figure 5. Wasted Power Consumption in Conjunction with School Timetable.

The principal questions of the questionnaire were as follows:

Q1.How many times did you browse the web site over the test period?

Q2.How many times did you turn off the lights in your classroom over the test period?

Q3.Have you ever studied how to conserve energy? If yes, when and where?

Q4.How likely do you think it would be that students would conduct energy conservation activity with the help of the visualization on the mobile phone?

Q5.What is the most useful content displayed on web site to encourage your energy-saving?

Q6.How useful do you find the overall system with regard to efficient energy usage?

Participants answered questions 5 and 6 using the Likert scale of 1-10, with 1 being the least positive and 10 the most positive.

4.2 Result of User Study

Figure 7 shows the performance factor, PF_{save}, over the week without notice. Average PF_{save} was around 80%. After notification of our test, average PF_{save} increased slightly. Participants had already formed habits for energy conservation. According to answers in the questionnaire, since the occurrence of the Great East Japan Earth Quake, 70% of participants had learned about power conservation as part of environmental education in junior high school (Q3). The number of students who voluntarily browsed the web site during the user test was less than 20% and only 10% of students turned off the light in the classroom (Q1,Q2). About 65% of students answered that the flashing light of LEDs was the most useful method to let students know aboutenergy conservation.

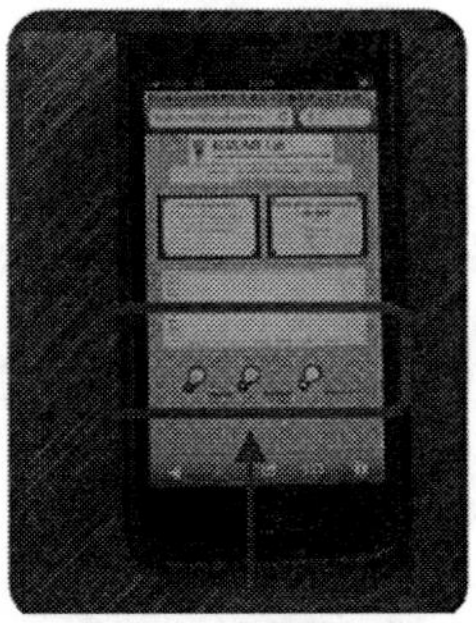

Figure 6. Notification via Mobile Phone.

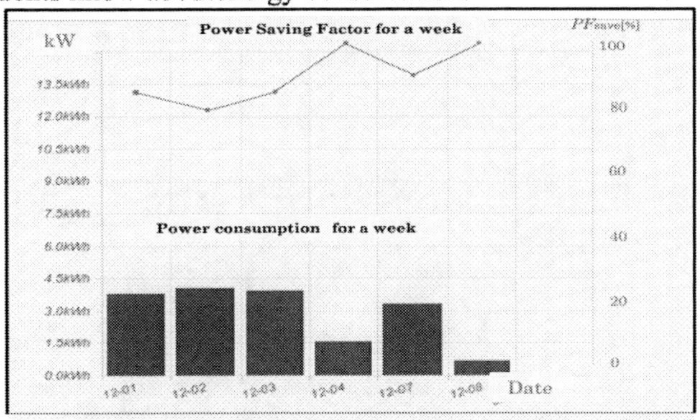

Figure 7. Report on Power Consumption over A Week.

(Likert value was over 7)(Q7).From content displayed on the web site, knowing the status of lights was utilized to prevent forgetting to turn off lights (over 80% of participant answered over 7 in Likert) (Q5,6).The 90% of answers to question7 with over 7 in Likert, showed our system had potential use in energy conservation activity (Q4).

5 Conclusions

In this research we proposed an electric power saving awareness system at school using ICT and constructed a prototype EAC. To evaluate the system we conducted a user test and questionnaire. We measured the wasted power consumption over a week and compared the performance factor in two cases. One involved no user notification; while for the other users were notified about the test. According to the questionnaire, our system had potential benefit when conducting energy conservation activity. To improve the performance factor, the user needed more opportunities to recognize energy saving activity. In other words, it would be more effective to add functions, e.g. the interaction between man and the EAC. It is important to note which methods or systems could assist people in changing their behaviour.

Acknowledgments

This research was partly supported by Grant-in-Aid for Scientific Research No.25330221.

References

1. Koichi Kuzume and Masakazu Okada, " Sensor Network System to Promote Energy Conservation --Realization of Energy Smart School--", The Proceedings of the IEEE International Conference on Pervasive Computer and Communications, Budapest (2013).
2. Shuichi Miura, "Practice and an effort of Environmental Education on Energy Conservation in an Elementary School", The Proceedings of the 22nd Energy and Resource Research Meeting, pp275-278, June, (2003) (in Japanese).
3. Florian Reinhart et.al,"Fostering Energy Awareness in Residential Homes using Mobile Devices", The Proceedings of the Fourth International Conference on Smart Grids, Green Communications and IT Energy-aware Technologies, IARIA, pp.35-43 (2014).
4. XBee®/XBee-PRO® ZB RF Modules, Digi International Inc.
5. http://www.arduino.cc/.

MATEC Web of Conferences **55**, 03003 (2016) DOI: 10.1051/matecconf/20165503003
ACPEE 2016

A Fault Estimation Method Based on UIO Robust Residual Generators

LIU Yongjin[1], FENG Gang[1], ZHANG Lin[1], WANG Wenfeng[1,2,3], JIA Zhonghui[1], ZHANG Bo[1,a]

[1]*Air Defense and Antimissile Institute, Air Force Engineering University, Xi'an, 710051, China*
[2]*The 302ed Institute of the Tenth Academy of China Aerospace Science & Industry Corporation; Guiyang 550009, China*
[3]*School of Automation Science and Electrical Engineering, Beihang University, Beijing 100191, China*

Abstract. This paper proposed a fault estimation method based on robust residual generators for a linear system. A system with additive actuator or component faults was considered in the case where the number of the independent faults was larger than that of the independent measurements. This method was achieved based on UIO robust residual generators and there was no need to design extra fault estimators. In this method, fault estimation was achieved via three steps. First, codingsets, which describe the sensitivity relationship between faults and generators, are designed. Second, a bank of robust residual generators are designed according to the coding sets. Finally, fault estimation is achieved by using the result of fault isolation and the output of robust residual generators. A sufficient condition on the application of the method was given and the asymptotic convergence property of the estimation error by using the method was proved. Simulation results demonstrate the effectiveness of the method.

1 Introduction

In recent years, with the rapid development of technologies, different kinds of systems are more complicated, and faults are easier to happen at the same time. Therefore, how to detect fault before it happens as soon as possible is the key to improve the reliability and safety of the system, it is also the hot topic that people research and there are series of research results [1-3]. The overall concept of fault diagnosis consists of fault detection, fault isolation, and fault analysis or identification. Fault diagnosis techniques are divided into model-based fault diagnosis techniques and data-driven design of fault diagnosis depending on the difference of character description and decision-making method between different systems. The former's core idea is to replace analytical redundancy with physical redundancy, and design observer to estimate the output of the system, then compare it with observed value to get information about fault [4-6].

As one of the model-based fault diagnosis techniques, the fault diagnosis technique based on robust residual generators is able to combine uncertainty of system and take full advantage of system's deep information, which is good for fault detection, fault identification, and fault isolation. The key of this method is to design a group of robust residual generators, and each robust residual generator is sensitive to a certain fault, but robust to the rest faults. Therefore, fault isolation can be achieved. Wang et al. designed robust observer for a linear system with additive actuator or component faults

[a]Corresponding author : dreamland_0628@163.com

and isolated the fault [7]. Chen et al. designed robust residual generators based on UIO, then diagnosed and isolated component faults of a linear system[8]. The dimension of the fault diagnosis techniques based on UIO robust residual generators is same as the diagnosed system.

How to estimate fault is also the research emphasis in fault diagnosis. Hou, et al. designed UIO-based fault estimation method in the case where the number of the independent faults was larger than that of the independent measurements, and it also needs derivative of output, but derivative of output is hard to get[9]. In order to estimate actuator faults, Hamayun et al. designed sliding mode observer, but it also requires the number of the independent faults was larger than that of the independent measurements[10]. These two methods are both have limitations, and hard to meet requirements. Researches on the case where the number of the independent faults less than that of the independent measurements seem to be less[11].

When the number of the input channel independent faults(mainly include additive actuator and component faults) less than that of the independent measurements in linear system, in this paper, a fault estimation method based on UIO robust residual generators and suitable conditions were given, and achieved fault isolation and estimation.

2 Problem Formulation

Consider a typical linear system with additive actuator or component faults described as[12,13]

$$\begin{cases} \dot{x}(t) = Ax(t) + Bu(t) + Df(t) \\ y(t) = Cx(t) \end{cases} \tag{1}$$

where, $x \in R^{n\times l}$ is the system state vector, $u \in R^{q\times l}$ is the system input vector, $y \in R^{p\times l}$ is the system output vector, $f \in R^{l\times 1}$ is the actuator or component fault signal. A is system matric, B is input matric, C is output matric; A, B, C are all with suitable dimensions and (A, C) is observable. D is fault direction vector, and $rank(C)=p>1$, $rank(D)>1$.

The fault diagnosis techniques based on *UIO* or sliding mode observer are able to achieve fault estimation in the case where $rank(CD)=rank(D)$,that is $l \le p$; these two methods will not be able to directly estimate fault when the number of the input channel independent faults less than that of the independent measurements. Therefore, in this paper, a fault estimation method based on *UIO* robust residual generators will be designed after fault isolation.

3 Fault Estimation

Considering there are many researches on fault detection and fault isolation when $l > p$, the basic idea is to detect fault by designing Luenberger observers, and isolate faults by designing code sets and robust residual generators. The specific process is as follows.

3.1 Fault Detection Strategy

In order to detect fault, at first, design Luenberger observer for the linear system described as [1]

$$\begin{cases} \dot{\hat{x}}(t) = A\hat{x}(t) + Bu(t) + L[y(t) - C\hat{x}(t)] \\ r(t) = y(t) - C\hat{x}(t) \end{cases} \tag{2}$$

With an observer gain L that ensures the observer stability.

Then design residual evaluation function $J(t)$ described as

$$J(t) = \|r(t)\| \tag{3}$$

The following inequalities are used to determine whether the fault occurs:

$$\begin{cases} J(t) > J_{th}, & \text{Fault occurs} \\ J(t) \le J_{th}, & \text{Fault doesn't occurs} \end{cases} \tag{4}$$

where J_{th} is threshold value, which is small and random.

3.2 Fault Isolation Strategy

(1) Encode fault signal. Suppose l={1,2,…,l}is the set composed of fault subscript, m={1,2,…,m} is the set composed of residual subscript. $\Omega_j \subseteq m$, $j \in l$ is the fault coding set composed of robust residual generator subscripts which is sensitive to fault number j, $\Gamma_i \subseteq l$, $i \in m$ is the fault set composed of fault subscripts which is sensitive to robust residual generator number i, that is to say, it is the set composed of fault subscripts that influence robust residual generator number i. Ω_j and Γ_i are both able to describe the corresponding relations between fault and residual, and achieve system fault isolation.

In order to isolate intercurrent fault as many as possible when $rank(C) = p$,fault coding set can be designed to let robust residual generator be robust to p-1faults, sensitive to l+1-p faults, so that C_l^{p-1} robust residual generators can achieve $p-1$intercurrent fault isolation.

(2) Design UIO-based robust residual generator

Suppose a typical linear system with additive unknown input are described as

$$\begin{cases} \dot{x}(t) = Ax(t) + Bu(t) + S\xi(t) \\ y(t) = Cx(t) \end{cases} \tag{5}$$

where $S\zeta(t)$ is unknown input. And satisfies

$$\begin{cases} rank(CS) = rankC \\ (C, A_1) is\ observable \end{cases} \tag{6}$$

where $A_1 = A - S[(CS)^T CS]^{-1}(CS)^T CA$.

When the system described as (5) has fault, it can be rewritten as (1) which is described as a system with single input fault vector.

$$\begin{cases} \dot{x}(t) = Ax(t) + Bu(t) + S\xi(t) \\ y(t) = Cx(t) \end{cases} \tag{7}$$

Proof. I_A is component fault identification matrix which describes component faults, and $f^1(t)$ describes actuator faults, so that we can obtain

$$\dot{x}(t) = Ax(t) + Bu(t) + I_A \Delta A_n x(t) + S\xi(t) + D^1 f^1(t) \tag{8}$$

By defining $D = [I_A \quad S \quad D^1]$, the system described as(5) can be rewritten as

$$\begin{cases} \dot{x}(t) = Ax(t) + Bu(t) + Df(t) \\ y(t) = Cx(t) \end{cases} \tag{9}$$

where fault vector $f(t)$ in the new state can be described as

$$f(t) = \begin{bmatrix} \Delta A_n x(t) \\ \xi(t) \\ f^1(t) \end{bmatrix} \tag{10}$$

UIO-based robust residual generator i can be described as

$$\begin{cases} \dot{z}_i(t) = F_i z_i(t) + G_i u(t) + E_i y(t) \\ r_i(t) = M_i z(t) + H_i y(t) \end{cases} \tag{11}$$

where $i \in m$, $z_i \in R^{s\times 1}$ is the state vector of generator *i*, $r_i \in R^{v\times 1}$ is residual output, $u \in R^{q\times 1}$ is the system input vector, $y \in R^{p\times 1}$ is the system output vector, F_i, G_i, E_i, M_i, H_i are unknown matrices with suitable dimensions to be determined, $T_i \in R^{s\times n}$ is transformation matrix, and T_i, F_i, E_i, M_i, H_i are determined by matric equation(12).

$$\begin{cases} E_i = E_{i1} + E_{i2} \\ T_i D^{(j)} = 0, j \in l - \Gamma_i \\ T_i D^{(k)} \neq 0, k \in \Gamma_i \\ F_i T_i = T_i A - E_i C \\ E_{i2} = F_i H \\ G_i = T_i B \\ M_i T_i + H_i C = 0 \end{cases} \tag{12}$$

where F_i is Hurwitz matrix, $l - \Gamma_i$ consists of elements in the l but not Γ_i.

In a similar way, fault detection of residual generator *i* is given by

$$J_i(t) = \|r_i(t)\| \tag{13}$$

Design one-to-one fault indicator function

$$RI_i = \begin{cases} 1, & J_i(t) > J_i^{th}(t) \\ 0, & J_i(t) \leq J_i^{th}(t) \end{cases} \tag{14}$$

where RI_i is 1 or 0, $J_i^{th}(t)$ is residual threshold value, which is small and random, and its value can be known through Monte Carlo experiment.

Suppose fault indicator function FI_j of fault j described as

$$FI_j = \bigcap_{k\in\Omega_j} RI_k, \quad j \in l \tag{15}$$

FI_j logical value is 1or 0, and it means

$$\begin{cases} FI_j = 1, & Fault\ j\ occurs \\ FI_j = 0, & Fault\ j\ doesn't\ occur \end{cases} \tag{16}$$

In this way, fault isolated can be achieved.

3.3 Fault Estimation Strategy

At presents, most of researches can only achieve fault detection and fault isolation when the number of the independent faults is larger than that of the independent measurements, researches on fault estimation seems to less. In this paper, UIO-based robust residual generator was designed at first to isolate fault when *l>p*, and then fault estimation method was designed after fault isolation. The specific process is as follows.

By defining

$$Q_i = M_i F_i^{-1} T_i D\,(Q_i \in R^{v\times l}),\ \ Q = [Q_1^T, Q_2^T, \cdots, Q_m^T]\,(Q_i \in R^{(vm)\times l}) \tag{17}$$

$$r(t) = [r_1^T(t), r_2^T(t), \cdots, r_m^T(t)]^T,\ \ f(t) = [f^{(1)}(t), f^{(2)}(t), \cdots, f^{(l)}(t)]^T \tag{18}$$

Fault estimation strategy is given by

$$\hat{f}(t) = \{Q[FI(t)]\}^+ r(t) \tag{19}$$

where $FI(t) = diag\{FI_1(t), FI_2(t), \cdots, FI_l(t)\}$ and $\{Q[FI(t)]\}^+$ is pseudo-inverse matrix.

Theorem. If the system described as (1)holds:

$$\begin{cases} rank(CD^{(\Gamma_i)}) = rank(D^{(\Gamma_i)}) \\ rank\begin{pmatrix} sI_n - A & D^{(\Gamma_i)} \\ C & 0 \end{pmatrix} = n + rank(D^{(\Gamma_i)}) \\ \forall i \in m, \forall s \in \mathrm{Re}(s) \geq 0 \end{cases} \tag{20}$$

where I_n is identity matrix with n dimension, s is complex number with real part equal or greater than 0, $D^{(\Gamma i)}$is matrix corresponding to Γ_i, we can achieve fault detection by (2)-(4) and fault isolation by(11)-(15). At the same time, asymptotically estimation of constant fault $\hat{f}(t)$ is given by

$$\lim_{t\to\infty} \hat{f}(t) = f(t) \tag{21}$$

Proof: It is easy to achieve fault detection and fault isolation by (2)~(4)and(11)~(15). We just prove that fault estimation is feasible.

By defining

$$e_i(t) = z_i(t) - T_i x(t), \qquad i \in m \tag{22}$$

It can be inferred from (11)and (12) that

$$\begin{aligned} \dot{e}_i(t) &= \dot{z}_i(t) - T_i \dot{x}(t) = F_i z_i(t) + G_i u(t) + E_i y(t) \\ &\quad - T_i A x(t) - T_i B u(t) - T_i D[FI(t)]f(t) \\ &= F_i e_i(t) - T_i D[FI(t)]f(t) \end{aligned} \tag{23}$$

And

$$r_i = M_i[z_i(t) - T_i x(t)] \tag{24}$$

Equation set (11) can be rewritten as

$$\begin{cases} \dot{e}(t) = F_i e_i(t) - T_i D[FI(t)]f(t) \\ r_i(t) = M_i e_i(t) \end{cases} \tag{25}$$

And we can obtain

$$\begin{aligned} \lim_{t\to\infty} r_i(t) &= \lim_{s\to 0} M_i (sI - F_i)^{-1} T_i D[FI(t)]f(t) \\ &= Q_i[FI(t)]f(t) \end{aligned} \tag{26}$$

It is because $r(t) = [r_1^T(t), r_1^T(t), \cdots, r_m^T(t)]^T$ that we obtain

$$\lim_{t\to\infty} r(t) = Q[FI(t)]f(t) \tag{27}$$

From equation set perspective, if we treat $\lim_{t\to\infty} r(t)$ as known quantity, $f(t)$ as matrix to be determined, (27) must have solution. Furthermore, there are p-1unknown quantities in (27)when p-1faults occur, and depending on the designing rule, only one robust residual generator is robust to this p-1faults in $\{r_1(t), r_2(t), \cdots, r_m(t)\}$, the rest of generators are all sensitive to this p-1faults. For the minimum row of matrices in equation is 1, the number of known quantities is $(C_l^{p-1} - 1) \times 1$ at least in (25). Furthermore, $(C_l^{p-1} - 1) \geq l - 1 > p - 1$, (27)has least square solution. The least square solution of (26) is the solution of equation based on the theory that equation must has solution. Therefore, we can obtain

$$\hat{f}(t) = \{Q[FI(t)]\}^{+} r(t) \tag{28}$$

4 Simulation Results

A linear system is employed with the following parameters:

$$A = \begin{bmatrix} -1 & -1 & -2 & 0 \\ 0 & -1 & 1 & 2 \\ 1 & 0 & -4 & 0 \\ 0 & -1 & 0 & -1 \end{bmatrix}, C = \begin{bmatrix} 1 & 0 & 1 & 0 \\ 0 & 0 & 2 & 0 \\ 0 & 1 & 0 & 1 \end{bmatrix}, B = D = \begin{bmatrix} 1 & 0 & 1 & 1 \\ 0 & 1 & 1 & 0 \\ 0 & 1 & 0 & 1 \\ 0 & 0 & 0 & 1 \end{bmatrix},$$

$$u=[1\ 1\ 1\ 1]^T, L=\begin{bmatrix}0.9091 & -1.4545 & -0.6364\\ -0.0909 & -0.5455 & 1.3636\\ 1 & -2 & 0\\ -0.1818 & 0.0909 & -0.2727\end{bmatrix}$$

There are four independent faults but three independent measurements in the system, so C_4^2 robust residual generators should be designed. Fault coding set is described as Table 1.

Table 1. Coding sets of simulation.

	$f^{(1)}$	$f^{(2)}$	$f^{(3)}$	$f^{(4)}$
r_1	0	0	1	1
r_2	0	1	0	1
r_3	0	1	1	0
r_4	1	0	0	1
r_5	1	0	1	0
r_6	1	1	0	0

In the Table 1, "0"means that residual is robust to fault, while"1"means that residual is sensitive to fault. The parameters of UIO-based robust residual generators described as (11) are set as

$$F_1=\begin{bmatrix}-0.6875 & 1.0877 & -3.8540 & 0.0271\\ 1.2540 & -1.7675 & -3.3496 & 1.2325\\ 0.1279 & -0.2762 & -0.1769 & -0.2762\\ 0.1073 & 0.1391 & -0.5029 & -1.3609\end{bmatrix}, F_2=\begin{bmatrix}-6 & 0 & 0 & 0\\ 0 & 0 & 0 & 0\\ 1 & 0 & 0 & 0\\ 0 & 0 & 0 & 0\end{bmatrix}, F_3=\begin{bmatrix}-0.9952 & 1.7166 & -3.4606 & -0.0940\\ 0.1829 & -1.4208 & 1.5985 & 1.8899\\ -0.0571 & -0.3318 & 0.0159 & -0.3318\\ 0.9362 & 0.0930 & -3.4584 & -1.4070\end{bmatrix},$$

$$F_4=\begin{bmatrix}-1.6667 & -0.6667 & 7 & 0.3333\\ -1 & -1.5 & -3 & 0\\ 0 & 0 & -1 & 0\\ 0.6667 & -0.3333 & -0.5 & -1.8333\end{bmatrix}, F_5=\begin{bmatrix}-0.3554 & 1.6699 & -2.4903 & 0.1317\\ 2.5674 & -0.4912 & 1.4922 & 1.8942\\ 0.0007 & -0.3082 & -0.0320 & -0.3082\\ -0.6567 & -0.1979 & -3.4815 & -1.6979\end{bmatrix}, F_6=\begin{bmatrix}-0.6348 & -0.6195 & 3.6172 & 0.5353\\ -2.4375 & -1.5411 & 3.8105 & -0.0411\\ 0.1073 & 0.0980 & -1.4114 & 0.0980\\ 1.9089 & -0.3908 & -4.9477 & -1.8908\end{bmatrix},$$

$$G_1=\begin{bmatrix}0 & 0 & -0.7071 & 0.7071\\ 0 & 0 & 0 & 1\end{bmatrix}, G_2=[0\ -1\ 0\ -1], G_3=\begin{bmatrix}0 & -0.2071 & -0.7071 & 0\\ 0 & -1.2071 & -0.7071 & 0\end{bmatrix},$$

$$G_4=[-0.5\ 0\ 0\ -0.5], G_5=\begin{bmatrix}-0.5236 & 0 & -1.1236 & 0\\ -0.5708 & 0 & -0.3708 & 0\end{bmatrix}, G_6=[-0.5774\ 0.5774\ 0\ 0],$$

$$E_1=\begin{bmatrix}0.7071 & -1.7015 & -0.6102\\ 0 & 0.0567 & 1.8271\\ 0 & -0.0452 & 0.1833\\ 0 & -0.2879 & -0.7816\end{bmatrix}, E_2=\begin{bmatrix}0.0635 & -1.0318 & 0.8939\\ -0.7580 & 0.3790 & -1.7988\\ 0.3333 & 0.3333 & -0.3333\\ 0.9587 & -0.4794 & -1.4963\end{bmatrix}, E_3=\begin{bmatrix}0.5 & -0.8738 & -1.3350\\ -0.5 & 1.4009 & 1.4161\\ 0 & -0.0687 & 0.1658\\ 0 & -0.4608 & -0.2411\end{bmatrix},$$

$$E_4=\begin{bmatrix}-0.8333 & 1.75 & 0.8333\\ -0.5 & 0.25 & 1.25\\ 0 & 0 & 0\\ 0.3333 & -0.25 & -1.0833\end{bmatrix}, E_5=\begin{bmatrix}1.8907 & -1.5172 & -0.4999\\ 1.4347 & -0.0732 & 1.8994\\ -0.1755 & 0.0568 & 0.1498\\ -0.4568 & 0.1556 & -0.8761\end{bmatrix}, E_6=\begin{bmatrix}0.2109 & 0.7607 & 0.3665\\ -1.4074 & 0.9537 & 2.1574\\ 0.0619 & -0.0310 & -0.0619\\ 1.1022 & -0.8011 & -1.8522\end{bmatrix},$$

$$M_1=\begin{bmatrix}0.75 & -0.3535 & 0 & 0\\ 0 & 0 & 0 & 0\\ -0.7071 & 0 & 0 & 0\\ 0.3536 & -0.5 & 0 & 0\end{bmatrix}, M_2=\begin{bmatrix}1 & 0 & 0 & 0\\ 0 & 0 & 0 & 0\\ 1 & 0 & 0 & 0\\ 0 & 0 & 0 & 0\end{bmatrix}, M_3=\begin{bmatrix}0.4893 & -0.4893 & 0 & 0\\ -0.1875 & 0.1875 & 0 & 0\\ -0.25 & 0.25 & 0 & 0\\ 0.3018 & -0.3018 & 0 & 0\end{bmatrix},$$

$$M_4=\begin{bmatrix}1.5&0&0&0\\-0.5&0&0&0\\0.5&0&0&0\\-0.5&0&0&0\end{bmatrix},M_5=\begin{bmatrix}1.3865&0.4214&0&0\\0.1888&0.3517&0&0\\-0.6&0.2&0&0\\0.3382&-0.4854&0&0\end{bmatrix},M_6=\begin{bmatrix}1.5773&0&0&0\\-0.5773&0&0&0\\0&0&0&0\\-0.5773&0&0&0\end{bmatrix},$$

$$H_1=\begin{bmatrix}1&-0.8536&0.3535\\0&0&0\\0&0.5&0\\0&0&0.5\end{bmatrix},H_2=\begin{bmatrix}1&0&0\\0&0&0.5\\0&0.5&0\\0&0&0.5\end{bmatrix},H_3=\begin{bmatrix}1&-0.75&0.6035\\0&0.25&-0.1035\\0&0.5&0\\0&0&0.5\end{bmatrix},$$

$$H_4=\begin{bmatrix}1.5&-0.5&-0.5\\0&0&0.5\\0&0.5&0\\0&0&0.5\end{bmatrix},H_5=\begin{bmatrix}1.5236&-1.0618&0.3382\\0.5708&-0.1854&0.0146\\0&0.5&0\\0&0&0.5\end{bmatrix},H_6=\begin{bmatrix}1.5774&-0.7887&-0.5774\\0&0&0.5\\0&0.5&0\\0&0&0.5\end{bmatrix}.$$

It is easily to find that the dimension of the robust residual generators based on UIO is four , and it is same as the original system. By Monte Carlo experiment, threshold values are given by

$$J_{th}=0.1,J_{th}^1=1\times10^{-8},J_{th}^2=5\times10^{-2},J_{th}^3=2\times10^{-3},J_{th}^4=5\times10^{-3},J_{th}^5=2\times10^{-3},J_{th}^6=2\times10^{-3}.$$

And the parameters of fault estimation strategy described as(13) are set as

$$T_1=\begin{bmatrix}0&-0.7071&0.7071&0\\0&0&0&1\end{bmatrix},T_2=\begin{bmatrix}0&0&-1&0\end{bmatrix},T_3=\begin{bmatrix}0&-0.7071&0.5&-0.5\\0&0.7071&-0.5&0.5\end{bmatrix},$$

T_4=[-0.5,0.5,-0.5,0.5], $T_5=\begin{bmatrix}-0.5236&-0.6&0.6&-0.0764\\-0.5708&0.2&-0.2&0.7708\end{bmatrix}$, T_6=[-0.5774 0.5774 0 0.5774],

$$Q_1=\begin{bmatrix}0.2368&0&0.9676&0.6022\\0&0&0&0\\-0.2869&0&-1.0860&-0.6865\\0.0480&0&0.2824&0.1652\end{bmatrix},Q_2=\begin{bmatrix}0.1622&0.1622&0&0\\0&0&0&0\\0.1622&0.1622&0&0\\0&0&0&0\end{bmatrix},$$

$$Q_3=\begin{bmatrix}0&0.0745&0.4345&0.1800\\0&-0.0286&-0.1665&-0.0690\\0&-0.0381&-0.2220&-0.0920\\0&0.0460&0.2680&0.1110\end{bmatrix},Q_4=\begin{bmatrix}0.7071&0&-1.4143&-0.7071\\-0.2357&0&0.4714&0.2357\\0.2357&0&-0.4714&-0.2357\\-0.2357&0&0.4714&0.2357\end{bmatrix}$$

$$,Q_5=\begin{bmatrix}0&0&-1.5652&0.0987\\0&0&-0.6364&-0.0579\\0&0&0.1296&-0.1354\\0&0&0.4608&0.1667\end{bmatrix},Q_6=\begin{bmatrix}0&0.8236&1.6472&0.8236\\0&-0.3014&-0.6029&-0.3014\\0&0&0&0\\0&-0.3014&-0.6029&-0.3014\end{bmatrix}.$$

5 Conclusion

In this paper, we proposed a fault estimation method based on UIO robust residual fenerators for a linear system with input channel independent faults when the number of the independent faults was larger than that of the independent measurements. A sufficient condition on the application of the method and specific process of fault detection, fault isolation, fault estimation was given. At the same time, asymptotic convergence of residual estimation was proved. This method can achieve fault estimation only after we obtain the results of fault isolation by residual generators. The simulation results demonstrate that this proposed approach is efficient.

References

1. Zhou D H, Hu Y Y. Fault diagnosis techniques for dynamic system. ActaAutomaticaSinica, 35(6): 748-758.(2009)
2. Ding S X. Model-based fault diagnosis techniques: design schemes, algorithms and tool. 2nd. London: Springer, 1-9.(2013)
3. Sari A H A.Data-driven design of fault diagnosis systems: nonlinear multimode processes. Fachmedien Wiesbaden. NY: Springer, 11-30.(2014)
4. Qin L, He X, Zhou D H. A survey of fault diagnosis for swarm systems. Systems Science&Control Engineering: An Open Acess Journal, 2(1):13-23.(2014)
5. Frank P M, Ding S X and Kppen-Seliger B. Current Developments in the Theory of FDI. Budapest, Hungary, 16-27.(2000)
6. Gentil S, Lesecq S, Barraud A. Improving decision making in fault detection and isolation using model validity. Engineering Applications of Artificial Intelligence, 22 (4-5): 534-545.(2009)
7. Wang X L, Wu P. Fault detection for control System based on unknown input observe. Lifting transportation machinery, (12):11-14.(2003)
8. Chen J, Patton R J. Robust model-based fault diagnosis for dynamic systems. US: Springer,65-108.(1999)
9. Hou M, Muller P C. Design of observers for linear systems with unknown inputs. Automatic Control, IEEE Transactions on, 37(6): 871- 875.(1992)
10. Hamayun M T, Edwards C, Alwi H. A fault tolerant control allocation scheme with output integral sliding modes. Automatica, 49(6): 1830-1837.(2013)
11. Zhu F. State estimation and unknown input reconstruction via both reduced-order and high-order sliding mode observers. Journal of Process Control, 22(1):296-302.(2012)
12. Zhang P. Design of Unknown Input Observer Based Fault Diagnosis Method for linear systems. Journal of ENGG College of Armed Pollce Force,24(6):29-32.(2008)
13. Hu Z K, Sun Y, Jiang B, He J, Zhang C F. An Optimal Unknown Input Observer Based Fault Diagnosis Method. ActaAutomaticaSinica,39(8):1225-1230.(2013)

MATEC Web of Conferences **55**, 03004 (2016) DOI: 10.1051/matecconf/20165503004
ACPEE 2016

A Fault Diagnosis Model of Surface to Air Missile Equipment Based on Wavelet Transformation and Support Vector Machine

ZHHENG Ni[1], ZHANG Lin[1], ZHANG Bo[1], WANG Wenfeng[1,a], SHI Wenjie[1], Si Wei[2]

[1] *Air Defense and Antimissile Institute, Air Force Engineering University, Xi'an710051, China*
[2] *The Research Institute on General Development and Evaluation of Equipment, EAAF of PLA, Beijing 100191, China*

Abstract. At present, the fault signals of surface to air missile equipment are hard to collect and the accuracy of fault diagnosis is very low. To solve the above problems, based on the superiority of wavelet transformation on processing non-stationary signals and the advantage of SVM on pattern classification, this paper proposes a fault diagnosis model and takes the typical analog circuit diagnosis of one power distribution system as an example to verify the fault diagnosis model based on Wavelet Transformation and SVM. The simulation results show that the model is able to achieve fault diagnosis based on a small amount of training samples, which improves the accuracy of fault diagnosis.

1 Introduction

The surface to air missile equipment is a kind of complex equipment with mechanical, electronic and hydraulic system. The operational effectiveness of the equipment will be affected or even disabled when any part of the equipment breaks down. The combat readiness intact rate of the equipment can be improved by applying condition-based maintenance to avoid burst faults [1]. Because the early fault signal of surface to air missile equipment is feeble which can be easily affected by noise and detection error, the accurate fault state of the equipment is unable to be acquired without special processing. Among many methods of fault diagnosis, Wavelet Transformation theory and SVM are widely used and both achieve good effect. The Wavelet Transformation is an important means for non-stationary signal processing which can easily separate the monitory point's original signal and noise background to achieve the purpose of signal denoising [2]. Taking the advantage of generalization performance of SVM, the low dimensional inseparable problem can be mapped to high-dimensional space to realize the separability through kernel function skills, which better solves the problems such as nonlinearity, small samples and local minimum points [3-4]. In this paper, a fault diagnosis model based on wavelet transformation and SVM is proposed by taking the power distribution system of surface to air missile equipment as the object of study. This model uses wavelet transformation to denoise fault signals and uses SVM to classify fault feature vector, which provides an effective method for fault diagnosis of surface to air missile equipment and lays the foundation of condition-based maintenance and PHM.

[a]Corresponding author : dreamland_0628@163.com

2 Design of Fault Diagnosis Model

2.1 The Basicprinciple of Wavelet Transformation

Wavelet Transformation is a local transformation on time and dimension, which can analyze signal from multi-dimension by flexing and translation to assign low-frequency signals lower time resolution and higher frequency resolution, in the meantime, to assign high-frequency signal higher time resolution and lower frequency resolution. The threshold-denoising method based on Wavelet Transformation is a common and effective method. The effective signal can be reconstructed by utilizing the huge difference feature between effective signal and noise signal, conversing noise signal by wavelet basis function, processing all resolved wavelet parameters by threshold and reverse-conversation.

In [5], the definition of continuous wavelet transformation and reverse-transformation is introduced in details. The continuous wavelet transformation is

$$(W_\psi f)(a,b) = \langle f, \psi_{a,b} \rangle = |a|^{-1/2} \int_R f(t)\psi^*\left(\frac{t-a}{b}\right)dt \tag{1}$$

The wavelet reverse-transformation is

$$f(t) = \frac{1}{C_\psi} \int_{-\infty}^{+\infty}\int_{-\infty}^{+\infty} (W_\psi f)(a,b)\psi_{a,b}(t)\frac{da}{a^2}db \tag{2}$$

The continuous wavelet transformation is mainly applied to the oretical study and analysis. The discrete format is generally used in numerical calculation. The discrete wavelet transformation is

$$(W_\psi f)(a,b) = \langle f, \psi_{a,b} \rangle = |a_0|^{-m/2} \int_R f(t)\bar{\psi}\left(a_0^{-m}t - nb_0\right)dt \tag{3}$$

The discrete wavelet re-transformation is

$$f(t) = c\sum_{-\infty}^{+\infty}\sum_{-\infty}^{+\infty}(W_\psi f)(a,b)\psi_{m,n}(t) \tag{4}$$

2.2 The Basic Principle of SVM

The main idea of SVM is to find an optimal hyper plane to separate different samples correctly [6]. SVM can be divided into linear separability and non-linear inseparability. Generally, actual operating system such as power distribution system belongs to the nonlinear inseparable problem. In order to realize SVM under the circumstance of non-linear inseparability, a non-linear transformation is needed to map the feature sample which has already been wavelet denoised and re-transformed to high feature space. Then the linear SVM is applied to fault classification.

According to the basic principle of SVM, the optimal linear classification function is needed to realize the fault classification and to achieve the purpose of diagnosis [7].

$$f(x) = sign[(w \bullet \theta(x) + b)] \tag{5}$$

In the formula, $w$is the weight vector and $b$is the deviation or classification threshold.

To meet the structural risk minimization, w and b need to be found to make the formula (6)hold.

$$\min R = \frac{1}{2}w^T w + C \bullet R_P \tag{6}$$

where, $w^T w$ is the complexity of the control model. C is the penalty factor. R_p is the error control function. The loss function which is chosen in the optimization goal is the 2-norm of error ζ, then the fault diagnosis model can be optimized as

$$\min Q(w,b,\xi)=\frac{1}{2}w^{T}w+\frac{1}{2}C\bullet\sum_{i=1}^{N}\xi_i^2$$
$$st.\quad y_i[w^{T}\bullet\theta(x_i)+b]=1-\xi_i, i=1,2,\ldots,N \tag{7}$$

Due to the condition that w may be infinite dimensional, it is very difficult to solve the above formula. Thus, formula(7) can be converted to following equations.

$$\begin{bmatrix} 0 & \boldsymbol{I}^{\mathrm{T}} \\ \boldsymbol{I} & K+C^{-1}I \end{bmatrix}\begin{bmatrix} \boldsymbol{b} \\ \boldsymbol{a} \end{bmatrix}=\begin{bmatrix} 0 \\ \boldsymbol{Y} \end{bmatrix} \tag{8}$$

where, $\boldsymbol{I}=[1,1,\ldots,1]^T$, $\boldsymbol{Y}=[y_1,y_2,\ldots,y_N]^T$, $\boldsymbol{a}=[a_1,a_2,\ldots,a_N]^T$, $K=\theta(x_i)^T\theta(x_j)$is the Kernel function.

Calculation difficulty will be increased when takes the polynomial kernel function at a higher order [8-9]. The precision of radial basis kernel function is higher than sigmoid kernel function, so the radial basis kernel function $K(x,x')=\exp(-\gamma\|x-x'\|^2)$ ($\gamma>0$) is chosen. $\gamma$is the coefficient of control radius. The least square method is used to get the value of a and b, then the fault classification function can be got after wavelet denoising.

$$f(x)=\mathrm{sgn}[\sum_{i=1}^{N}a_iK(x,x_i)+b] \tag{9}$$

2.3 The Design of Fault Diagnosis Model based on Wavelet Transformation and SVM

The fault diagnosis schematic based on Wavelet Transformation and SVM is shown in Figure 1. Firstly the Wavelet Transformation is used to denoise the extractive fault signal and to reconstruct the signal. The fault information is extracted to build fault feature vector set. The SVM classifier is constructed by using the method of one to one. At last, the sample data is inputted into the fault diagnosis model to forecast and analyze to output the diagnosis result.

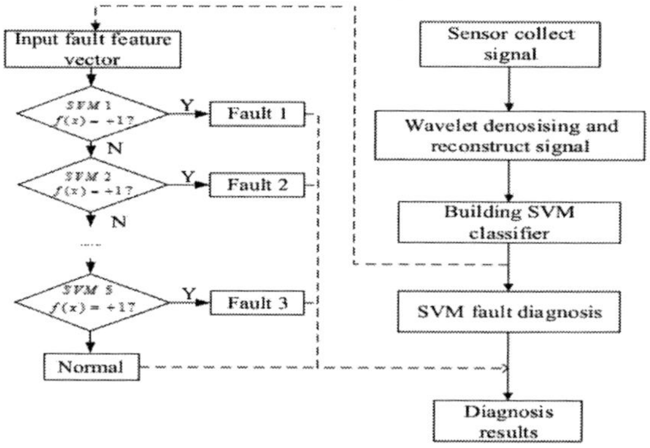

Figure 1. The schematic diagram of fault diagnosis model based on Wavelet Transformation and SVM.

3 Simulation Analysis

According to the dimension of circuit fault feature vector and the circuit fault mode, combining the expert experience and the characteristics of the equipment, the method of fault diagnosis based on the bottom node monitoring model is selected. After the extraction of the test node's signal, denoising technique will be applied to construct the fault diagnosis model based on SVM, so as to achieve fault diagnosis.

Taking the typical analog circuit diagnosis of one power distribution system as an example to verify the fault diagnosis model based on Wavelet Transformation and SVM, as shown in Figure2. The circuit is composed of four filters and an adder, which is divided into five parts H_1, H_2, L_1, A_1 and

L_2 respectively. Monitoring nodes are N_1, N_2, N_3, N_4 and N_5. Fault types are fault H_1, fault H_2, fault L_1, fault A_1 and fault L_2. According to the characteristics of the hypothesis circuit, the sample data can be trained by 5 degrees SVM fault classifier to judge the circuit state.

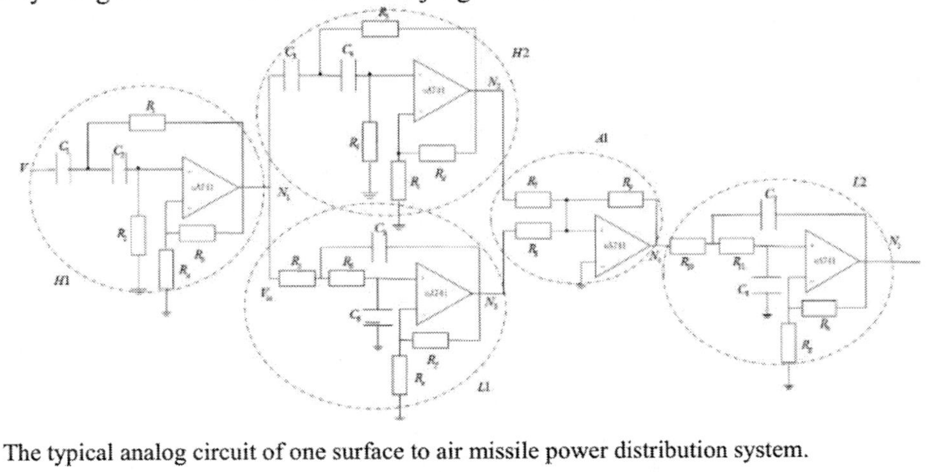

Figure 2. The typical analog circuit of one surface to air missile power distribution system.

3.1 Parameter Selection

3.1.1 The performance compare of wavelet denoising

Taking N_1 as the monitor node and the voltage signal as the feature vector, different denoising methods are analyzed separately based on the same voltage signal and the results are shown in Figure3. It indicates that the soft threshold method is better than the other two methods. Hence, soft threshold method is chosen in the paper. The wavelet basis has a determining influence on the effect of wavelet transformation. Test the Synm N wavelet class and the DbN wavelet class based on the same test signal [10].Comparing the results, the former one is proved to be better than the latter one. Simulating and analyzing the Sym wavelet basis and the Db wavelet basis, the results are shown in Figure 4.

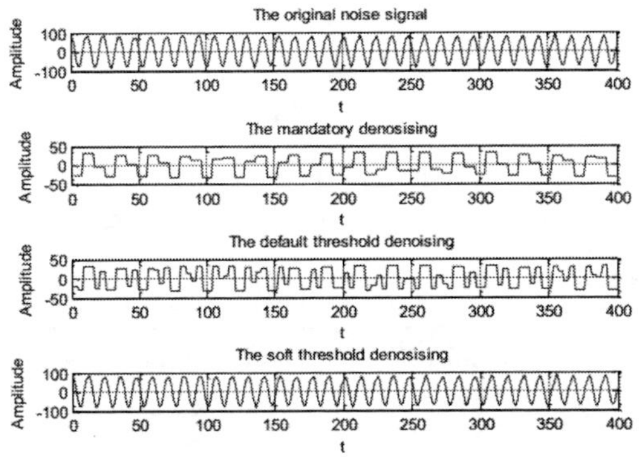

Figure 3.Denoising effect figures of three modes.

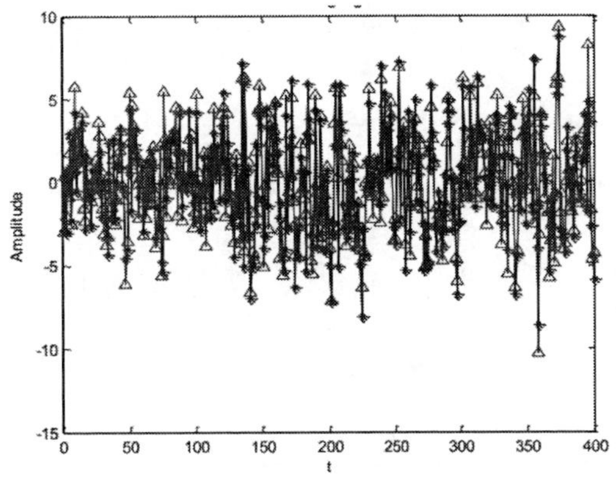

Figure 4.Denosising curve.

Using Sym wavelet basis to denoise:

$$SNR_{sym} = 18.5476(\text{dB}),\ A_{sym} = 1.0842,\ MSE_{sym}=6.3896$$

Using Db wavelet basis to denoise:

$$SNR_{db} = 17.6567(\text{dB}),\ A_{db} = 1.0584,\ MSE_{db}=6.9897$$

By comparison and analysis, the effect of Sym is slightly better. Hence, Sym wavelet basis is selected in the paper.

 DOI: 10.1051/matecconf/20165503004

3.1.2 The selection of parameter γ in SVM

The value of the constant Cwhich was pre-determined in radial basis function has no influence on the result, while the value of the parameter γ does have a great one. In this paper Grid-Search is applied to the optimization of parameter γin SVM kernel function. The specific steps are:

Step1:When$C$=10, according to the SVM principle, the maximum range of the parameter $\gamma$is $\gamma\in[0:0.1:1]$.

Step2:Create 1*100 network planes and 100 parameter pairs(C_i, γ_i) within this range, use all node parameters as learning samples and acquire learning errors after putting them in SVM, the pair $(C_i, \gamma_i)_{min}$ corresponds to the node which has the minimum error will be adopted.

Step3: When the learning accuracy cannot meet the requirements, regard $(C_i, \gamma_i)_{min}$ as a center, and wide the grid width with 0.01～5 times to reconstruct parameter pairs(C_i, γ_i). Finally the required accuracy will be achieved via optimizing the parameter γ in SVM.

3.2 Fault Simulation

Taking node $N_1$for example, the voltage signal is extracted by the sensor acquisition. The signal is shown in Figure 5 after applying the Sym wavelet basis and the soft threshold method to denoise.

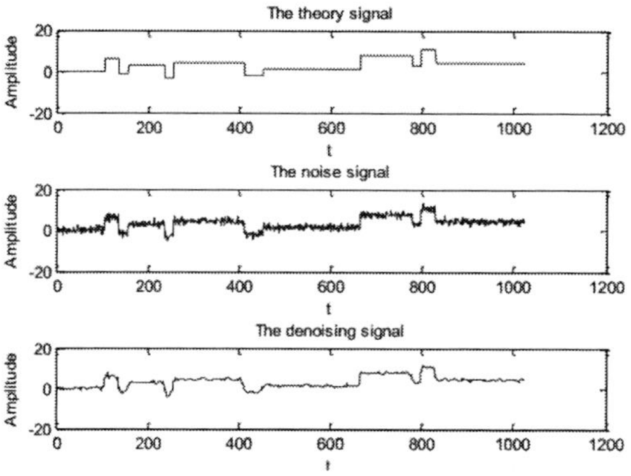

Figure 5. The collection and denosising of node N_1.

3.3 Results analysis

Apply the fault diagnosis model and the parameter was confirmed by using Grid-Search optimization method. The results of training and testing are shown in Table 1.

Table.1 The Results of Training and Testing.

SVM	Circuit state	Testing samples	Correct numbers	Accuracy %	Operation time (s)	Optimal parameter (C_i,γ_i)
SVM_1	Normal	25	22	88	23.5	(56.172, 0.1854)
	Fault	13	13	100		
SVM_2	Normal	15	14	93.3	25.3	(38.543, 0.6787)
	Fault	25	22	88		
SVM_3	Normal	10	8	80	22.8	(251.3, 1.0583)
	Fault	30	28	93.3		
SVM_4	Normal	10	9	90	25.3	(267.4, 0.1864)
	Fault	30	28	93.3		
SVM_5	Normal	15	13	86.7	24.7	(156.8, 0.8756)
	Fault	25	23	92		

Monte-Carloanalysis was carried out on the circuit and the sampling time was 10s. Each kind of fault classification is analysed for 120 times and the number of fault response data samples was total 600. Two-thirds samples were used for training and one third for testing.After the training of two-thirds sample data, the accuracy rate of the circuit's SVM fault diagnosis model is about 90%, which can meet the requirement of fault diagnosis for power distribution system.

4 Conclusions

To solve the problem of fault diagnosis of the surface to air missile equipment, taking the power distribution system as the study object, a fault diagnosis model based on Wavelet Transformation and SVM is established. Different wavelet basises and denoising method are compared. According to the characteristics of the power distribution system, the parameter of SVM's kernel function is optimized and five kinds of fault feature vector sets are trained based on the feature vectors from the monitoring node N_1. Simulations are made on the typical analog circuit of one surface to air missile power distribution system. Results show that the accuracy of the fault diagnosis model based on Wavelet Transformation and SVM is about 90%, which shows a good prospect in the fault diagnosis of the surface to air missile equipment.

References

1. ZHANG Lin,SUN Anquan,WANG Tianyi,et al. Intelligent fault diagnosis of certain missile equipment. Journal of Central South University(Science and Technology), 216-220.(2013)
2. LI Xue-liang,LIU Shi-bin,CHENdong,QIU Song-song. Denoising of LAPS signal based on wavelet transform. Journal of Optoelectronics • Laser,25(5):835-839. (2014)
3. GU Wen-cheng;CHAIBao-ren;TENG Yan-ping. Research on Support Vector Machine Based on Particle Swarm Optimization.Transactions of Beijing Institute of Technology, 34(7):705-709.(2014)
4. Lin Zhang, Liu Tao, etc. Research on Fault Diagnosis of Weapon Equipment Based on SVM.2012International Conference on Intelligent System and Applied Material GSAM,1242-1245. (2012)
5. ZENG Zuo-qin. Application of Wavelet Transform for Data Processing in Making and Breaking Test.Electrotechnics Electric.2:42-47.(2012)
6. Hu Liang-mou. Support Vector Machine fault diagnosis and control technology.Beijing: Academic Press.(2011)
7. An-quan Sun, Lin Zhang, Wen-feng Wang, et al. Robust FDI for a Class of Nonlinear Networked Systems with ROQs. Mathematical Problems in Engineering, 1-8.(2014)
8. WANG Chun-lin,ZHOUhao,LIGuo-neng.etc. Combining Support Vector Machine and Genetic Algorithm to Predict Ash Fusion Temperature.Proceedings of the CSEE, 08:11-15.(2007)
9. Lin H T, Lin C J. A study on sigmoid kernels forSVM and the training of non-PSD kernels by SMO-type methods[EB]. http:/www.csie.ntu. edu.Tw/~cilin/papers.Html.(2003)
10. ZHENG Jun,HOU Rui-feng. Selection of Wavelet Base in Denoising of Wavelet Transform. Journal of Shenyang University,21(2):108-110.(2009)

MATEC Web of Conferences **55**, 03005 (2016) DOI: 10.1051/matecconf/20165503005
ACPEE 2016

Research on De-noising of Fault Signal Based on Improved Multi-wavelets

ZHANG Jiali[1], WANG Jie[1], WANG Wenfeng[1,2,3], ZHANG Lin[1,a], ZHAO Haizhou[1], LEI Rongqiang[1]

[1.] *Air Defense and Antimissile Institute, Air Force Engineering University, Xi'an,710051, China*
[2.] *The 302ed Institute of the Tenth Academy of China Aerospace Science & Industry Corporation; Guiyang 550009, China*
[3.] *School of Automation Science and Electrical Engineering, Beihang University, Beijing 100191,China*

Abstract. Aiming at signal de-noising of fault diagnosis, we need obtain fault signal of high signal to noise ratio with multi-wavelets which need little prior knowledge and are uninfluenced by waveform matching. Here, this paper determines the wavelet basis through contrast experiment, and the decomposition level is confirmed by the principles of making noise fraction from details and making low energy from general picture, threshold is ensured by the method of midpoint optimization. The simulation shows that this method can enhance signal to noise ratio obviously.

1 Introduction

Fault diagnosis is to find out the reasons of object's triggering some operation function being not consistent with normal condition in the case of given working environment, judge the elements of arousing the occurrence of this condition, and even forecast the tendency and the loss of the development of the condition[1-5]. Basic flow path can be divided into four procedures commonly as Figure 1 shows:

1. Signal collection. Collect the representative fault signal in different conditions.
2. Signal de-noising. Dislodge the interfering signal in the collection signal.
3. Feature extraction. Extract the signal of reflecting fault's feature in the time-domain, frequency-domain or time- frequency domain of de-noising signal.
4. Fault diagnosis. According to relevant algorithm, judge the fault of characteristic signal.

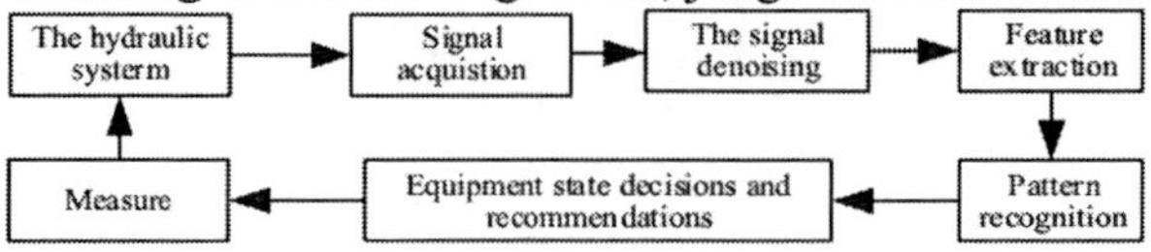

Figure 1.The flow chart of fault diagnosis.

As the key of fault diagnosis, fault de-noising normally has some methods at present, such as repeating average de-noising, multipoint smooth de-noising, fourier de-noising, wavelet de-noising

[a] Corresponding author : dreamland_0628@163.com

and so on [6-9].As a new method of signal de-noising in recent years, wavelet transform is a method of time-frequency analysis [10]. It can analyze a signal's details, and at the same time, can see the whole outline of signal through transforming time and frequency partly.

The scale function of multi-wavelet is generated by two or more functions, so it is less affected by wave form matching and has little requirement on the signal's prior knowledge. What's more, it can combine the smooth, symmetry, tight branch, perpendicularity and vanishing moment well. This paper obtains fault signal of high signal to noise ratio through improving multi-wavelet determination method in wavelet basis, wavelet decomposition level and various detail thresholds.

2 Improving Multi-wavelet Algorithm

2.1 The Brief Introduction of Multi-wavelet

When conducting signal de-noising process, multi-wavelet follows the steps as Figure 2 shows. Firstly, pre-filter noise signal. Secondly, perform threshold to fault signal that has experienced multi-wavelet decomposition. Thirdly, obtain the new reconfiguration signal through multi-wavelet inverse transformation. Finally, achieve target value by using post-filtering to dispose fault signal.

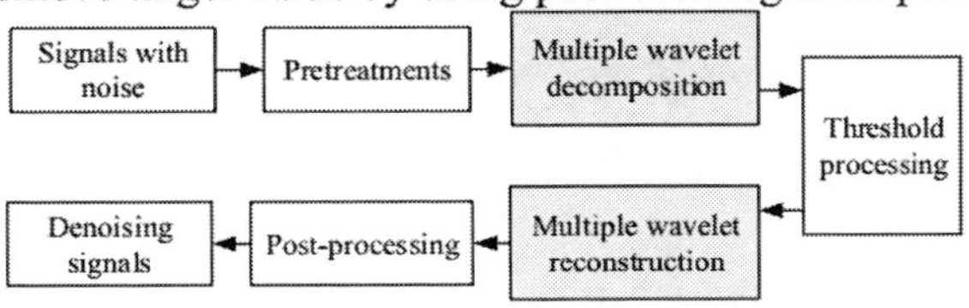

Figure 2. The process of wavelet de-noising.

The multi-resolution analysis of multi-wavelet de-noising consists of multi-scale function, but $L^2(R)$ space base consists of many traditional and flexible wavelet functions. Suppose $\Phi(t)=[\varphi_1(t),\varphi_2(t),\cdots,\varphi_r(t)]$ as r dimension scale function, $\Psi(t)=[\psi_1(t),\psi_2(t),\cdots,\psi_r(t)]$ as wavelet function, h_k and g_k as $r\times r$ dimension filter array. Therefore, the equation is

$$\Phi(t)=\sum_k h_k\Phi(2t-k) \quad k\in Z$$
$$\Psi(t)=\sum_k g_k\Phi(2t-k) \quad k\in Z \tag{1}$$

Set $s_{j\text{-}1,n}$ as r dimension low-frequency component, $d_{j\text{-}1,n}$ as r dimension high-frequency component, h_k^*、g_k^* as the conjugate transpose of h_k and g_k. The decomposition and reconfiguration of multi-wavelet follow the equation

$$s_{j\text{-}1,n}=\sum_k h_{k\text{-}2n}s_{j,k}, d_{j\text{-}1,n}=\sum_k g_{k\text{-}2n}s_{j,k} \tag{2}$$

$$s_{j,n}=\sum_n (h^*_{k\text{-}2n}s_{j\text{-}1,n}+g^*_{k+2n}d_{j\text{-}1,n}) \tag{3}$$

2.2 The Choice of Multi-wavelet Index

In the procedure of multi-wavelet signal de-noising, three prime factors of affecting the effect of signal de-noising are the structure of multi-wavelet, the number of piles choice of wavelet decomposition, and the determination of detail threshold.

2.2.1 The structure of multi-wavelet

Wavelet transformation is different from Laplace transform and Fourier transform that have only one fixed kernel function. Not only does it not have fixed wavelet base function, it also has great difference in systematic ness, support length and wave path .However, the three conditions as follows must be fitted for wavelet base function:

Allowance condition: $\int_{-\infty}^{+\infty}\frac{|\omega|^2}{\omega}d\omega<\infty$ which guarantees the inversion geometry of wavelet transform. And volatility: $\int_{-\infty}^{+\infty}\psi(t)d\omega=0$.

Time-frequency attenuating property: the frequency spectrum of wavelet base function must be committed in a relevant narrow frequency ban. And at the same time, in time threshold, it also must be concentrated on stated small interval.

Besides, when choosing wavelet base function for the practical apply, it is necessary to consider perpendicularity, symmetry, tight branch, vanishing moment and similarity synthetically.

In addition to the conditions above, the choice of multi-wavelet base should follow the principles of high time-frequency resolution ratio, small decomposition index ratio and restructure error, estimated variance approaching practice .Among the three conditions, the prior two conditions are used to improve the effect of signal de-noising and the third condition is used to simplify calculating steps.GHM multi-wavelet is symmetrical, double orthogonal multi-wavelet, obtained by Geronimo Hardin and Massopust through complicated and fractal interpolation method. Commonly, we use GHM multi-wavelet and Db2 wavelet to process the simulation signal of the pitting fault of bearing outer ring, and prove the superiority of GHM in the aspect of signal de-noising .Therefore, this paper choose GHM to conduct de-noising.

2.2.2 The choice of number plies of multi-wavelet decomposition

Set the number of plies of multi-wavelet decomposition of collection frequency as f_s, the maximal frequency of the prime energy interval of fault signal as $f_m(f_m>0)$.if prime energy is in the frequency scale of the latest plié of general picture, then

$$\frac{f_s}{2^{n+1}}>f_m \tag{4}$$

Signal experienced multi-wavelet decomposition at least exists a detail component ($n\geq 1$), so that we can obtain equation (5), what's more, we obtain equation(6)

$$\frac{f_s}{2^{n+1}}\leq\frac{f_s}{2^2} \tag{5}$$

$$f_s>2^2f_m \tag{6}$$

The more the number of plies of wavelet is, the more detail component will be generated and the more amount of noise will be wiped out. We can obtain the most detail components based on equation (7).Because the number of plies of multi-wavelet decomposition is a non-negative positive integer, the final certain equation of the greatest number of plies of wavelet decomposition of wavelet de-noising can be indicated as equation (8).Literature has detailed discussion.

$${f_s}/{2^{n+2}}\leq f_m\leq {f_s}/{2^{n+1}} \tag{7}$$

$$n=ent(lb\frac{f_s}{2^2f_m}) \tag{8}$$

2.2.3 The determination of various detail thresholds of multi-wavelet

In the common methods of wavelet threshold processing, necessary characteristic component energy only fixes on some limited wavelet coefficient, however, the noise energy of fault signal almost scatters to the whole wavelet domain, so it needs to choose appropriate threshold .Set $\lambda=\sigma\sqrt{2lnN}$ (N is the length of signal), σ as the standard deviation of noise. It is regular to use equation (9) to estimate threshold in engineering.

$$\sigma=\frac{median|w_{i,j}|}{0.6745} \tag{9}$$

So the definition of hard threshold method is:

$$\widehat{w}_{i,j}=\begin{cases} w_{j,k}, & |w_{j,k}|\geq\lambda \\ 0 & |w_{j,k}|<\lambda \end{cases} \tag{10}$$

In hard threshold function, wavelet coefficient $\widehat{w}_{i,j}$ is in continuous between λ λ and $-\lambda$.

$$\widehat{w}_{i,j}=\begin{cases} \operatorname{sgn}(w_{j,k})(|w_{j,k}|-\lambda), & |w_{j,k}|\geq\lambda \\ 0 & |w_{j,k}|<\lambda \end{cases} \tag{11}$$

In soft threshold function, because of the contract processing of wavelet coefficient, there is unavoidable error between $\widehat{w}_{i,j}$ and $w_{j,k}$. Meanwhile, it is not consistent, which generates error hard to be eliminated in hard threshold method and soft threshold method. In order to make up for the inconsistence of hard threshold method and reduce the constant error of soft threshold method, this paper uses a new threshold function, the function is as equation (12) shows:

$$\widehat{w}_{i,j}=\begin{cases} \operatorname{sgn}(w_{j,k})\left(|w_{j,k}|-\alpha\lambda-\frac{\lambda-\alpha\lambda}{\exp(w_{j,k}-\lambda)}\right), & |w_{j,k}|\geq\lambda, 0\leq\alpha\leq1 \\ 0 & |w_{j,k}|<\lambda \end{cases} \tag{12}$$

In the equation, α is between 0 and 1.above equation, ifα=0, threshold curve will become soft threshold curve; if α=1, the asymptotic line of threshold curve is hard threshold curve. Although, using the strike an average method, in which set α=0.5, increases signal to noise ratio, α is not most one. This paper search optimal α(set α as α_{mid})by the means of mid-point method, and the flow path of the algorithm is as Figure 3 shows.

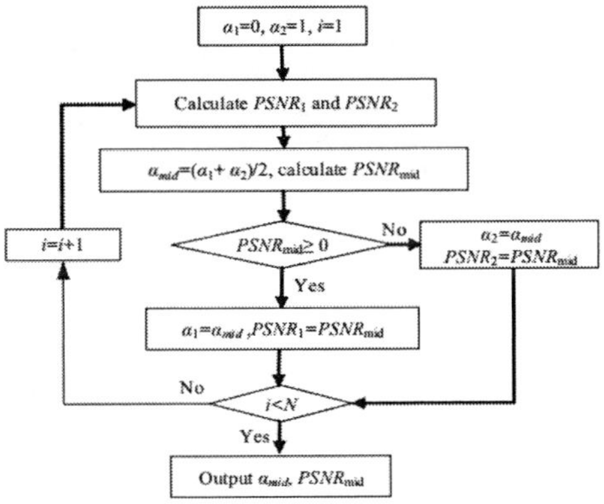

Figure 3. The algorithm calculate flow

3 Simulation Confirmation

In order to check the effectiveness of improving multi-wavelet algorithm, This paper chooses simulation signal$f(t)=\sin(10\pi t)+0.7\sin(24\pi t)$to check it. After adding 3dB signal to noise ratio, collection frequency is set to be 1024Hz, and the time of lasting collection is 1s. And we use the method in chapter 2.2 to determine Multi-wavelet base, decomposition and threshold.

3.1 The pretreatment and reconfiguration of GHM multi-wavelet

As GHM is a multi-wavelet, where r=2, for one-dimensional fault signal $x[n]$,we can process it and obtain two-dimensional input signal $s_{0,n}$. In order to keep the trait of system first-order vanishing moment trait of system, we usually choose $c=\sqrt{2}$.

$$s_{0,n}=\begin{pmatrix} x[n] \\ c\cdot x[n] \end{pmatrix} \tag{13}$$

The coefficient of low-pass of GHM multi-wavelet can be indicated as

$$h_0=\begin{bmatrix}\frac{3}{10} & \frac{2\sqrt{2}}{5}\\ -\frac{\sqrt{2}}{40} & -\frac{3}{20}\end{bmatrix} h_1=\begin{bmatrix}0 & 0\\ -\frac{9\sqrt{2}}{40} & -\frac{3}{20}\end{bmatrix} h_1=\begin{bmatrix}0 & 0\\ -\frac{9\sqrt{2}}{40} & -\frac{3}{20}\end{bmatrix} h_1=\begin{bmatrix}0 & 0\\ -\frac{\sqrt{2}}{40} & 0\end{bmatrix} \tag{14}$$

The coefficient of high-pass of CHM multi-wavelet can be described as

$$g_0=\begin{bmatrix}-\frac{\sqrt{2}}{40} & -\frac{3}{20}\\ -\frac{1}{20} & -\frac{3\sqrt{2}}{20}\end{bmatrix} g_1=\begin{bmatrix}\frac{9\sqrt{2}}{40} & -\frac{1}{2}\\ \frac{9}{20} & 0\end{bmatrix} h_1=\begin{bmatrix}\frac{9\sqrt{2}}{40} & -\frac{3}{20}\\ -\frac{9}{20} & \frac{3\sqrt{2}}{20}\end{bmatrix} h_1=\begin{bmatrix}-\frac{\sqrt{2}}{40} & 0\\ \frac{1}{20} & 0\end{bmatrix} \tag{15}$$

When retreating GHM multi-wavelet, we can retreat it longitudinally, and retreat it breadthwise, then process post-filtering .Post-filter $Q(k)(k=0,1)$ can be indicated as

$$Q_0=\begin{bmatrix}0 & 0\\ \frac{4\sqrt{2}}{5} & -\frac{3}{10}\end{bmatrix} Q_1=\begin{bmatrix}0 & 0\\ 0 & -\frac{3}{10}\end{bmatrix} \tag{16}$$

3.2 Simulation Experiment

Simulation original signal and noise signal are as Figure 4 shows. Bringf_s=1024 and f_m=12 into equation (8), and obtain the optimal number of plies of plies of multi-wavelet decomposition, 4 and α, the value of threshold function is 0.6213 based on Figure 3.In order to judge the veracity of the number of plies of wavelet decomposition in this paper, set the number of plies of wavelet decomposition as, n=3, n=4 and n=5, under the same condition, and compare their effects of signal de-noising. The effect of simulation signal effect is as Figure 5 shows, we can see when n=3, the effect wave form of signal de-noising is obviously variant, and when n=4or n=5, the effect is similar.

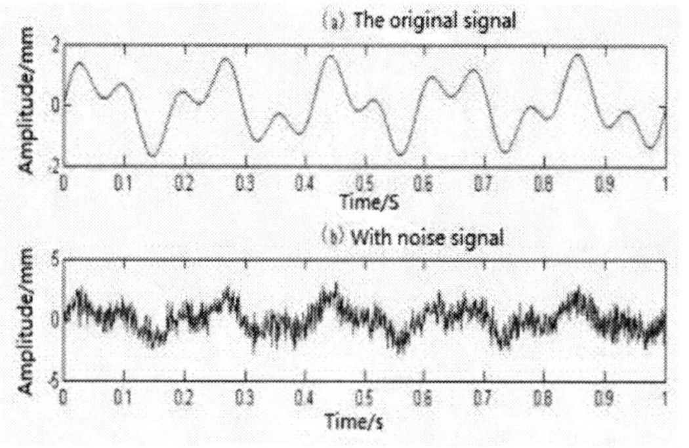

Figure 4. The simulation signal.

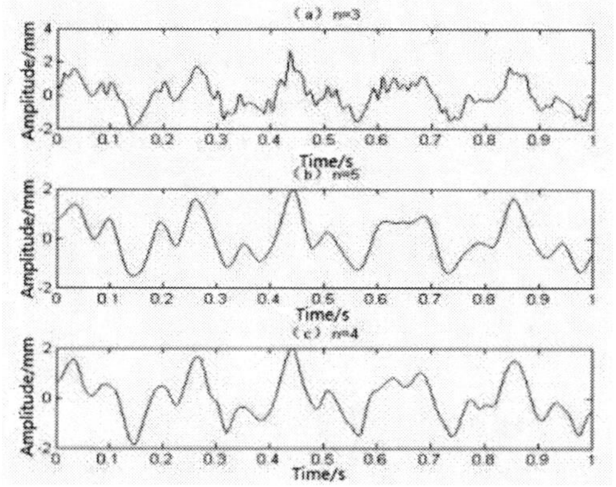

Figure 5. The simulation signal de-noising effect.

Figure 6 provides the de-noising effect frequency spectrum of simulation signal, we can see that when n=5,frequency component changes greatly besides 20Hz,which is different from original signal. When n=3, de-noising signal exists error in high frequency. However, when n=5, the prime frequency trait of simulation signal is the same as original signal. Therefore, we can draw a conclusion that when n=4 the effect is best. To sum up, the proposed method of wavelet decomposition can resolve low frequency into general picture effectively.

In order to check the effectiveness of detail threshold determination method proposed in this paper, the effects of hard threshold de-noising, soft threshold de-noising and threshold de-noising given by this paper are as Figure 7 shows in the same condition, the effect of threshold de-noising proposed in this paper is obviously better than the effect of soft threshold de-noising and hard threshold de-noising.

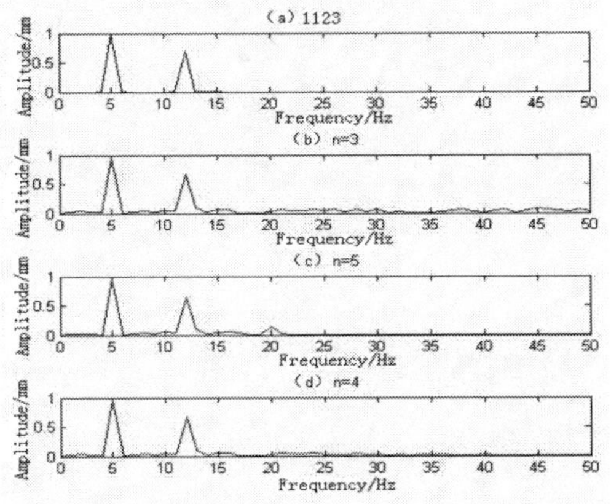

Figure 6. The signal de-noising effects of frequency domain.

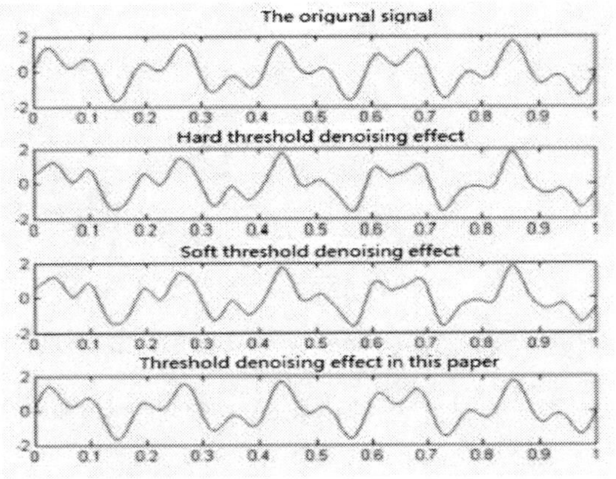

Figure 7. De-noising effect with different thresholds.

4 Conclusion

According to the principles of reflecting the low-frequency component decomposed into general picture and decomposing noise component into details, we choose the method of the number of plies of wavelet decomposition in fault signal de-noising. Through a midpoint optimization method, we determine the optimal in wavelet de-noising threshold parameter, ensuring the fault signal with high signal-to-noise ratio. Meanwhile, simulation experiments have verified the correctness of this method.

References

1. Wu Jinpei, Xiao Jianhua. Intelligent fault diagnosis and expert system. Science Press.(1997)
2. Wang Yilong, Yang Shouzhi. The wavelet denoising method of based consistent threshold function. Shan Tou University Academic Journal (Natural Science), 28(4):66-74.(2013)
3. Mallat S,Hwang W L. Singularity detection and processing with wavelet. IEEE Transaction Information Theory, 38(2):617-643.(1992)
4. Lavrik I, Young Jung Y, Ruggeri F, Vidakovic B. Bayesian false discovery rate wavelet shrinkage: theory and applications. Communications in Statistics-Simulation and Computation, 37(6): 1086-1100.(2008)
5. Xu, Y., Weaver, J. B., Healy, Jr., D. M., etal. Wavelet transform domain filters a spatially selective noise filtration technique. IEEE Trans Image Processing, 3(6): 747-758.(1994)
6. Banham M R, Katsaggelos A K. Spatially adaptive wavelet-based multiscale image restoration. IEEE Trans. Image Processing, 5(4): 619-634.(1996)
7. Mariantonia Cotronei, Laura B Mntefusco, Luigia Puccio. Multiwavelet analysis and signal processing. IEEE Transactions on Circuits and Systems- II : Analog and Digital Signal Processing, 45(8): 970-987.(1998)
8. Gao Taichang, Han Xiaodong, Qu Shi. The electric field variability signal denoising of multi-wavelet transform. The PLA university of science and technology Academic Journal, 10(12):90-94.(2009)
9. Xiang Dongyang, Wu Zhengguo, Hou Xinguo, Zhao Hua. Improving the denosing algorithm of multi-wavelet transform coefficient. High voltage technology, 37(7):1728-1733.(2011)
10. Zhang Wenjuan, Zhou Dandan,Wang Ling. A improvement project of wavelet threshold denoising. Computer Development and Applying, 20(9):28-29.(2007)

MATEC Web of Conferences **55**, 03006 (2016) DOI: 10.1051/matecconf/20165503006
ACPEE 2016

Research for the Influence of Distribution Network Line Reclosing Current on Line Protection

YU Kansheng[1,a], WU Ying[2], FAN Ruixiang[1],PAN Benren[1],CAO Bei[1]

[1] *Jiangxi Electric Power Research Institute, 330096 Nanchang ,China*

[2] *Jiangxi Electric Power CO. Maintenance Branch,330096 Nanchang,China*

Abstract. According to the distribution network line structure and reclosing control strategy, the system simulation model of distribution network lines has established based on the real-time digital simulation RTDS. Based on this, distribution network switching impulse current characteristic has researched with different capacity, different distribution and different load power factor under the different voltage switching angles. The results of the study provide a scientific basis for distribution network line protection setting, in order to further lay the foundation for improvement the validity and reliability of distribution network line protection action.

1 Introduction

With the continuous random access of residential load and commercial unit into distribution network, the load current of distribution network increases constantly. Meanwhile, grid-connected operation of new distribution transformers which come into service recently has to some extent changed the system structure and power flow distribution. Therefore, a higher demand of safe and reliable operation of power network protection and control equipment is proposed. For example, the reclosing impact current of distribution network line will change, and the amplitude and fall time of impact current will change to a certain extent[1,2]. Post- acceleration protection setting of distribution network was studied deeply and comprehensively in literatures [3-5], however, only the determination of post-acceleration protection set point is studied from the single aspect of setting calculation, but the calculation did not involve the distribution situation of distribution transformer capacity of distribution line, as well as the parameters and load character of distribution line. Obviously, the accuracy level of post-acceleration protection set point of setting is poor, which then lead to the reduction of the success rate of reclosing of distribution line.

The establishment of technology test platform for distribution network line based on real-time digital simulation technology and line protective device is an efficient way to solve above problem[6]. In this paper, 10kV system distribution line is taken as research object, a distribution line real-time simulation platform based on RTDS is established, and on this bases, the characteristics of closing impact current of distribution line under different distribution of distribution transformer capacity, different load power factor of low voltage side of the distribution transformer and different voltage

[a] Corresponding author : 249245208@qq.com

DOI: 10.1051/matecconf/20165503006

closing angle is studied, and the main factors which influence the characteristics of closing impact current is sorted out.

2 Real-time Simulation Model of Distribution Line Simulation

2.1 Modular Design of Real-time Simulation

RTDS model includes 110kV infinity system model, 110kV-10kV step-down transformer model, 10kV switch model, 10kV distribution network line model, 10kV transformer model and 0.4kV low voltage load model. When modeling, simulation precision requirement of single transformer's magnetizing rush current should be met, meanwhile, limits on the number of compute nodes by hardware of simulation system should be considered. Therefore, it is best to use the modularized design method, and to optimize the allocation of RTDS processor cards.

2.2 Modeling Progress of RTDS

Real-time simulation modeling of distribution network lines can be divided into four parts: main loop of the distribution network wiring system, breaker analog output model, breaker digital input and output model, and fault control strategy model.Main loop of the distribution network wiring system uses infinite power to replace 110kV power network system. The control strategy model is established according to corresponding expressions and control structure diagram and based on the component model library of RTDS control system.

3 Field Measurement Experiment of Distribution Transformer

3.1 Modular Design of Real-time Simulation

In the analysis of the influence from distribution network line reclosing impact current to the line protection, two important parameters should be mainly considered, which are over current ration in acceleration section and delay time ration of independent setting in acceleration section. In simulation modeling, the lines and transformer model which determine the current current ration can be built correctly, but fall time of distribution transformer magnetizing rush current, which is related to time setting of protection in acceleration section, is unable to be obtained from known parameters of distribution transformer. It is required to measure the distribution transformer on field to obtain the fall time of magnetizing rush current and achieve the parameter identification of distribution transformer in simulation experiment. In the experiment, the field measurement is according to capacity of 4MW, 8MW and 10MW. The constant identification[7] of fall time of distribution transformer is approximated as 0.3.

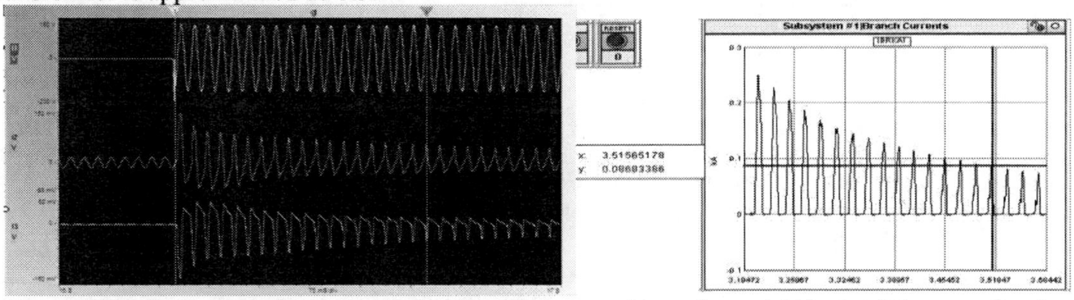

Figure 1. Distribution transformer no-load test(Left is measured waveform, right is simulation waveform)

4 Research on Real-time Simulation Experiment

4.1 Scheme on Real-time Simulation Test

To research the influence of distribution network line's transformer capacity, distribution, load power factor and closing angel to distribution network line closing impact current, bracketing method is used in the real-time simulation experiment, as shown in Table 1, where:

Table 1. Table of experimental factors

Closing angle (°)	0	30	60	90
Load power factor	0.5	0.6	0.75	No-load
Total capacity of distribution transformer(MVA)	4	8		
Distribution of distribution transformers	"136"	"334"	"631"	

(1) "136" denotes that the distribution of transformers of distribution network line is roughly that the transformer capacity close to exit section (forepart) of transformer substation is low, the capacity of distribution transform at end of line is high, and the capacity of distribution transform at mid-piece of line is between two parties.

(2) "334" denotes that the distribution of capacity of distribution transform at exit section of transformer substation, mid-piece and end of line is uniform.

(3) "631" denotes that the distribution of transformers of distribution network line is roughly that the transformer capacity close to exit section (forepart) of transformer substation is high, the capacity of distribution transform at end of line is low, and the capacity of distribution transform at mid-piece of line is between two parties.

4.2 Real-time Simulation Analysis

4.2.1 Characteristic analysis of distribution network line no-load closing impact current

In the simulation, the total capacity of distribution transformer is 8MVA, and after cutting of the switch while short circuit fault happens, the residual magnetic of distribution transformer is kept in a low level, the experiment data are shown in Table 2. Table 2 shows that under same closing angle, different distribution will lead to different closing impact current, and the higher capacity away from power side (outlet switch of transformer substation), the higher closing impact current is. For same distribution, the amplitude of closing impact current is different under different closing angle. The amplitude of closing impact current decreases with increase of closing angle, and the amplitude of closing impact current is maximum when 0°.

Table 2. The maximum peak of no-load current

Impact current(A)	"136"	"334"	"631"
0°	447.77	687.64	1017.9
30°	437.87	723.79	997.12
60°	356.16	596.61	849.52
90°	153.98	267.87	368.86

4.2.2 Characteristic analysis of impact current under different power factor and load feature

In the simulation, the total capacity of distribution transformer is 8MVA, and the distribution is "334", the experiment data are shown in Table 3. Table 3 shows that, considering the error of simulation, under same closing angle, the amplitude of impact current under different distribution, same capacity load and different power factor keeps essentially constant. It can be confirmed that under same load capacity, different power factor will not cause large change of amplitude of closing impact current.

Table 3. The maximum peak of current under different power factor

Impact current (A)	Power factor is 0.5	Power factor is 0.6	Power factor is 0.75
0°	649.12	618.67	608.98
30°	569.92	566.46	529.41
60°	437.12	434.58	405.92
90°	363.37	367.67	338.48

4.2.3 Characteristic analysis of impact current under different distribution of distribution transformers

In the simulation, the total capacity of distribution transformer is 8MVA, and the power factor is 0.5, the experiment data are shown in Table 4. Table 4 shows that under same closing angle and different distribution of distribution transformers, the closing impact current is different. The characteristic is represented as that under the same closing angle, the higher distribution transformer capacity away from power side, the higher closing impact current is.

Table 4. The maximum peak of current under different distribution

Impact current(A)	"136"	"334"	"631"
0°	649.12	1000.34	1357.55
30°	569.92	848.22	1224.90
60°	437.12	668.86	955.85
90°	363.37	450.80	857.41

4.2.4 Characteristic analysis of impact current under different capacity of distribution transforms

In the simulation, the power factor of low voltage of distribution transformer is 0.5, and the experiment data are shown in Table 5. Table 5 shows that, under same closing angle and same distribution of distribution transforms, for different total distribution transform capacity, the closing impact current is different. The characteristic is represented as that under same closing angle, the higher total distribution transform capacity, the higher closing impact current is.

Table 5. The maximum peak of current under different capacity

Impact current(A)	Closing angle(°)	"136"	"334"	"631"
8MVA	0	649.12	1000.3	1357.55
4MVA	0	408.28	744.18	972.05
8MVA	30	569.92	848.22	1224.90
4MVA	30	347.25	735.37	910.70
8MVA	60	437.12	668.86	955.85
4MVA	60	288.42	513.63	409.52
8MVA	90	363.37	450.80	857.41
4MVA	90	236.00	352.97	405.91

4.3 Analysis of Fall Time of Closing Impact Current

On the basis of above studies, we select four typical lines with different distribution transformer capacity, under different closing angles, to analyze the needed time that closing impact current falls to selected current value.

(1)"631" distribution:

Table 6. Fall time of impact current under "631" distribution

Time(s)		4M	6M	8M	10M
165A	231A	0.14	0.14	0.14	0.14
220A	308A	0.1	0.1	0.1	0.1
247A	346A	—	0.1	0.1	0.1
330A	462A	—	0.08	0.06	0.06
440A	616A	—	—	0.06	0.06
412A	577A	—	—	—	0.06
550A	770A	—	—	—	0.04

(2)"136" distribution:

Table 7. Fall time of impact current under "136" distribution

Time(s)		4M	6M	8M	10M
165A	231A	0.04	0.06	0.06	0.08
220A	308A	0.02	0.04	0.04	0.06
247A	346A	—	0.04	0.04	0.06
330A	462A	—	0.04	0.02	0.04
440A	616A	—	—	0.02	0.02
412A	577A	—	—	—	0.02
550A	770A	—	—	—	0.02

(3)"334" distribution:

Table 8. Fall time of impact current under "334" distribution

Time(s)		4M	6M	8M	10M
165A	231A	0.12	0.12	0.10	0.14
220A	308A	0.08	0.10	0.06	0.10
247A	346A	—	0.08	0.04	0.08
330A	462A	—	0.06	0.04	0.06
440A	616A	—	—	0.04	0.04
412A	577A	—	—	—	0.04
550A	770A	—	—	—	0.04

(4)"361" distribution:

Table 9. Fall time of impact current under "361" distribution

Time(s)		4M	6M	8M	10M
165A	231A	0.12	0.12	0.12	0.14
220A	308A	0.08	0.10	0.06	0.10
247A	346A	—	0.08	0.06	0.06
330A	462A	—	0.06	0.04	0.06
440A	616A	—	—	0.04	0.04
412A	577A	—	—	—	0.04
550A	770A	—	—	—	0.04

Explanation: in the above four tables (Table 6-9), the first column denotes the maximum load current value of the line; the second column denotes post-acceleration protection setting value, which is 1.4 times of maximum load current; the first row denotes different typical distribution transformer total capacity, where 4M denotes the total capacity of distribution transformer is 4MW.

Through the simulation analysis of four typical lines, we find that there are some regular pattern of fall time of impact current under same capacity but different distribution, which is represented as that the higher capacity away from power source, the longer fall time of impact current is.

5 Real-time Simulation Experiment of Relay Protection Device

5.1 Structure of Closed-Loop Experiment System

The closing post-acceleration protection of distribution network line is mainly influenced by closing impact current, and different distribution of distribution transformer capacity will influence the closing impact current.

The structure of test system is as shown in Figure 2:

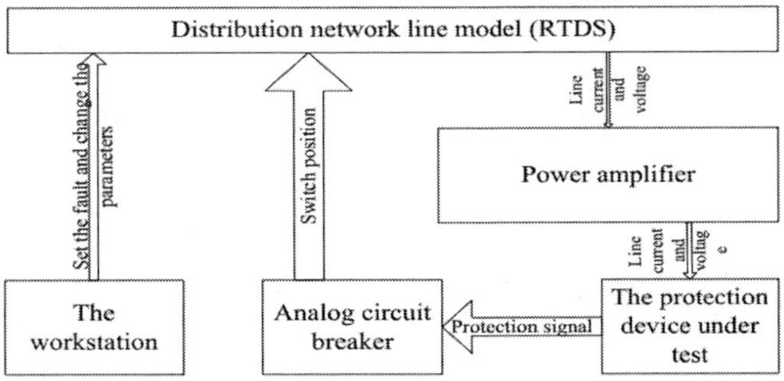

Figure 2. Structure of testing system

The real-time simulation model of distribution network line is established via parameters of unsuccessful reclosing to simulate the scene of unsuccessful reclosing when fault tripping. 10kV fault current (as shown in Figure 2) and voltage is output from RTDS, through the power amplifier and input to the protection device to be measured. Via simulating real-time change of CT and PT sampled values, fixed value setting is provided to the protection device, and a closed-loop test platform[8] of relay protection device for distribution network line is formed.

5.2 Model Experimental Verification

Real-time in-loop simulation system is built via the case line. Before the experiment, Setting parameters of line protective device are as follows: where the transformation ratio of protective device is 79.3.

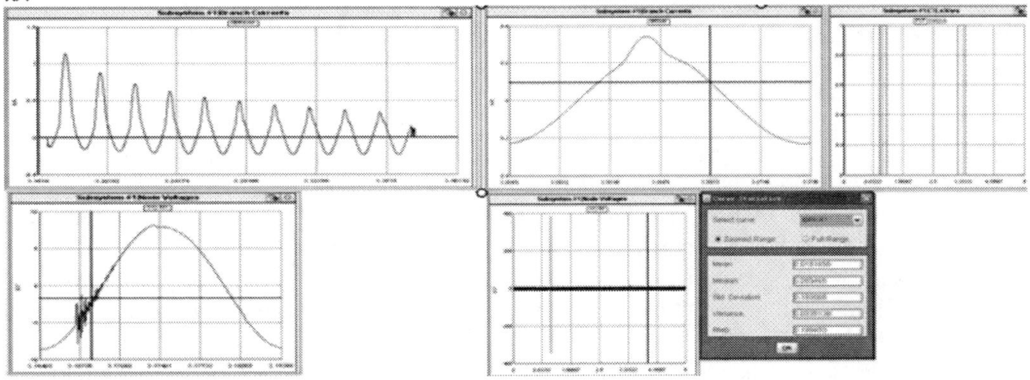

Figure 3. Structure of validation testing system

The field of unsuccessful closing of distribution network line in the case is simulated via the validation experiment of real-time in-loop simulation system model (Figure 3), and the model's reliability is verified.

5.3 Post-acceleration setting time reference table

When it is difficult to define which distribution the transmission line is, combined with analysis in section 3, the post-acceleration current setting is set as the maximum load current setting, the Post-acceleration setting time reference table is as shown in Table 10.

Table 10. Fall time of impact current under different setting value and different capacity

Time(s)		4M	6M	8M	10M
165A	231A	0.14	0.14	0.14	0.14
220A	308A	0.10	0.10	0.10	0.10
247A	346A	—	0.10	0.10	0.10
330A	462A	—	0.08	0.06	0.06
440A	616A	—	—	0.06	0.06
412A	577A	—	—	—	0.06
550A	770A	—	—	—	0.04

6 Conclusions

In this paper, a distribution network line reclosing simulation model is established based on real-time digital simulation technology. Through packaging design of submodule, the advantages of RTDS modeling, such as expansibility and easy to be used in closed-loop test, are made good use. The real-time simulation experiment platform is established, then on this base, characteristics of reclosing impact current under different distribution transformer capacity distribution and different operation mode are researched, and magnetizing rush current fall time reference table under four typical capacity distribution is proposed, which provide the basis for post-acceleration setting of distribution network line relay protection device, ensure the protection device to avoid the magnetizing rush current and mistrip, and improve the success rate of distribution network line reclosing and the power supply reliability.

References

1. Lü Li-ting, Luo Jian, Huang Zheng-xuan, Ou Zu-guo, Simulation and analysis of the inrush of power transformer, *Relay* , **05**(2007)
2. Hao Zhi-guo, Zhang Bao-hui, Chu Yun-long, Trend and Situation of Distinguish Technology of Transformer Magnetizing Inrush Current J, *Transformer*, **07**(2005)
3. Huang Fengpin, Exploration on Protection Setting Coordination of 10 kV Distribution Lines J, *Guangdong science &technology*, **20**(2011)
4. Ma Enda, Computation of Protection Setting Coordination of 10 kV Distribution Lines, *Science &technology information*, **09**(2012)
5. Chen Jia-lin, Brief analysis setting calculation and application of 10kV overhead line protection,*Science &technology information* , **28**(2006)
6. Liang Zhi-chen, Ma Xian-dong, Wang Like, KuffelR, Jiang Xiao, Realtime digital simulator RTDS and its application, *Automation of electric power Systems*, **10**(1997)
7. Suonan Jia-le, Kang Xiao-ning, Song Guo-bing, Jiao Zai-bin, Yun Bao-ji, Survey on Relay Protection Using Parameter Identification, *Proceedings of the electric power system and automation,***01**(2007)
8. Zhou Wei, Zhang Pei-chao, Yang Xing-xing, A real-time closed-loop digital simulation system for computer protections based on RTDS, *Power system protection and control*,**16**(2010)

MATEC Web of Conferences **55**, 03007 (2016) DOI: 10.1051/matecconf/20165503007
ACPEE 2016

Deep Learning for Intelligent Substation Device Infrared Fault Image Analysis

Ying Lin[1,a], Jianjun Sun[1], Chengqi Li[1], Yan Ma[1], Yujie Geng[1] and Yufeng Chen[1]
[1]*State Grid Shandong Electric Power Research Institute, Jinan, China*

Abstract. As an important kind of data for device status evaluation, the increasing infrared image data in electrical system puts forward a new challenge to traditional manually processing mode. To overcome this problem, this paper proposes a feasible way to automatically process massive infrared fault images. We take advantage of the imaging characteristics of infrared fault images and detect fault regions together with its belonging device part by our proposed algorithm, which first segment images into superpixels, and then adopt the state-of-the-art convolutional and recursive neural network for intelligent object recognition. In the experiment, we compare several unsupervised pre-training methods considering the importance of a pre-train procedure, and discuss the proper parameters for the proposed network. The experimental results show the good performance of our algorithm, and its efficiency for infrared analysis.

1 Introduction

Infrared fault image is an important way to evaluate the situation of substation device. According to previous study, some fault of substation device can be diagnosed by analysing its infrared image. Currently, infrared fault image analysis is mainly based on a manual way, which has high requirements for the professional quality of an operator. To improve the efficiency, studies turn to automatic extraction and analysis of the infrared images by artificial intelligent ways.

Currently, main manufacturers of thermal imaging cameras such as FLIR Systems Inc. provide some closed tools to show their specific data. These tools help the operators to get internal information of the thermal image. It seems that analyse the thermal information in signal ways is possible. Unfortunately, without the specific port and for other manufacturers, the tools cannot be used.

Most studies adopted basic imaging processing algorithms and did not make further analysis. Considering the difficulty to get internal data and drawback of previous woks, we turn to automatically detect the fault region and recognize its belonging device part in a novel image processing way. The former is an image segmentation process which takes advantage of superpixel segmentation method in this paper. Different from other segmentation methods aiming at edges or textures, a superpixel one emphasizes the homogeneous property of small superpixel regions. The latter is a typical pattern recognition problem which is applied by a machine learning method through exacted features. Different from typical feature description methods for visual images include SIFT[1], SURF[2], HOG[3] et al, we prefer to the deep learning method[4]. The popular deep learning methods

[a] Corresponding author : lysgwork@163.com

include deep belief networks (DBN) [5], convolutional neural network (CNN)[6], deconvolutional network (DNN)[7]and other algorithms [8-9].With the concept local receptive field, CNN method is more suitable for image classification. The performance of CNN for image recognition has already be proved by several practical applications [10-12].

Hence, the first contribution of our work is taking advantage of the imaging characteristics for infrared image segmentation. The use of superpixel segmentation makes the method suitable for various shapes of fault region. Our second contribution is incorporating the deep learning algorithms into unsupervised device part recognition. Not considering the selection of feature description, this way makes the classification result more reliable and robust.

2 Fault Region Detection

An infrared fault image generated from a thermal imaging camera is created from the temperature matrix and the given palette. The image can be divided into three parts: a) the background region; b) the device region; c) the fault region. Generally, the fault region is inside the device region and of great saliency in contrast to its neighbors. Different from the traditional visible light image, structured features such as textures, labels are lost in a regular infrared thermal image. Besides, pixels inside each kind of region are homogeneous with similar properties. Therefore, we adopt the concept of superpixel segmentation. The original infrared image is firstly divided into several homogeneous regions. The insight property of the fault region is further used to detect the demanded region.

2.1 Fault Region Segmentation

Superpixel algorithms group pixels into perceptually meaningful atomic regions, which can be used to replace the rigid structure of the pixel grid. We here choose three popular superpixel algorithms for compare: Turbopixel [13], SLIC [14] and Entropy rate superpixel (ERS) [15]. The segmentation results of a sample image are shown in Figure 1. We can see that SLIC and ERS algorithms are not suitable for our demands with their irregular boundaries. The regular lattice generated by Turbopixel maybe more suitable. The segmented regions are labeled as R_i; $i = 1; 2; \dots n$, n is the predefined segmentation number.

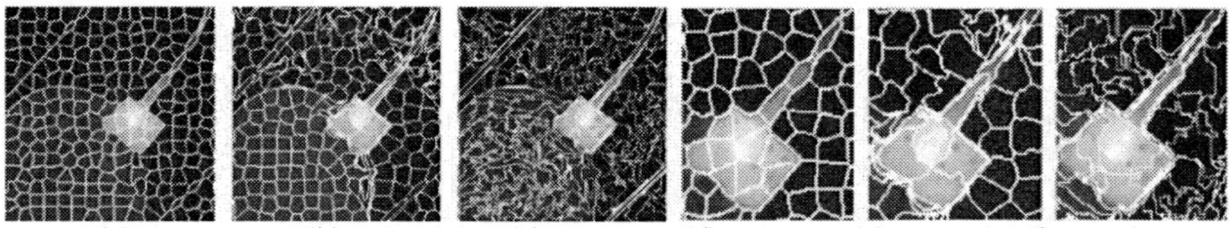

(a) (b) (c) (d) (e) (f)

Figure 1. Results of Different Superpixel Segmentation Algorithms. (a)(d) are the results of Turbopixel, (b)(e) are the results of SLIC and (c)(f) are the results of ERS.

There are several color palettes in a thermal imaging camera for the operator to select, such as Ironbow, BlueRed, Grayscale, Amber. Among these palettes, the Ironbow palette is proved to be suitable for the infrared fault analysis procedure. Based on this commonly used palette, we further choose the HSV color space for fault region detection. The focused fault region is more often displayed with nearly yellow and white color, which is:

$$white: r \approx g \approx b; yellow: r \approx g \gg b \tag{1}$$

Here, r, g, b are three channels in the RGB color space. To highlight these regions, we modify the standard hue value(H) equation as shown in (2). With (2), each R_i can be represented by its regional H value where taking the mean value $[\bar{r}_i, \bar{g}_i, \bar{b}_i]$ of the region's color as the regional color value as shown in Figure 3(b). With threshold H > 0.65 the binary image can be further obtained.

DOI: 10.1051/matecconf/20165503007

$$H=\begin{cases}1 & r,g,b>180\\ \frac{1}{6}(4+\frac{g-b}{\max-\min}) & \max=r\\ \frac{1}{6}(2+\frac{b-r}{\max-\min}) & \max=g\\ \frac{1}{6}|\frac{r-g}{\max-\min}| & \max=b\end{cases} \tag{2}$$

2.2 Fault and Device Region Validation

As mentioned above, the device region is the main object in the image. To diagnose the fault, we should know the location where the fault occurs and its belonging device part. We here choose three properties to find the most likely fault region: Average brightness $\overline{I}_{R_i}$: the average gray value of R_i ;Solidity S_{R_i} : the ratio between the region's area and area of its minimum convex hull; Center distance d_{R_i} : the distance from the image center to the region.

By maximizing the objective function (3), the fault region can be extracted as shown in Figure 3(c), and labelled as R_s.

$$\max_{R_i} \overline{I}_{R_i} S_{R_i} / d_{R_i} \tag{3}$$

The binarized images are shown in Figure 3 (d). As the fault region is inside the device region, the approximate device fault part can be greedily searched by taking the fault region as the searching center as shown in Figure 3 (e).

3 Deep Learning for Device Part Recognition

After the fault and device region are extracted, we should further recognize the type of the belonging device part. The type of the device part could show the most possible causes of heating and other properties. Here, we turn to choose the state-of-the-art deep learning algorithms for device parts recognition.

As shown in Figure 2, we change the network in [16] called a convolutional-recursive network (CRNN). The network is combined by two stages. The first stage is a CNN-layer with convolution and subsampling, while the second stage is a multi-layer RNN(recursive neural network) [9]. The features generated by the last two RNN layers are fully-connected to a softmax classifier.

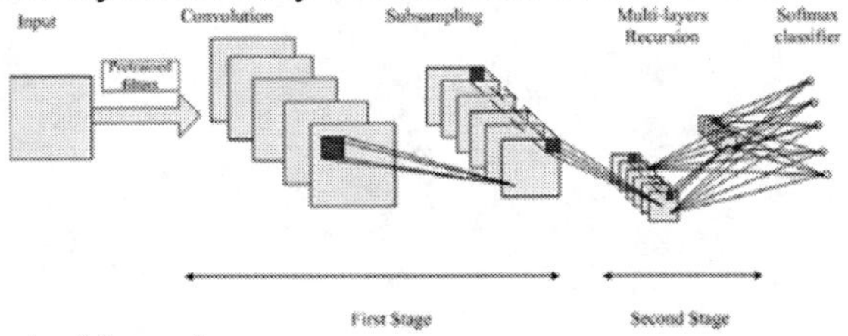

Figure 2. Convolutional-Recursive Network

3.1 Pre-train

Before trained by a CRNN network, the filters for convolution should better first pre-trained by unsupervised learning algorithms as discussed in [17]. As mentioned in [18] and [19], they found out that K-means algorithm attains best performance with only one parameter to tune. On the other hand, [20] proposed a new feature learning algorithm called sparse filtering. In this paper, we choose K-means, Sparse Filtering and RICA[21]algorithms for comparison. The comparison results are shown in the experiments part.

3.2 Training

The main idea of CNNs is to convolve filters over the input image in order to extract features. First of all, we resize all the input samples to size $p \times p$. Then convolve each sample with K_1 filters of size $r \times r$. When the convolutional stride is s, the output feature size is $p_1 \times p_1, p_1 = (p-r)/s+1$.With a subsampling block size $r_1 \times r_1$ and stride size s_1, we further generate the feature of size $p_2 \times p_2, p_2 = (p_1 - r_1)/s_1 + 1$.After the first CNN stage, the output feature size of each sample is $K_1 \times p_2 \times p_2$.In the RNN stage, all features in one layer are connected with features in the next layer. Each layer shares the same weights, and the feature size is gradually reduced until the final response size of each feature map is 1×1. Assuming the size of local receptive field is $l \times l$, the number of RNNs in each layer is K_2. The size of final feature of each sample when combing the last two layers' output is $K_2 \times K_1 \times (l^2+1)$.

4 Experiments

The performance of the analysis result mainly depends on the fault region detection, device part recognition with known device type. We have collected a large amount of infrared samples containing various device parts for evaluation.

4.1 Image Segmentation

Three images in which faults occur in different part to display the segmentation process are shown in Figure 3. Here, (a) is the input infrared images, (b) is the *H* value gray image, (c) is the validated fault region, (d) is the binarized image for device region, and (e) is the detected fault region and its located device part. The solid box represents the bounding box of fault region, and the dashed box stands for the device part needed to be classified in the deep learning step. The superpixel segmentation ensures pixels inside homogenous region to be smooth, contrast between fault region and the background region are strong enough in the *H* value gray image. Therefore, the fault region and the device region are easy to be extracted. From the results, we can see that the segmentation method is applicable for fault with different locations, shapes and sizes.

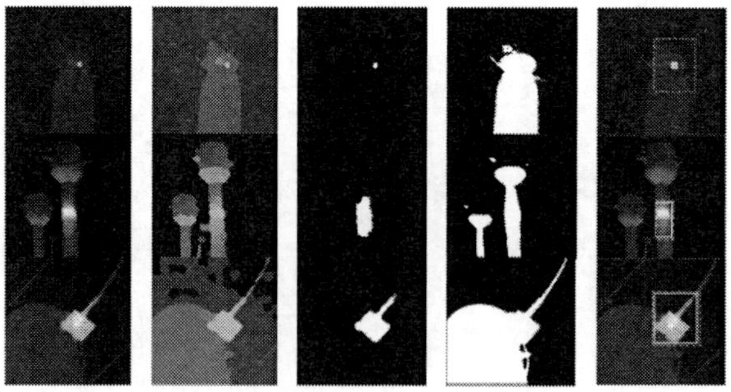

(a) (b) (c) (d) (e)

Figure 3. Image segmentation Procedure. (a) is the input infrared image, (b) is the H value gray image, (c) is the validated fault region, (d) is the binarized image for device region, and (e) is the detected fault region and its located device part.

4.2 Parameters Comparison

In this section, we take the analysis of a current transformer for example. According to controllable appearance and fault, the current transformer is divided into three parts: joint, inner join and bushing. Each class has 3000 samples for train, and the total number is 9000. 1500 images are chosen for test.

with enough samples of different part appearance, the classification algorithm could get comparable performance for other kind of devices such as voltage transformer, arrester and so on.

According to Sect.3, we here define the sample size as 67×67, the filter size as 8×8 with stride 1. After convolution, the output feature map size is 60×60. With pooling stride 2, the feature map size after the $1st$ stage is 27×27. We take the receptive field in RNN stage as 3×3, after 3 layers the size becomes 1×1. With the defined multi-layer feature output, the final feature dimension generated by each sample is $K_2 \times K_1 \times 10$.

When taking K_1 as 64, the optimized filters of the three pre-train algorithms can be shown in Figure 4. In this process, when taking advantage of sparse filtering, we find the objective function is hard to converge. The filters generated by sparse filtering are a little chaotic.

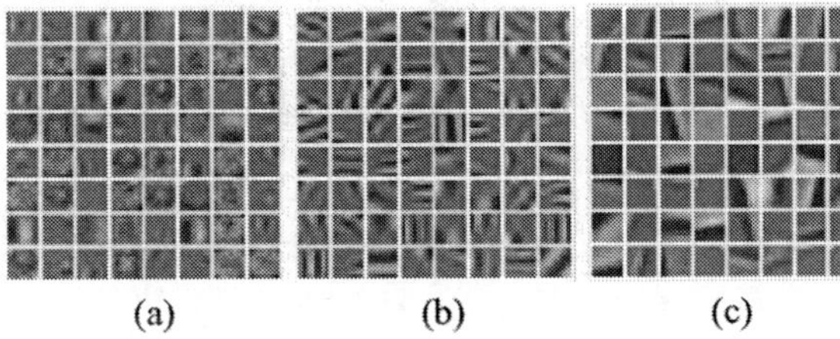

(a) (b) (c)

Figure 4. Pre-train Results. (a) is the filter from sparse filtering, (b) is the filter generated by RICA, (c) is generated by K-means.

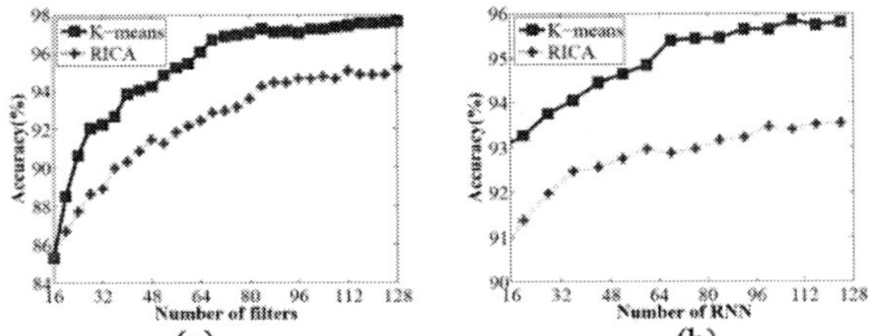

(a) (b)

Figure 5. Classification Accuracy with different parameters.

Therefore, our pre-train method and parameters decision is focus on RICA and K-means. K_1 is changed from 16 to 128 with step 4. The classification precision is shown in Fig.5. Then fix K_1 as 64, change K_2 from 16 to 128 with step 8, the result is shown in Fig.5. Similarly, we also find the performance of K-means is much better than that of RICA. When K_1 is larger than 64 and K_2 larger than 64, the number of filters and RNN does not influence much. Therefore, we take the parameters with $K_1 = 64, K_2 = 64$.

4.3 Discussion and Future Work

Our processing object is a purely infrared image generated by a thermal imaging camera without knowing its original temperature matrix. In this case, after region detection and recognition, we could know the area of fault region, its belonging device part, its location in the device and the temperature rise and difference in pseudo-color distance.The recognition of the temperature label is a typical digit recognition problem and already has mature solutions. With the recognized temperature range, we could get the precise temperature distribution by the look-up table of the selected Ironbow palette. This demand requires further study and is not the main purpose in this paper which would be realized in our future work. Meanwhile, further and deeply analysis would be practical based on the results of this paper.

5 Conclusions

In this paper, a novel method for intelligent infrared fault image analysis of substation device is presented. We first use a Turbopixel segmentation method to approximately segment an infrared image into multiple homogeneous regions. The regions are further checked by their hue property based on the most commonly used palette. To deeply analyze the fault, its belonging device part is classified by the CRNN network, which is the key point of the automatic analysis. From the experimental results, the algorithm of this paper is of high accuracy, and can be adaptable to infrared analysis of different substation devices. Due to the lake of temperature range, we do not make further statistics in this paper. And this is our aim of further study to get an entire automatic study of infrared fault image analysis for substation device.

Acknowledgement

This research is supported by the National High-Tech Research and Development Program of China under Grant No. 2015AA050204.

References

1. D. G. Lowe, *Proceedings of the International Conference on Computer Vision*, 1150-1157(1999).
2. H. Bay, A. Ess, T. Tuytelaars, L. V. Gool, Computer Vision and Image Understanding, **110**, 3, 346-359(2008).
3. N. Dalal, B. Triggs, *IEEE Conference on Computer Vision and Pattern Recognition*, **1**, 12, 886-893(2005).
4. Y. Bengio, Foundations and Trends in Machine Learning, **2**, 1(2009).
5. H. G. E., O. S., T. Y., Neural Computation, **18**, 7, 1527-1554(2006).
6. Y. Lecun, L. Bottou, Y. Bengio, P. Haffner, *Proceedings of the IEEE*, **86**, 11, 2278-2324(1998).
7. M. D. Zeiler, D. Krishnan, G. W. Taylor, R. Fergus, *IEEE Conference on Computer Vision and Pattern Recognition*, 2528-2535(2010).
8. H. Lee, C. Ekanadham, A. Y. Ng, *Advances in Neural Information Processing Systems 20*, 873-880(2008).
9. R. Socher, C. C. Lin, A. Y. Ng, C. D. Manning, *Proceedings of the 26th International Conference on Machine Learning*,129-136(2011).
10. A. Krizhevsky, I. Sutskever, G. E. Hinton, *Advances in Neural Information Processing Systems 25*, 1097-1105(2012).
11. S. Y., W. X., T. X., *IEEE Conference on Computer Vision and Pattern Recognition*, 3476-3483(2013).
12. O. Abdel-Hamid, A. Mohamed, H. Jiang, G. Penn, *IEEE International Conference on Acoustics, Speech, and Signal Processing*, **22**,10, 4277-4280(2012).
13. A. Levinshtein, A. Stere, K. N. Kutulakos, D. J. Fleet, S. J. Dickinson, K. Siddiqi, *IEEE Transactions on Pattern Analysis and Machine Intelligence*, **31**, 12, 2290-2297(2009).
14. R. Achanta, A. Shaji, K. Smith, A. Lucchi, P. Fua, S. Susstrunk, IEEE Transactions on Pattern Analysis and Machine Intelligence, **34**, 11, 2274-2282(2012).
15. M. Y. Liu, S. Ramalingam, O. Tuzel, R. Chellappa, *IEEE Conference on Computer Vision and Pattern Recognition*,2097-2104(2011).
16. R. Socher, B. Huval, B. Bath, C. D. Manning, A. Y. Ng, *Advances in Neural Information* Processing *Systems 25*, 656-664(2012).
17. E. D., B. Y., C. A., M. P., V. P., Journal of Machine Learning Research, **11**, 3, 625-660(2010).
18. A. Coates, H. Lee, A. Y. Ng, Journal of Machine Learning Research **15**,215-223(2011).
19. A. Coates A. Y. Ng, Neural Networks: Tricks of the Trade, 561-580(2012).
20. J. Ngiam, Z. Chen, S. A. Bhaskar, W. K. Pang, A. Y. Ng, *Advances in Neural Information Processing Systems*, 1125-1133(2011).
21. Q. V. Le, A. Karpenko, J. Ngiam, A. Y. Ng, *Advances in Neural Information Processing Systems 24*, 1017-1025(2011).

MATEC Web of Conferences **55**, 03008 (2016) DOI: 10.1051/matecconf/20165503008
ACPEE 2016

Reliability Evaluation of Distribution System Considering Sequential Characteristics of Distributed Generation

Wanxing Sheng[1], Xiaoli Meng[1], Tingting Fan[2] and Songhuai Du[2,a]

[1]*China Electric Power Research Institute, Power Distribution Department, 100192, Beijing*
[2]*China Agricultural University, College of Information and Electrical Engineering, 100083, Beijing*

Abstract. In allusion to the randomness of output power of distributed generation (DG), a reliability evaluation model based on sequential Monte Carlo simulation (SMCS) for distribution system with DG is proposed. Operating states of the distribution system can be sampled by SMCS in chronological order thus the corresponding output power of DG can be generated. The proposed method has been tested on feeder F4 of IEEE-RBTS Bus 6. The results show that reliability evaluation of distribution system considering the uncertainty of output power of DG can be effectively implemented by SMCS.

1 Introduction

With the development of social economy, the requirement of power quality and reliability becomes higher and higher. Distribution network is connected directly with consumers, and it has great impact on the reliability of power supply[1-3]. Through analyzing the reliability level of distribution system, reliability evaluation can help find weaknesses in the system and put forward effective improvement measures.

Distributed generations have greatly increase the complexity of reliability analysis of distribution network. Some investigations have been made. In [4], the operating characteristics and reliability of DG operated in isolated island mode based on interval analysis had been proposed. With the consideration of the stochastic characteristics of power output of photovoltaic generation and wind power generation, a reliability evaluation algorithm for distribution network containing micro-grid had been investigated in [5]. The formation probability of isolated island had been computed based on analytical method in [6-7].

This paper describes a method to identify the faults' effect on load points' reliability index based on equivalent simplification of distribution network. The partition simplification approach can simplify the process of fault analysis and improve the calculation efficiency. According to the randomness of output power of DG and the time series characteristic of load, a reliability evaluation model based on sequential Monte Carlo simulation for distribution system is established. Then the algorithm process based on sequential simulation method is proposed to make the reliability evaluation of feeder F4 of IEEE-RBTS Bus 6.

[a] Corresponding author : songhuaidu@cau.edu.cn

2 Complex Distribution Network Reliability Evaluation

2.1 Reliability Indices of Distribution System

The reliability level of one distribution system can be evaluated from two aspects. One is consumer level, another is system level. Two frequently used reliability indices for users of load points are as follows. Fault rate λ_i (number of outages/year) and annual outage duration U_i (sum of the outages time/year).

The reliability level of the system is evaluated according to the outage frequency and the average outage time of load points. The commonly used indices are system average interruption frequency index (SAIFI) and system average power interruption duration index (SAIDI). Their respective formula is expressed as:

$$SAIFI = \frac{\sum \lambda_i N_i}{\sum N_i} \tag{1}$$

$$SAIDI = \frac{\sum U_i N_i}{\sum N_i} \tag{2}$$

where λ_i, N_i and U_i stand for fault rate, total number of users and annual average outage duration of load point i, respectively.

2.2 Partitions of Distribution System

Distribution system reliability is not only related to network topology as well as switch configuration which decides the range of failure influence and power supply recovery. The process can be divided into two phases after fault occurred in distribution network: fault isolation and fault recovery. In fault isolation stage, quick removal of failure is completed by circuit breaker; in fault recovery stage, sectionalizers isolate fault and circuit breaker close in order to restore the power supply for blackout zone which is non-fault. Therefore, the switches can be seen as boundaries which can divide the whole network into several region units which can be defined as feeder line areas. The components belonging to the same region have the same fault effect on load points. Similarly, the load points located in the same region are under the same influence of any fault. Therefore, they have the same failure rate and the same outage duration and can be equivalent to one load point. The partition approach can simplify the process of fault analysis and improve the calculation efficiency.

A feeder line area does not contain switches and it is the smallest analysis unit of fault. Equivalent reliability parameters of feeder line area can be calculated as follows:

$$f_{el} = \sum_{i=1}^{N_{el}} f_i \tag{3}$$

$$t_{el} = \sum_{i=1}^{N_{el}} f_i t_i \Big/ f_{el} \tag{4}$$

where f_{el} and t_{el} are the equivalent failure rate and equivalent repair time of feeder line area l; f_i and t_i are the failure rate and repair time of component i; N_{el} is the number of components in feeder line area l.

2.3 Area Classification for Distribution Network with DGs based on Fault Effect

When a specific failure occurs in distribution system, areas of different locations will be affected by different degrees. Therefore, corresponding to the fault occurred, areas in distribution system can be divided into the following categories:

(1) Fault area

The area where the failed component located is defined as fault area. The outage time for load points in fault area will be the repair time of feeder line area.

(2) No-effect area

Load points in this area will not be affected by failure, so there is no power outage.

(3) Isolate area

The area is located in the upper reaches of the fault and it can be connected with primary substation after the fault is isolated. Another case is that the area is located in the lower reaches of the fault and load can be transferred to other line by loop switches. The outage time for load points in isolate area will be the switching time of sectionalizer.

(4) Island area

When fault occurs, if there is a switch installed between the fault and some DGs, the area downstream the switch may operate in island mode. Whether the load points are outage or not depends on the ability of DG supply to load demand in island [8].

3 Reliability Assessment Model for Distribution System

3.1 Distribution Network Components Modelling

The main components of distribution network, such as feeders, transformers, circuit breakers and sectionalizers can be seen as repairable components. Their reliability parameters are failure rate (λ) and repair rate (μ). It is assumed that the time to failure (*TTF*) and the time to repair (*TTR*) obey exponential distribution.

Taking *TTF* for example, the probability density function is:

$$f(t)=\lambda e^{-\lambda t} \tag{5}$$

where $f(t)$ representsthe probability of failure of components in the time of t, its distribution function is:

$$F(t)=1-e^{-\lambda t}=\delta \tag{6}$$

$F(t)$ represents the probability of the components work properly until time is t, and δ is a random number between 0 and 1.The *TTF* of components can be obtained based on the following equations:

$$TTF=-ln\delta/\lambda \tag{7}$$

Through Monte Carlo sampling to simulate the evaluation and fault status of components, then to simulate the entire distribution system operation condition.

3.2 Distributed generation modelling

3.2.1 Diesel generator

Traditional distributed generation is not affected by the factors of natural environment and its output can be controlled at a constant value. Therefore, traditional distributed generation can be represented by constant power model. Availability rate ρ_A is used to reflect generating units' performance [9].

3.2.2 Wind turbine

It is assumed that the wind speed of each hour obeys two parameter Weibull distribution. Its cumulative probability density function is:

$$F(v) = p(x < v) = 1 - \exp\left[-\left(\frac{v}{c}\right)^k\right] \tag{8}$$

The parameters k and c can be calculated by parametric estimation according to historical data [10-11]:

$$k = \left(\frac{\sigma}{v_m}\right)^{-1.086} \tag{9}$$

$$c = \frac{v_m}{\Gamma(1+1/k)} \tag{10}$$

where v_m and σ is respectively the mean value and variance of wind speed.
In the Monte Carlo simulation, Weibull distributed random number can be generated by inverse transformation:

$$v = c\left[-\ln(1-R)\right]^{1/k} = c\left[-\ln(R)\right]^{1/k} \tag{11}$$

The output power of a wind turbine is dependent on the wind speed as well as the parameters of the power performance curve. Therefore, once the Weibull parameters k and c were got, the output power can be calculated as follows [12]:

$$P(v) = \begin{cases} 0 & (0 \le v \le v_i) \\ P_r * \dfrac{(v - v_i)}{(v_r - v_i)} & (v_i \le v \le v_r) \\ P_r & (v_r \le v \le v_o) \\ 0 & (v_o \le v) \end{cases} \tag{12}$$

where v_i, v_r and v_o are the cut-in speed, rated speed and cut off speed of the wind turbine, respectively. P_r is the rated output power of the wind turbine.

3.3 Load Modelling

The load will be assumed to be constant during an hour and changes discretely for every hour. The load value is determined from the sequential load data of IEEE-RTS 79 reliability test system. The system provides weekly peak load as a percentage of annual peak load; daily peak load cycle as a percentage of the weekly peak load; and hourly peak load as a percentage of the daily peak load.

4 Sequential Simulation Process of Reliability Evaluation

When analysing reliability of distribution system including DGs, two different operating modes should be taken into consideration. Through the above analysis, the reliability evaluation algorithm processes was proposed based on sequential Monte Carlo simulation.

(1) Data initialization. Determine the simulation time and initialize the system. Set $t = 0$.

(2) Simplify system and divide it into several partitions according to topological structure.

(3) Determine the number of partitions and randomly generate the same account of random numbers δ_1, $\delta_2 \ldots \delta_n$. Calculate *TTF* of each area according to the random numbers: $TTF = -ln\delta/\lambda$.

(4) Select the area that has the minimum *TTF* as the fault area. Treat *TTF* as the minimum working time of system and add up simulation time. Set $t = TTF + t$.

(5) After sampling to get the fault area, calculate the repair time of it. Choose the equivalent repair time as its *TTR*.

(6) According to the fault that occurs, analyse the fault effect types of load points respectively. The load points are divided into four categories: ① load points in fault area; ② load points in no-effect area; ③ load points in isolate area; ④ load points in island area.

(7) For the first type, load points will not be outage; for the second type and third type, add up the outage duration and the number of blackouts of load points respectively. Treat the fourth type as follows:

① Calculate the total output power of DGs $P_{DG}(t)$ and the total power of load demand $P_L(t)$ within island in the time of t.

② Determine whether or not it meet that: $P_{DG}(t) >= P_L(t)$. If it meets the condition, load points will not be outage; if not, record the blackout and add up the outage duration of load points in island.

③ Determine whether or not t equals to the time when fault recovery has been completed. If so, output the fault records of load points and go to the next step; if not, set $t = t+1$ and go to step ①.

(8) Determine whether or not the present time t achieves the desired simulation Years. If so, go to the next step; if not, go to step (3) and continue to simulation.

(9) Calculate the failure rate and annual average outage duration of each load point according to the recorded data.

(10) Calculate the reliability of overall system according to load points' index.

The above algorithm process is shown in Figure 1.

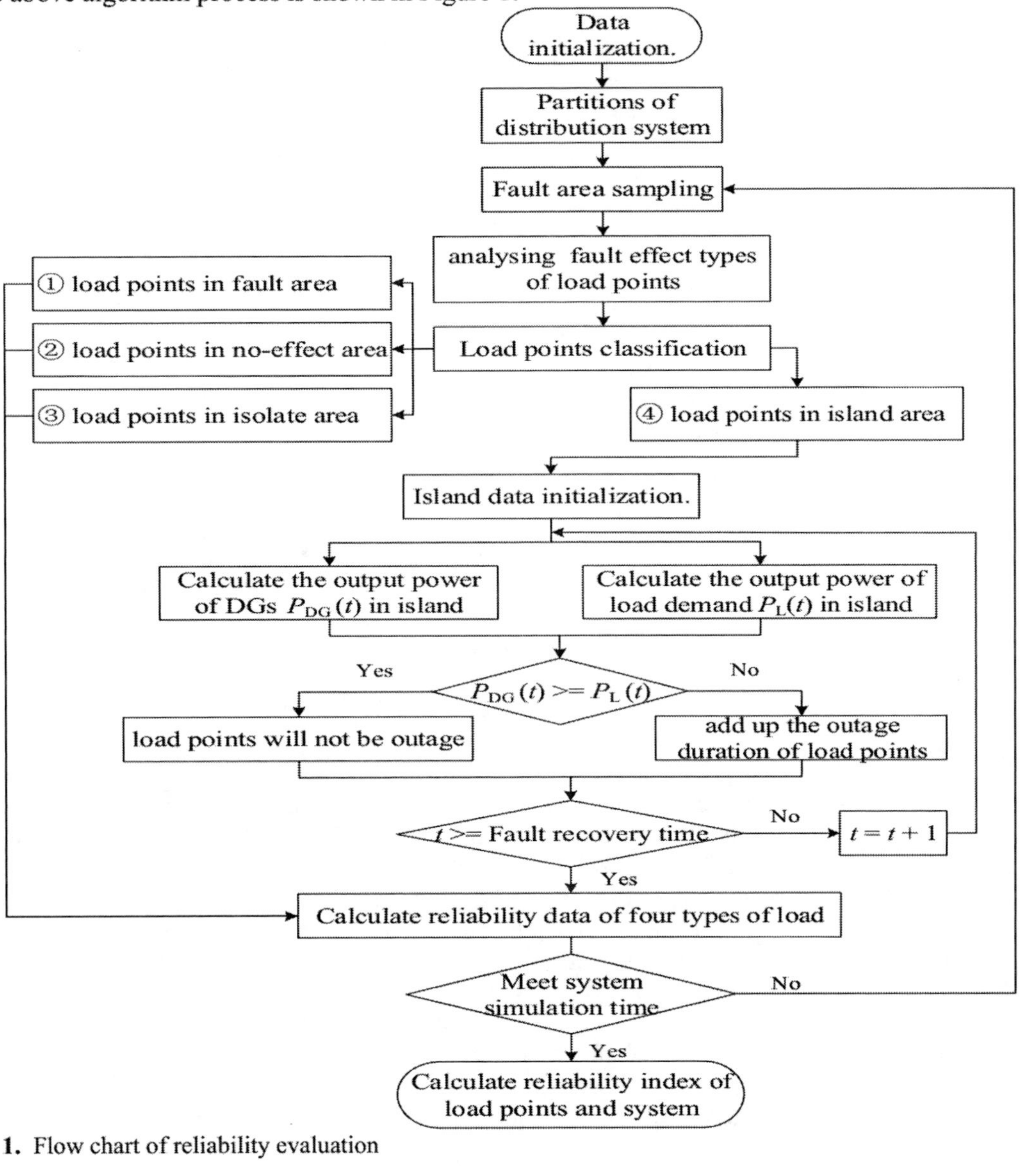

Figure 1. Flow chart of reliability evaluation

5 Case Study

The paper has analysed reliability of feeder F4 of IEEE RBTS-BUS6 network, which is shown in Figure 2. Renewable distributed generations are integrated to feeder 19 and feeder 25, which constitute island 1 and island 2. The peak load of island 1 and island 2 are 2.4 MW and 2.6 MW. Wind power and photovoltaic power generation operate in parallel with grid. Diesel generating sets are running as standby. Cut in, rated and cut out speed of wind turbine are respectively 4, 15 and 25 m/s. The mean value and variance of wind speed are respectively 7 and 13.4. It is assumed that there are only circuit breakers and sectionalizers in the feeder which are completely reliable. The parameters of lines can be found in the references [13-14]. The availability rate of DG is 0.95 and conversion time is 0.1 h [9]. The operation time for sectionalizer is 2 h.

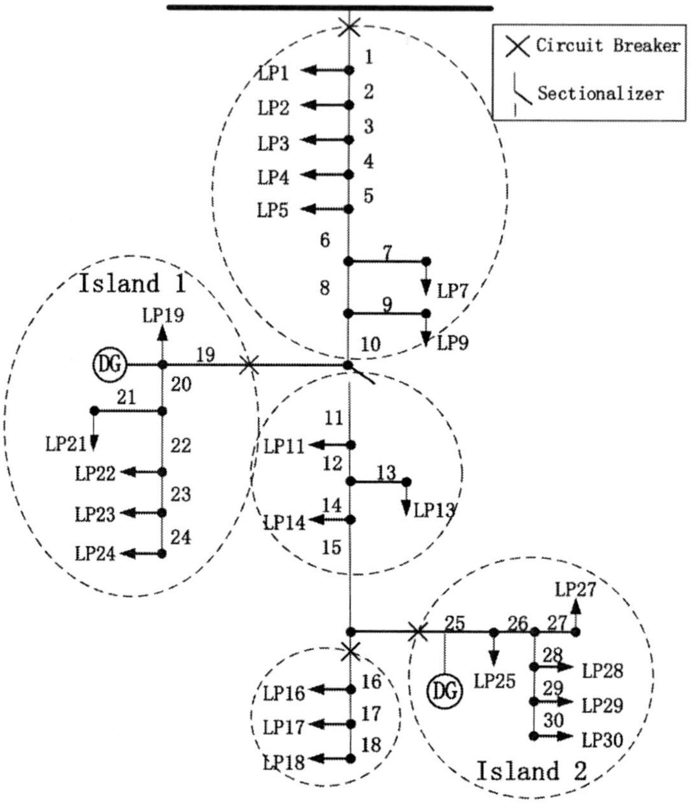

Figure 2. Test system

According to the proposed partition method in this paper to simplify the test system, as shown in Figure 2, it can be seen that branches 1-10 belong to region 1 and branches 11-15 belong to region 11, branches 16-18 belongs to region 16, branches 19-24 belong to region 19, branches 25-30 belong to region 25. The load points are numbered according to a certain sequence number, and the corresponding numbers of the upstream branch feeders are used to name.

According to the flow chart of algorithm proposed in this paper to carry out reliability evaluation of distribution system containing distributed generation:

Scenario 1: No distributed generation is contained in distribution system. Analysis method and Monte Carlo method were used to calculate reliability of system respectively.

Scenario 2: Distributed generation is installed in island 1. The capacity of distributed power supply is 1.15 times the load demand in island 1. Among them, the capacity of diesel engine accounts for

50% of total power generation capacity. Analysis method and Monte Carlo method were used respectively.

Scenario 3: Distributed generation is installed in island 2. The capacity of distributed power supply is 1.15 times the load demand in island 2. Among them, the capacity of diesel engine accounts for 50% of total power generation capacity. Analysis method and Monte Carlo method were used respectively.

In the above scenarios, the reliability indices of 5 typical load points and system reliability indices, as shown in Table 1-Table 3.

Table 1. Failure rates for typical load points

Scenario		LP 1	LP 11	LP 16	LP 19	LP 25
1	Analytical method	1.2569	1.2557	1.6491	1.9006	1.8872
	SMCS	1.2569	1.2569	1.6488	1.9006	1.8870
2	Analytical method	1.2569	1.2557	1.6491	1.9006	1.8872
	SMCS	1.2569	1.2569	1.6488	1.9006	1.8870
3	Analytical method	1.2569	1.2557	1.6491	1.9006	1.8872
	SMCS	1.2569	1.2569	1.6488	1.9006	1.8870

Table 2. Annual average outage duration for typical load points

Scenario		LP 1	LP 11	LP 16	LP 19	LP 25
1	Analytical method	6.8622	10.0547	13.1911	12.0168	15.0959
	SMCS	6.8616	10.0550	13.1906	12.0177	15.0962
2	Analytical method	6.8622	10.0547	13.1911	6.3884	15.0959
	SMCS	6.8616	10.0550	13.1906	6.3885	15.0962
3	Analytical method	6.8622	10.0547	13.1911	12.0168	6.7002
	SMCS	6.8616	10.0550	13.1906	12.0177	6.7004

Table 3. System reliability indices

Scenario		SAIFI	SAIDI
1	Analytical method	1.4785	9.9048
	SMCS	1.4787	9.9048
2	Analytical method	1.4785	9.0539
	SMCS	1.4787	9.0539
3	Analytical method	1.4785	8.7836
	SMCS	1.4787	8.7836

According to the data of three scenarios, it can be seen that the results of Monte Carlo method are very close to the results of analytical method. In addition, it can be seen from table 2, the failure rate of load point 19 and 25 is higher, which belong to relatively weak part in system. Therefore, distributed generation is connected to feeder line area 19 and feeder line area 25 respectively in scenario 2 and scenario 3.

Compare reliability assessment results of scenario 2, 3 with scenario 1, it can be seen that when distributed generation access to distribution system, the reliability indicators of load points which are out of the island do not change, however, the annual outage time of load points that are in the island has been significantly improved. Reliability of the whole system has also improved. The reason for this is that, when the primary grid fails, if distributed generation output is greater than load demand in island, the load points' interruption duration only equals to the start time of distributed generation, which is far less than fault repair and isolation time.

The reliability assessment results of scenario 2 and scenario 3 show that, when distributed generation is connected to the island 2, the reliability level of system is better than that distributed generation connected to island 1. The main reason is that, island 2 is in the feeder terminal, which is far from the superior substation. When there is no distributed power in island 2, load points will be outage until the fault is repaired. However, the fault interruption duration of load points in island 1 may be the fault repair time or isolation time. Therefore, distributed power installed far from the location of primary substation can improve the reliability of the whole system better.

6 Conclusions

Considering the randomness of distributed generation output power and the sequential characteristic of load, the reliability evaluation model based on Monte Carlo simulation for distribution system is established. In the case study, the application of the proposed method has been verified on the feeder F4 of IEEE RBTS-BUS6. The results show that reliability evaluation of distribution system considering the uncertainty of DG and load can be effectively implemented by SMCS and islanding operation can positively improve the reliability of distribution network.

References

1. R. Billinton, R.N. Allan, "Reliability Evaluation of Power Systems", New York: Premium Press, **14**(1), pp. 51-57, (1996).
2. W. Huang, C.H. Sun, Z.P. Wu, "A Review on Microgrid Technology Containing Distributed Generation System", *Power System Technology*, **33**(9), pp. 14-18, (2009).
3. L.Y. Kang, H.X. Guo, J. Wu, "Characteristics of Distributed Generation System and Related Research Issues Caused by Connecting It to Power System", *Power System Technology*, **34**(11), pp. 43-47, (2010).
4. Y. Sun, M. Bollen, G. Ault, "Reliability Analysis of Islanded Distribution Systems with Distributed Energy Resources", *Power System Technology*, **32**(23), pp. 77-81, (2013).
5. X.D. Wang, J.K. Lin, "Reliability Evaluation Based on Network Simplification for the Distribution System with Distributed Generation", *Automation of Electric Power Systems*, **34**(4), pp. 38-43, (2010).
6. Y.M. Atwa, E.F. El-Saadany, A. Guise, "Supply Adequacy Assessment of Distribution System Including Wind-based DG during Different Modes of Operation", *IEEE Trans on Power Systems*, **19**(25), pp. 78-86, (2011).
7. M. Atwa, E.F. El-Saadany, M.A. Salama, "Adequacy Evaluation of Distribution System Including Wind/Solar DG during Different Modes of Operation", *IEEE Trans on Power Systems*, **4**(26), pp. 1945-1952, (2011).
8. G. Zini, C. Mangeant, J. Merten, "Reliability of Large-scale Grid-connected Photovoltaic System", *Renewable Energy*, **7**(36), pp. 1670-1676, (2011).
9. S. Conti, R. Nicolosi, S. A. Rizzo, "Generalized Systematic Approach to Assess Distribution

System Reliability with Renewable Distributed Generators and Microgrids", *IEEE Trans. Power Delivery*, **27**, pp. 261-270, (2012).
10. M. Al-Muhaini, G. T. Heydt, A. Huynh, "The Reliability of Power Distribution Systems as Calculated Using System Theoretic Concepts", *IEEE Power Energy Society General Meeting*, pp, 1-8, (2010).
11. X. Xie, Y. Yuan, W.Z. Zou, "Reliability Evaluation of Microgrid Power Supply in Emergency Non-scheduled Isolated Mode", *Automation of Electric Power Systems*. **36**(9), pp. 8-22, (2012).
12. Y. Xu, Y.C. Wu, "Reliability Evaluation for Distribution System Connected with Wind-turbine Generators", *Power System Technology*, **35**(4), pp. 154-158, (2011).
13. Y. M. Atwa, E. F. El-Saadany, M. M. A. Salama, R. Seethapathy, M. Essam, and S. Conti, "Adequacy evaluation of distribution system including wind/solar DG during different modes of operation", *IEEE Trans. Power System.*, **26**(4), pp. 1945–1952, (2011).
14. R.N. Allan, R. Billinton, I. Sjarief, "A Reliability Test System for Educational Purposes: Basic Distribution System Data and Results", *IEEE Trans on Power Systems*, **6**(2), pp.813-820, (1991).
15. R. Billinton, S. Johnnavithula, "A Test System for Teaching Overall Power System Reliability Assessment", *IEEE Trans on Power System*, **11**(4), pp.1670-1676, (1996).

MATEC Web of Conferences **55**, 03009 (2016) DOI: 10.1051/matecconf/20165503009
ACPEE 2016

Risk-based Security Assessment in Distribution Network with the Integration of Photovoltaic

Wanxing Sheng[1], Keyan Liu[1], Yating Cai[2] and Songhuai Du[2,a]

[1]*China Electric Power Research Institute, Power Distribution Department, 100192, Beijing*
[2]*China Agricultural University, College of Information and Electrical Engineering, 100083, Beijing*

Abstract. With the development of distribution network, distributed generation such as wind and photovoltaic (PV) power will become increasingly prominent in the near future. PV is widely constructed because of advantages it has. However, the volatility and randomness of PV makes it more complex than traditional energy in the security assessment of distribution network. Based on risk theory, considering the randomness of PV, node low voltage risk index and line overload risk index are established in this paper. Also, K (N - 1 + 1) principle for distribution network which is developed from traditional (N-1) deterministic principle is applied to reflect the flexible structure of distribution network. IEEE three-feeder example system is utilized to investigate the influence of PV power on the security assessment of distribution network.

1 Introduction

With the increasing demand for energy, development and utilization of renewable energy have become more and more important. Secure and reliable wind power, PV power and other distributed power ought to be developed to adjust to smart distribution network. PV is widely favoured because of its safety, short construction period, no noise, no pollution and other advantages. However, the output of PV power is influenced by its volatility and uncertainty. These features make it different from the traditional standby power in the reliability and risk assessment [1]. Therefore, it's necessary to study the security assessment based on risk theory of distribution network containing PV power.

Line overload, node low voltage and load points blackout risk indices are established and distribution network risk assessment system are proposed based on utility theory in document [2]. Node low voltage and line overload are selected as risk indices analysing the security of distribution network in the case of different solar access capacity in document [3]. Combined with distribution network reliability index, energy loss rate of accident and customers' loss rate of accident risk indices are established and corrected in the access of distributed power in document [4].

The structure of distribution network varies flexibly. When the component (branch) switches off, branches along the direction of the flow will be completely de-energized and non-fault power failure region occurs. In this case, traditional safety indicators and analysis methods are no longer suitable for

[a] Corresponding author : songhuaidu@cau.edu.cn

the network [5]. Referring to the conventional (N-1) security criterion used in bulk power network, K (N-1+1) principle is proposed for distribution network analysis in document [6], according to the characteristics of structure and operation of the distribution network. The static security based on risk theory is evaluated in the absence of distributed power in document [7].

Based on documents mentioned above, node low voltage index and line overload index are established in this paper to analyse the security of distribution network by combining risk theory and K (N-1+1) deterministic principle.

2 Models for Security Assessment based on Risk Theory

Risk is a quantization of uncertain factors of the system. Risk index, usually defined as the product of probability and consequences [8], is computed as equation (1).

$$Risk\left(X_f\right)=\sum P_r\left(E\right)Sev\left(E,X_f\right) \tag{1}$$

where, X_f is the operating state of the system before accidents; E is the accident; P_r is the probability of accident E ; $Sev\left(E,X_f\right)$ is the severity caused by accident E under the operating state X_f ; $Risk\left(X_f\right)$ is the risk index.

In addition to probability and severity model of accidents occurring, the power outputs of PV system and probability model of different light intensity are also needed in the paper. The following parts will introduce these models in detail.

2.1 Probability model of the accident

In traditional power system reliability assessment, the failure rate of components generally is modelled as Markov state, which is averaged from historical statistical data. Reliability assessment, usually involved the long term assessment, therefore, mostly adopt the forced shutdown rate as the failure rate of the component. But for security assessment based on risk theory, as its short assessment cycle, time factors must be considered.

Assuming real-time failure rate of the component as λ , components' failure times obey Poisson distribution in interval t . According to the probability formula of Poisson distribution [9], failure rate of the component in interval t can be derived from equation (2).

$$P\left(F_i\right)=1-e^{-\lambda_i} \tag{2}$$

where, F_i is the failure state of components; λ_i is the failure rate of components.

Assuming that components are independent from each other, fault probability of the system caused by a component fault is computed as equation (3).

$$\begin{aligned}P\left(E_i\right)&=P\left(F_1\cap F_2\cdots F_{i-1}\cap F_i\cap F_{i+1}\cdots F_{N_c}\right)\\&=P\left(F_i\right)\prod_{j\neq i}P\left(F_j\right)\\&=\left(1-e^{-\lambda_i}\right)e^{-\sum_{j\neq i}\lambda_j}\end{aligned} \tag{3}$$

where E_i is the system accident, F_j is the normal state of components.

2.2 Severity function of the accident

The fundamental principles of severity functions defined in risk assessment [10]: (1) not involved in operation process after failures occur; (2) easier for the dispatcher to understand; (3) associated with deterministic criteria; (4) the expression should be intuitive and concise; (5) reflect the characteristics of different problems; (6) reflect the severity of the components. Based on these principles, a series of simple linear functions are designed to characterize the severity of power flow overload, low voltage and etc. in document [11].The application of severity function will be introduced in the following part.

2.2.1 Severity of node low voltage

For each node in the distribution system, according to the degree of exceeding voltage, the severity function of low voltage is defined. The node low voltage severity function is shown in Fig. 1. Assuming the limit of low voltage as 0.95, define the severity as equation (4). Where, V is the voltage magnitude.

$$Sev^{lv}(V)=\begin{cases}\dfrac{1-V}{1-0.95}, & V\le 1.0\\ 0, & V>1.0\end{cases} \tag{4}$$

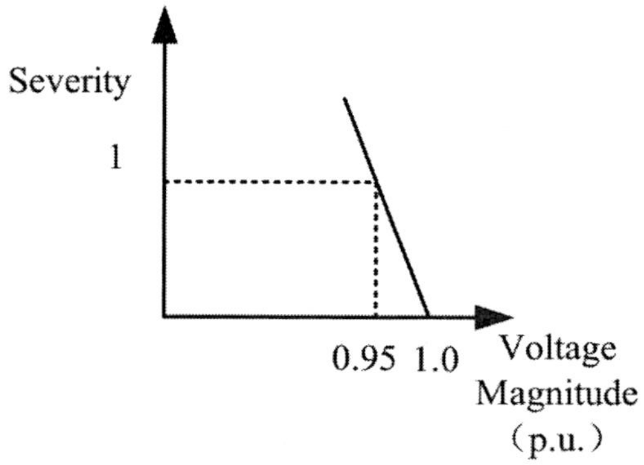

Figure 1. Node low voltage severity function

From Fig. 1, we can see that the node low voltage severity is 1 when the node voltage is 0.95; the severity is between 0 and 1 when the node voltage is in the interval (0.95, 1.0) and the severity increases as the node voltage decreases; the severity is over 1 when the node voltage is lower than 0.95.

2.2.1 Severity of line overload

For each line in distribution system, assuming the limit of L is 0.8, the overload severity function is defined as equation (5). Where, L is the ratio of the real line power and rated line power.

$$Sev^{ol}(L)=\begin{cases}1-\dfrac{1-L}{1-0.8}, & L\ge 0.8\\ 0, & L<0.8\end{cases} \tag{5}$$

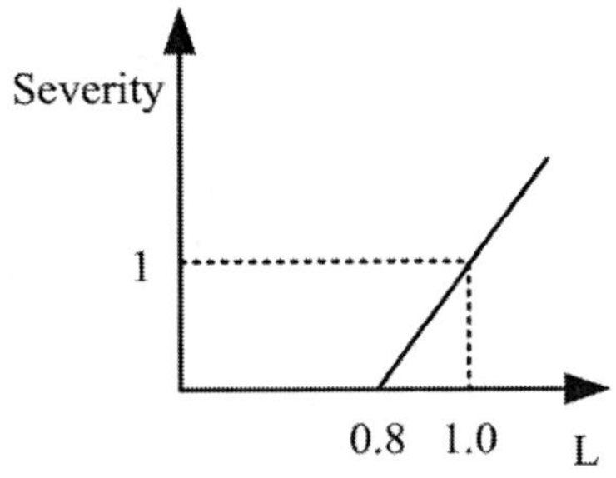

Figure 2. Line overload severity function

It is shown in Fig.2 that the line overload severity is 0 when L is lower than 0.8; the severity is between 0 and 1 when L is in the interval (0.8, 1.0) and the severity increases as L increases; the severity is higher than 1 when L is higher than 1.0.

2.3 Output model of PV power

The output of photovoltaic power generation system is closely related to the light intensity. The output model of photovoltaic system under varying light intensity is described as equation (6) [12] and the relation between light intensity and output of PV power is shown in Figure 3. With the increasing of the light intensity, the output curve is divided into three sections. In the first two sections, the outputs of PV system keep increasing .When the light intensity reaches some degree, the outputs stay constant.

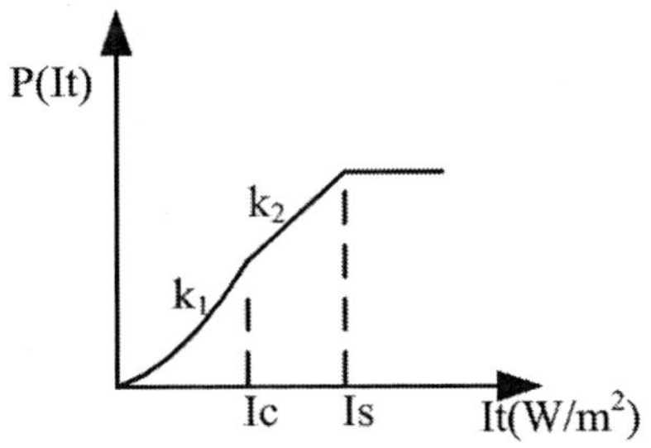

Figure 3. PV System Power Output Curve

$$P(I_t)=\begin{cases}\dfrac{P_r}{I_s I_c} I_t^2 & 0 \le I_t \le I_c \\ \dfrac{P_r}{I_s} I_t & I_c \le I_t \le I_s \\ P_r & I_t \ge I_s \end{cases} \tag{6}$$

where $P(I_t)$ is photovoltaic system power output; I_s is the standard light intensity, $I_s = 1000W/m^2$; I_c is the light intensity when the photoelectric conversion efficiency reaches maximum, $I_c = 150W/m^2$; P_r is the rated capacity of photovoltaic power generation system.

2.4 Light intensity probability model

The year average light intensity is used instead of the real light intensity to reduce the computational complexity. Light intensity value and the probability in this paper are referred to statistical data in document [13] in which light intensity is averagely divided into 12 groups from 0 W/m^2 to 1100 W/m^2. Table 1 shows the different light intensity and its corresponding probability.

Table 1. Power Output and Probability of PV

Light intensity /W m^{-2}	Output/ MW	Probability for light intensity
0	0	0.55017
100	0.4	0.11519
200	1.2	0.06532
300	1.8	0.05290
400	2.4	0.04636
500	3.0	0.04479
600	3.6	0.03930
700	4.2	0.03355
800	4.8	0.02756
900	5.4	0.01933
1000	6.0	0.00537
1100	6.0	0.00014

3 Definition of Risk Indices

Considering the randomness of photovoltaic power, appropriate correction is made to the risk index of node low voltage and line overload. Define the node low voltage risk index as equation (7) and line overload risk index as equation (8).

$$R_{lv}\left(X_f\right)=\sum_{k=1}^{n}\left[\omega_k P_r\left(E\right)Sev_k^{lv}\left(E,X_f\right)\right] \tag{7}$$

$$R_{ol}\left(X_f\right)=\sum_{k=1}^{n}\left[\omega_k P_r\left(E\right)Sev_k^{ol}\left(E,X_f\right)\right] \tag{8}$$

where, n is the total of light intensity groups; ω_k is the probability of group k for light intensity; $Sev_k^{lv}\left(E,X_f\right)$ is node low voltage severity of group k ; $Sev_k^{ol}\left(E,X_f\right)$ is line overload severity of group k .

4 Case Study

To verify the feasibility of indices introduced, IEEE three-feeder distribution system is applied in this paper. Table 2 shows the branch attributes of the distribution system, referring to document [7] for other specific parameters. The capacity of PV power is set as 6 MW. The topology diagram of the distribution network is shown in Figure 4.

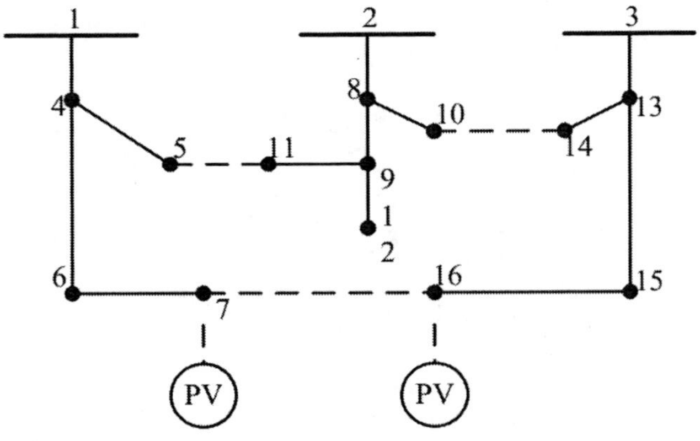

Figure 4. EEE Three-feeder System

Table 2. Branch attributes

branch	Length /km	Failure rate %	Rated current /kA
1-4	1	0.2	0.8
4-5	1	0.027	2.2
4-6	1	0.027	2.2
6-7	1	0.2	0.8
2-8	1	0.027	2.2
8-9	1	0.027	2.2
8-10	1	0.027	2.2
9-11	1	0.027	2.2
9-12	1	0.027	2.2
3-13	1	0.2	0.8
13-14	1	0.027	2.2
13-15	1	0.2	0.8
15-16	1	0.2	0.8

Two cases are proposed to achieve risk based security assessment on distribution network: (1) no access of PV power; (2) accessing PV power at node 7 and node 16. When disconnecting line 1-4, 2-8, 3-13, there are two transfer solutions. To reduce the complexity of calculation, for these three lines, choose one transfer solution respectively by connecting switch 5-11, 5 -11, 10-14.

Case (1): no access of PV power.

The test results are shown in table 3. It can be seen that the line overload risk is the highest when the line 1-4 is disconnected. This is mainly due to that the line 1-4 is directly connected to the substation and load supplied by substation 1 has to be supplied by substation 2 and 3 when line 1-4 is disconnected and the failure rate of the line1-4 is higher than other lines. For node low voltage risk, we can see that when any of line 1-4, 6-7, 3-13, 13-15 and 15-16 is disconnected ,the risk is much higher which is also due to the higher failure rate of these branches shown in Table 2.

Table 3. Results without PV power

Failure line	Node low voltage Risk index	Line overload Risk index
1-4	0.1119	0.04115
4-5	0.004443	0.001491
4-6	0.005292	0.01064
6-7	0.029745	0.02748
2-8	0.008132	0.01653

Failure line	Node low voltage Risk index	Line overload Risk index
8-9	0.005089	0.01141
8-10	0.003675	0.003527
9-11	0.003516	0.004181
9-12	0.002695	0.003025
3-13	0.051219	0.02304
13-14	0.003988	0.002899
13-15	0.037013	0.05478
15-16	0.03116	0.04054

Case (2): accessing PV power at node 7 and node 16.

The test results are shown in table 4. From table 4, it's clearly shown that the risk of the distribution system tends to be decreasing. Though the line overload risk is still high when disconnecting line 1-4, it has been reduced by accessing PV at node 7 and node 16 which is because the access of PV can help supply part of load on feeder 1. Also, we can see that accessing PV power at the end of the line, in some degree, can help reduce node low voltage risk.

Table 4. Results with PV generation at node 7

Failure line	Node low voltage Risk index	Line overload Risk index
1-4	0.09327	0.03015
4-5	0.003323	0.001287
4-6	0.004167	0.009127
6-7	0.027364	0.02634
2-8	0.005194	0.008231
8-9	0.004569	0.009234
8-10	0.003562	0.002865
9-11	0.002865	0.003259
9-12	0.002573	0.002914
3-13	0.04931	0.02086
13-14	0.003564	0.002682
13-15	0.03259	0.04383
15-16	0.02768	0.03928

By comparing Table 3 and Table 4, when accessing PV power at node 7 and node 16, changes of risk indices tend to be consistent. It is shown that the access of PV power can effectively reduce the risk of the system regardless of load changes although the randomness of PV power is involved. Through table 3 and table 4, it can be seen that not only the consequences of accidents but also probability of accidents occurring is worth to be considered in the security assessment.

5 Conclusions

In this paper, by adding uncertain factors to risk indices, the randomness of PV power outputs reflected. Considering the flexibility of distribution network structure after failures occur, the security based on risk theory of distribution network is assessed. Test results show that in the end of the branch accessing PV power supply can effectively reduce the risk of system. However, effects caused by different capacity of PV power, different access site and time-varying characteristics of load haven't been analyzed. If these influence factors are appropriately considered , the assessment will be more reasonable.

References

1. C.Q. Liu, Y. Zhang, "Distribution Network Reliability Considering Distribution Generation", *Automation of Electric power systems*, **31**(22), pp. 46-49,(2007).
2. Y.Y. Yang, J.Y. Yang, Z. Huang. " Risk Assessment System of Distribution Network and Its Application", *Modern Electric Power*, **28**(4), pp. 20-23, (2011).
3. W.P. Zhou. " Risk Analysis of Distributed Photovoltaic Power Network"[D], North China Electric Power University, (2014).
4. Y. Xu, K. Wang, B. Chen, "Reliability and Risk Assessment Techniques for Distribution Networks with Distributed generations", *Proceedings of the CSU-EPSA*, **25**(4), pp. 117-121,(2013).
5. Z.Z. Guo, W. Liu. "Research on Security Indices of Distribution Networks", *Proceedings of the CSEE*, **23**(8), pp. 85-90, (2003).
6. T.H. Wang, P.Y. Wang, M.T. Fan. "Optimal Feeder Automation Planning Using Binary Programming", Proceedings of the CSEE, **20**(5), pp. 54-58, (2000).
7. R.X. Liu, J.H. Zhang, D. Wu. " Research on Static Security Index of Distribution Network Based on Risk Theory", *Power System Protection and Control*, **39**(15), pp. 89-95, (2011).
8. J.D. McCalley, V. Vittal, N. Abi-Samra, "An Overview of Risk Based Security Assessment". Edmonton: *Power engineering Society Summer Meeting*,1999.
9. H. Wan, J.D. McCalley, V. Vittal, "Risk Based Voltage Security Assessment", *IEEE Transactions on Power Systems*, **15**(4), pp. 1247-1254, (2000).
10. M. Ni, J.D. Mccalley, V. Vittal, et al, "Online Risk-Based Security Assessment", *Power Engineering Review IEEE*, **22**(1), pp. 258 – 265, (2002).
11. J.D. McCalley, B.L. Agrawal, V. Vittal, et al, "A Risk-based Security Index for Determining Operating Limits in Stability-limited Electric Power System", *IEEE Transaction on power systems*, **12**(3), pp. 1210-1219, (1997).
12. J. Park, L. Wu, J. Choi, et al. "A Probabilistic Reliability Evaluation of a Power System Including Solar/Photovoltaic cell generator". *Power & Energy Society General Meeting*, (2009).
13. Cha, Seung Tea, et al. "Reliability Evaluation of Distribution System Connected Photovoltaic Generation Considering Weather Effects", *Probabilistic Methods Applied to Power Systems*, 2004 International Conference on IEEE, pp. 451 – 456, (2004).

MATEC Web of Conferences **55**, 03010 (2016) DOI: 10.1051/matecconf/20165503010
ACPEE 2016

Reliability Analysis of Distributed Grid-connected Photovoltaic System Monitoring Network

FU Zhixin[a], LI Qiaomu, YUAN Yue

College of Energy and Electrical Engineering, Hohai University, 211100 Nanjing, China

Abstract. A large amount of distributed grid-connected Photovoltaic systems have brought new challenges to the dispatching of power network. Real-time monitoring the PV system can efficiently help improve the ability of power network to accept and control the distributed PV systems, and thus mitigate the impulse on the power network imposed by the uncertainty of its power output. To study the reliability of distributed PV monitoring network, it is of great significance to look for a method to build a highly reliable monitoring system, analyze the weak links and key nodes of its monitoring performance in improving the performance of the monitoring network. Firstly a reliability model of PV system was constructed based on WSN technology. Then, in view of the dynamic characteristics of the network's reliability, fault tree analysis was used to judge any possible reasons that cause the failure of the network and logical relationship between them. Finally, the reliability of the monitoring network was analyzed to figure out the weak links and key nodes. This paper provides guidance to build a stable and reliable monitoring network of a distributed PV system.

1 Introduction

In recent years, Photovoltaic (PV) power generation in the form of distributed power dominates the PV market, and has the trend of large-scale development. Affected by the uncertainty of light intensity and environment temperature changes. Adjustable control of distributed PV power is poor, at the same time, with the growing number of access points, the interconnection of distributed PV power plants to the grid has brought new challenges to the grid scheduling, the corresponding monitoring problems have also been paid more and more attention[1,2].

Aiming at the shortcomings of traditional monitoring network, many scholars at home and abroad apply sensor technology to the monitoring filed. Because of its good timeliness and flexibility, Wireless Sensor Network (WSN) technology provides new ideas to the construction of PV monitoring network[3,4]. While the studies on reliability analysis of these monitoring networks are still rare. The existing studies have mainly focused on its software, hardware and communication network's topology design[5,6], which have taken the operating characteristics of the monitoring network's constituent elements in less consideration. However, the reliability of the monitoring network based on the WSN technology is closely related to the operating characteristics of its sensor nodes: On the one hand, the monitoring network based on WSN technology is constructed by energy limited sensor nodes. So the energy issue has become one of most important factors restricting the reliability of

[a] Corresponding author : zhixinfu@hhu.edu.cn

monitoring network. On the other hand, because WSN uses wireless communication method, its channel bandwidth is narrow. Moreover, it's susceptible to the external interference from the environment and the other electromagnetic signals, which resulting in information latency, packet loss and other issues. So the communication quality has become another important factor restricting the reliability of the monitoring network[7,8]. Based on the above two points, in view of the insufficient research on the reliability of distributed PV monitoring network. This paper took fully consideration of the topological structure and the operating characteristics of its constituent elements, when analyzing the reliability of the monitoring network. It explored methods to improve the reliability of distributed PV monitoring network in order to improve its operating performance.

Given the characteristic that the reliability of the monitoring network changes with time. Firstly, this paper established a reliability model of a distributed PV monitoring network. Secondly, the fault tree analysis method was applied to determine possible causes leading to the failure of the monitoring network, and logical relationships between them. Finally, according to the simulation data, it looked for weak links and key nodes of the monitoring network, while analyzing its reliability.

2 Topological structure of PV monitoring network and its reliability evaluation indicators

2.1 Construction of the monitoring network based on WSN technology

In order to adapt to the characteristics that the number of access points of the future distributed PV systems is huge, and its access points are dispersed, the construction of the monitoring network should be simple and easy to maintain. The clustering topology of WSN has good scalability, and it is easy to manage and maintain[9], which make it suitable for the wide and scattered distributed PV systems. In this paper, based on WSN the classical triangle mesh method is used to deploy light intensity sensor nodes which are located on solar panels. This method can ensure the coverage degree of solar panels, at the same time, effectively reduce the cost and the number of sensor nodes.

Taking a $75 \times 60m^2$ solar panels for example. The monitoring network uses the light intensity sensor node with the perception radius of 10m. Given that the tree topology of WSN has advantages of simple structure, little energy consumption for information transmission and high efficiency. In this paper, light intensity sensor nodes use tree topology to transmit information between them. According to the energy consumption model of the sensor node, and the principle of minimum energy consumption, the cluster head node should be located at the geometric center of the solar panels, and the transmission of information uses the principle of the minimum number of hops to maximize the life of the monitoring network[10]. The deployment of light intensity sensor nodes on solar panels is shown in figure 1. In this figure, the Arabic number represents the number of each light intensity sensor node (Such as: 3 represents the No.3 light intensity sensor node). L represents the transmission path(Such as: L8 represents the No.8 transmission path).

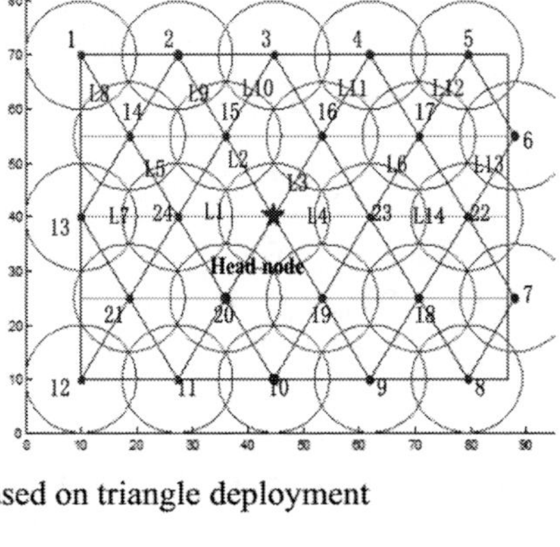

Figure 1. PV monitoring system based on triangle deployment

Figure 1 shows that, the light intensity sensor nodes are uniformly distributed on the solar panel. What's more, all sensor nodes and communication links are in a symmetrical state. Therefore, this paper considers only the light intensity sensor nodes and communication links situated on the upper part of the solar panel. The transmission paths between sensor nodes are:

$1 \to 14 \to 24 \to$ head node	$14 \to 24 \to$ head node	$2 \to 15 \to$ head node
$15 \to$ head node	$3 \to 15 \to$ head node	$16 \to$ head node
$4 \to 16 \to$ head node	$17 \to 23 \to$ head node	$5 \to 17 \to 23 \to$ head node
$22 \to 23 \to$ head node	$6 \to 22 \to 23 \to$ head node	$23 \to$ head node
$13 \to 24 \to$ head node	$24 \to$ head node	

2.2 Reliability evaluation indicators of the monitoring network

The coverage degree and the connectivity degree are two basic evaluation indicators to evaluate the reliability of the distributed PV monitoring network. The coverage degree reflects the collection reliability of light intensity information and other physical information of the monitored area. And the connectivity degree reflects the information transmission reliability between sensor nodes. Therefore, the coverage degree and the connectivity degree affect directly the monitoring quality of the monitoring network. So they are important evaluation indicators to evaluate the reliability of the monitoring network. According to the reliability evaluation indicators proposed in research [11], this paper makes the following definitions for the factors which affect the reliability of the distributed PV monitoring network: (1)Failure rate of light intensity sensor nodes. It is the ratio of the number of failure nodes to the total number of sensor nodes. It reflects the impact of the failure nodes to the performance of the monitoring network. (2)Coverage degree. It is the ratio of the effective coverage area to the total monitored area. It reflects the effective coverage ratio in the case of the failure of sensor nodes. (3)Reliability of the communication links. It is used to judge if the light intensity information or other information can be transmitted correctly between sensor nodes. It reflects the reliability of the transmission of information when the nodes are in normal working condition. If any one of these three indicators can't achieve the monitoring network's reliability evaluation indicators, it means that the monitoring network fails.

According to the monitoring network's requirement to the reliability of data, set the reliability evaluation indicators as follows: When the failure rate of light intensity sensor nodes reaches 20%, the coverage degree of the solar panel decreases to 80%, or 20% or more light intensity information can't be transmitted correctly to the cluster head node, it means that the monitoring network fails.

3 Fault tree model of PV monitoring network

3.1 Advantages of fault tree analysis method

The Fault Tree Analysis (FTA) method has the characteristics of strong logical, analytical simple, computationally efficient, etc[12], which make it have obvious advantages in assessing risks and tracking weak links of systems with complex structure. Thus, it has been widely used in the actual projects. This paper set the failure of the monitoring network as the Top Event in the FTA method. Then according to the topology of the monitoring network and the information transmission paths between sensor nodes, it analyzed any possible causes leading to the failure of the monitoring network layer by layer. After that, it established the fault tree model of the monitoring network with the help of specific event symbols, logical gates and transfer symbols. Table 1 shows the basic logical symbols and the corresponding relationships with the monitoring network's elements when analyzing the reliability of the distributed PV monitoring network using the FTA method.

Depending on the operating characteristics of the distributed PV monitoring network, the steps to analyze the reliability of the monitoring network with FTA method are as follows: (1)Determine the top event of the fault tree, which means the failure of the monitoring network. (2)Making logical analysis to determine the possible causes of the failure of the monitoring network. (3)Establish the fault tree model of the failure of the monitoring networt. (4)Making qualitative and quantitative analysis of the fault tree model.

Table 1. Symbols of monitoring system's fault tree model

Symbol	Symbol	Name	Explanation	Correspondence with monitoring network
Event Symbol		Basic Event	Event which can not or don't Need to break down	Failure of sensor nodes or communication links
		Top Event	Target Event of Fault Tree	Failure of monitoring network
	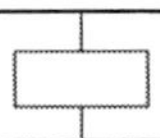	Intermediate Event	Located between Top Event and Basic Event	Possible causes of the failure of the monitoring network
Logic Gates	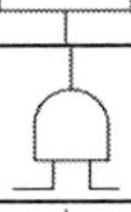	AND gate	Output event occurs only when all input events occur	
	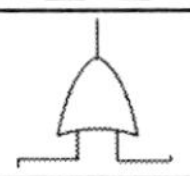	Or gate	Output event occurs as long as one input event occurs	
	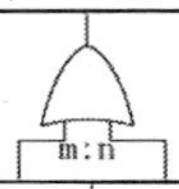	Vote gate	Output event occurs when the *m* of *n* input events occur	
Transfer Symbol	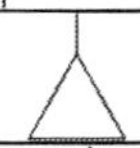	Transfer-in		
	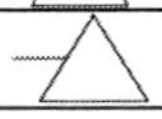	Transfer-out		

3.2 Qualitative analysis of the monitoring network's fault tree model

On the basis of the previous analysis, it needs to make further qualitative analysis of the fault tree model of the monitoring network. Before making the qualitative analysis, this paper firstly made the following assumptions: (1)The Top Event and Basic Events of the monitoring network's fault tree model only have two states: normal state and abnormal state. And the failures of these events are independent. (2)"1" means abnormal state, "0" means normal state. (3)The malicious destructions of the human beings and the interferences from the external environment to the performance of the monitoring network are not taken into consideration. (4)E represents Basic Events of the fault tree model, F represents its Top Event. Then, $X_i = \phi(E_i)$, $F = \phi(X_1, X_2 \cdots X_n)$.

$$X_i = \begin{cases} 1, \text{ Basic Event occurs } (i = 1,2,\cdots n) \\ 0, \text{Basic Event doesn't occur } (i = 1,2,\cdots n) \end{cases} \tag{1}$$

$$F = \begin{cases} 1, \text{ Top Event occurs} \\ 0, \text{ Top Event doesn't occur} \end{cases} \tag{2}$$

The state of the Top Event F is determined by the states of Basic Events X, $F = \phi(X_i)$, ϕ represents the structure function of the fault tree model. If the number of Basic Events is n, then with AND-gate, the structure function F can be expressed as:

$$F = \phi(X) = \bigcap_{i=1}^{n} X_i = \prod_{i=1}^{n} X_i \tag{3}$$

With Or-gate, it can be expressed as:

$$F = \phi(X) = \bigcup_{i=1}^{n} X_i = 1 - \prod_{i=1}^{n}(1 - X_i) \tag{4}$$

If all the events in one Basic Event occur will lead to the occurrence of the Top Event, this Basic Event is called a cut set. If any one event in this Basic Event is removed, this Top Event will no longer occur, then this cut set is called the smallest cut set. The process to make qualitative analysis of the monitoring network's fault tree model is the process to find the smallest cut set. The common methods to find the smallest cut set of the fault tree model are: upward method, downward method, Boolean cut set method, ect. This paper used the downward method to find out the smallest cut set of the monitoring network's fault tree model.

3.3 Establishment of the monitoring network's fault tree model

This paper analyzed the reliability of the distributed PV monitoring network running for one year. The Top Event of its fault tree model is the failure of the monitoring network. The external interference is not taken into consideration, according to the reliability evaluation indicators proposed previously, the failure of the monitoring network is caused by three aspects: (1)The rising of the failure rate of light intensity sensor nodes. (2)The decreasing of the coverage degree of the solar panels' monitored area. (3)The decreasing of the reliability of information transmission due to the failure of the communication links. The main fault tree model of the failure of the monitoring network is shown in figure 2. Then, this paper analyzed respectively the failure of the monitoring network caused by these three aspects.

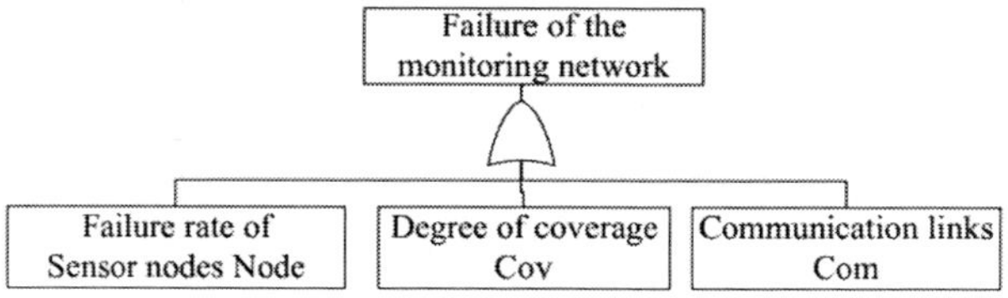

Figure 2. Main fault tree model of the failure of the monitoring network

3.4 Reliability of communication links

According to the information transmission paths established in section 1, the fault tree model of communication links is shown in figure 3.

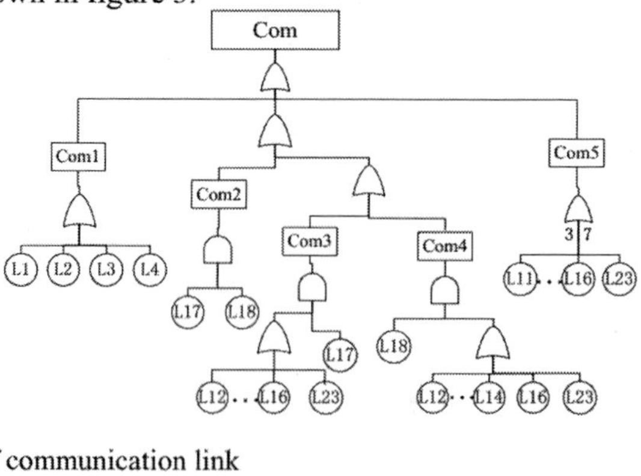

Figure 3. Fault tree model of communication link

3.5 Coverage degree

The reasons which may cause the decrease of the solar panel's coverage degree, can be divided into the following three groups: (1)The decreasing of coverage degree is caused only by boundary nodes. (2)The decreasing of coverage degree is caused only by internal nodes. (3)The decreasing of coverage degree is caused by both boundary nodes and internal nodes. The triangular grid method proposed in [13] is utilized to calculate the effective coverage area of the monitored area. The fault tree model of the solar panel's coverage degree is shown in figure 4.

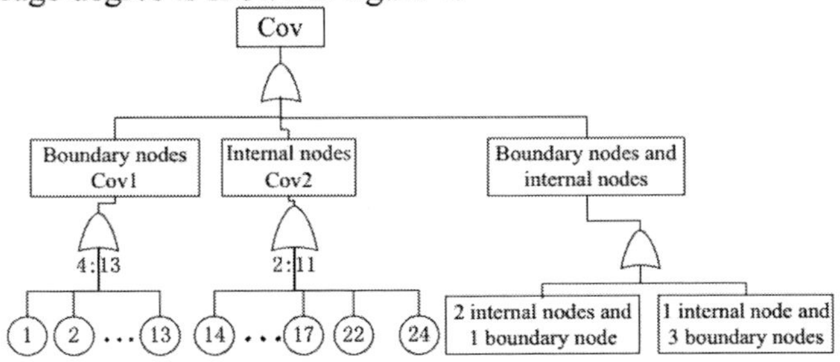

Figure 4. Fault tree model of the coverage degree

3.6 Failure rate of light intensity sensor nodes

For the sensor node *i*, who is responsible for receiving a number of *n* nodes' information, its abnormal working states can be presented as: (1)The node *i* fails itself. (2)The number of information received successfully by the node *i* is less than *n*. For example: For the node No.23, when it is under normal working state, it should receive 7 light intensity messages transmitted respectively from node No.5, 6, 7, 8, 17, 18 and 22. If one or more in these seven nodes fail, then the node No.23 can't receive 7 messages successfully, which means it is in an abnormal working state. Else, if the node No.23 fails itself, of cause, it is also under an abnormal working state. According to the information transmission paths established earlier, by analyzing the logical relationships between the monitoring network's sensor nodes, the fault tree model of the sensor nodes' failure rate is as shown in figure 5.

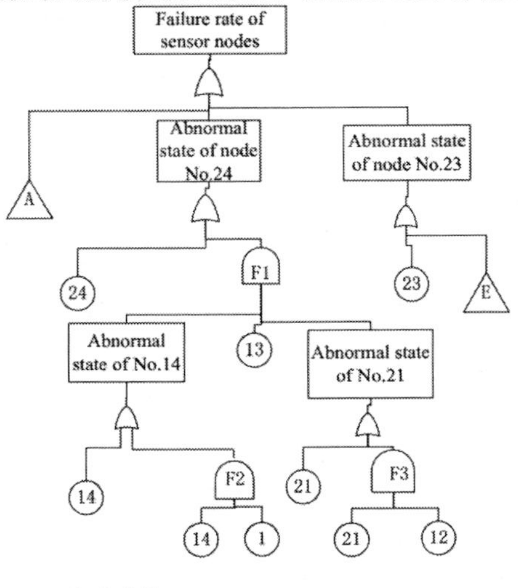

Figure 5. Fault tree model of sensor nodes' failure rate

In this fault tree model, E represents: except node No.23 fails itself, all the other causes leading to its abnormal working state. A represents: except node No.23 and No.24 themselves are in abnormal states, all the other causes causing the increasing of the sensor nodes' failure rate. For the PV monitoring network constructed in section 1, through analyzing the logical relationships between sensor nodes, the sub-fault tree E and A can be shown respectively in figure 6 and figure 7.

In the sub-fault tree A. B represents any one node inNo.14, 16, 17 or 22 is in an abnormal working state. C represents two or more nodes in No.14, 16, 17 and 22 fail. D represents three or more nodes in No.1, 2, 3, 4, 5, 6 and 13 fail. When establishing the sub-fault trees B, C, and D, it should use the Or-gate and Vote-gate. This paper no longer gives specific presentations of them.

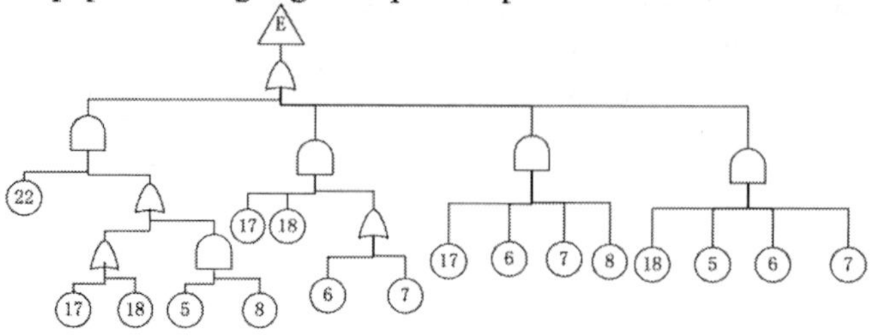

Figure 6. Sub-fault tree E

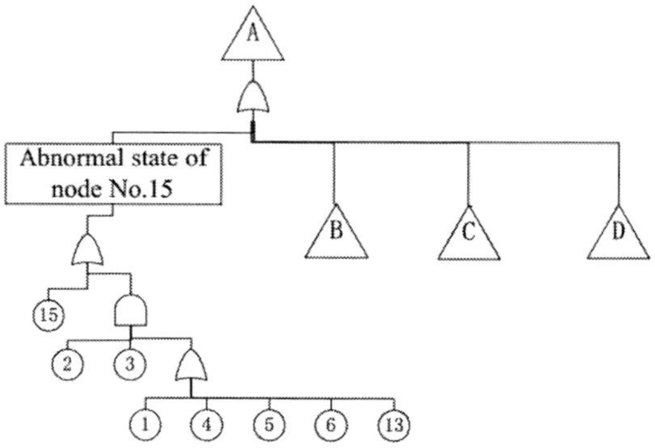

Figure 7. Sub-fault tree A

3.7 Finding the smallest cut sets with the downward method

The basic theory of the downward method is: From the Top Event of the fault tree model, according to the established logical relationships, analyze layer by layer from the top to the bottom until that the Top Event can be expressed by Basic Events. Then find the smallest cut sets of fault tree models with Boolean algebra rules.

Took the node No.24 as an example, searches for the smallest cut set which may cause the abnormal working state of the monitoring network. The solving process is as shown in table 2. In this table, *G* represents abnormal working state of sensor nodes, *E* represents Basic Event, *F* represents the collection of Basic Events. For example, G_{24} represents the abnormal working state of the node No.24. E_{24} represents the node No.24 fails itself. F_2 represents nodes No.1 and No.14 fail at the same time. It is the collection of E_1 and E_{14}, written as E_1E_{14}.

Table 2. Solving Process of the smallest cut-set by the downward method

G_{24}	E_{24}	E_{24}	E_{24}	E_{24}	E_{24}
	F_1	$E_{13}G_{14}G_{21}$	$E_{13}E_{14}G_{21}$	$E_{13}E_{14}E_{21}$	$E_{13}E_{14}E_{21}$
			$E_{13}F_2G_{21}$	$E_{13}E_{14}F_3$	$E_{13}E_{14}E_{12}E_{21}$
			$E_{13}G_{14}E_{21}$	$E_{13}F_2E_{21}$	$E_{13}E_1E_{14}E_{21}$
			$E_{13}G_{14}F_3$	$E_{13}E_2E_3$	$E_{13}E_1E_{14}E_{12}E_{21}$
				$E_{13}E_{14}E_{21}$	$E_{13}E_{14}E_{21}$
				$E_{13}E_2E_{21}$	$E_{13}E_1E_{14}E_{21}$
				$E_{13}E_{14}F_3$	$E_{13}E_{14}E_{12}E_{21}$
				$E_{13}F_2F_3$	$E_{13}E_1E_{14}E_{12}E_{21}$

 DOI: 10.1051/matecconf/20165503010

The last column of this table shows all cut sets that may cause the abnormal working state of the node No.24. After simplifying with Boolean algebra rules, the smallest cut sets X_1 and X_2 can be obtained respectively.

$$\begin{cases} X_1 = E_{24} \\ X_2 = E_{13}E_{14}E_{21} \end{cases} \tag{5}$$

$$G_{24} = X_1 + X_2 \tag{6}$$

In the same way, the smallest cut set which may lead to the abnormal working state of the node No.23 can be written as:

$$\begin{aligned} G23 = {} & E_{23} + E_{17}E_{22} + E_{18}E_{22} + E_5E_8E_{22} + \\ & E_6E_{17}E_{18} + E_7E_{17}E_{18} + E_6E_7E_8E_{17} + E_5E_6E_7E_{18} \end{aligned} \tag{7}$$

It can use the same method to solve the smallest cut sets of the other fault tree models. This paper no longer show them one by one.

4 Simulation analysis

In the simulation, the expectancy of life of the light intensity sensor nodes is $10^6 h$, then $\lambda = 10^{-6}$. The bit error rate is $p_{bit} = 0.2 \times 10^{-3}$. The length of data frame of light intensity information is $l_n = 65 \times 8 bit$. The simulation time is $T_{max} = 24 \times 365 h$.

The times that Basic Events occur caused by the failures of communication links and sensor nodes, are shown respectively in figure 8 and figure 9. Comprehensively consider the figure 1, figure 3 and figure 8, it can be known that the communication links connected directly with the cluster head node have the most important influence on the operation of the monitoring network. It could take the retransmission mechanism or increase the number of communication paths between sensor nodes to improve the reliability of the information transmission. From figure 1 and figure 9, it can be known that the failures of node No.23 and node No.24, which are connected directly to the cluster head node, are the main causes causing the failure of the monitoring network. Thus, in the operation of the monitoring network, it should focus on the working states of node No.23 and No.24. Once their working states are abnormal, they should be repaired immediately.

According to the simulation results showed in figure 8 and figure 9, in 10^5 times of simulation, different causes causing the failure of the monitoring network are shown in table 3. In this table, Node represents the failure rate of sensor nodes, Cov represents the coverage degree, Com represents the reliability of communication links. According to table 3, the decreasing of the reliability of communication links is the main cause which may lead to the failure of the monitoring network. Therefore, to improve the reliability of the monitoring network, it should make efforts to improve the reliability of the communication links, and to ensure the precise information transmission.

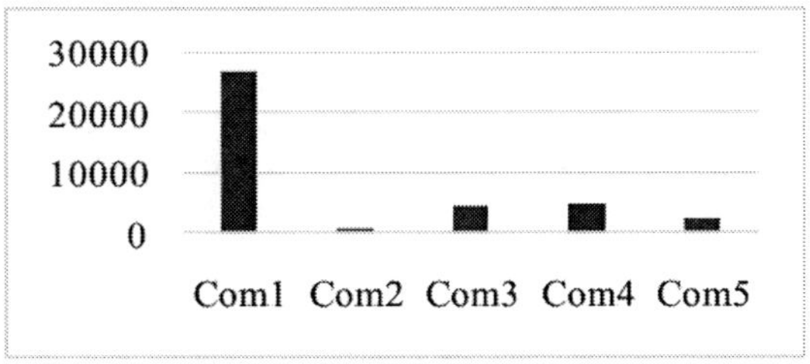

Figure 8. Times that basic events occur caused by the failure of communication links

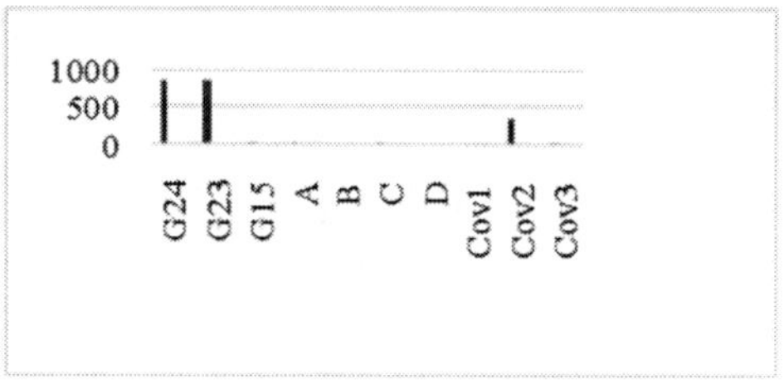

Figure 9. Times that basic events occur caused by the failure of sensor nodes

Table 3. Occurrence of different failure causes

Node	Cov	Com
1871	416	39621

5 Conclusion

This paper built a reliability model of a distributed PV monitoring network based on WSN technology. On the basis of fully considered the dynamic changes of its failure, the FTA method was used to determine any possible causes which may cause the failure of the network. The results of this paper offer the thought for further research on the reliability of the distributed PV monitoring network.

Acknowledgment

This work is supported by the National Natural Science Foundation of China (no.51307044) and Six Talent Peaks Project of Jiangsu Province of China (no.2014-XNY-008).

References

1. S. Pelland, D. Turcotte, G. Colgate, et al. IEEE Trans. on Sust. Energ., **3**, 1 (2012).
2. T. Mukai, M. Tomasella, A.K. Parlikad, et al. IEEE Trans. on Sust. Energ., **5**, 4 (2014)
3. B. Ando, S. Baglio, A. Pistorio, et al. IEEE Int. Workshop on Meas. and Net. Proc., (2013)
4. X. L. Xu, D. E. Qiao. Int. Conf. on Elec. Info. and Ctrl. Eng., (2011)
5. F. Y. Ma, X. R. Zhang. IEEE 30th Chinese Ctrl. Conf., (2011)
6. M. Jiang, Y. P. Liu, X. Z. Gu. 3th IEEE Int. Conf. on Secure Softw. Int. and Reliab. Impr., (2009)
7. V. D. Nitika, K. M. Durgesh, T. Sanjeev, et al. 3th Int. Conf. on Comput. Int., Com. Sys. and Net., (2011)
8. H. Yousefi, K. Mizanian, A. H. Jahangir. 24th IEEE Int. Conf. on Adv. Info. Net. Appl., (2010)
9. W. Heinzelman, A. Chandrakasan, H. Balakrishnan. IEEE Trans. on Wireless Com., **1**, 4 (2002)
10. J. Jia, J. Chen, G.R. Chang, et al. Comput. and Maths. with Appl., **57**, 11 (2009)
11. N. Mohamed, J. Al-Jaroodi, I. Jawhar, et al. Int. Symp. on Perf. Eval. of Comput. and Telecom. Sys., (2013)
12. R. X. Duan, G. C. Wan, D. C. Dong. Int. Conf. on Int. Comput. Technol. and Autom., (2010)
13. S. Kumar, M. Dave, S. Dahiya. IEEE Int. Conf. on Adv. Net. and Telecom. Sys., (2013)

MATEC Web of Conferences **55**, 03011 (2016) DOI: 10.1051/matecconf/20165503011
ACPEE 2016

Development and Uncertainty Evaluation of Calibrating System for Digital Energy Setting

Sha Zhao[1,a], Jingfen Bai[1], Fantao Lin[1], Xiangjiang Yang[1] and Chunping Yu[1]

[1] *China Electrical Power Research Institute, 100192 Beijing, China*

Abstract. The power measurement mode has made fundamental changes with the development of smart substation, traditional energy metering analog devices are replaced by digital devices according to the requirements of digitalized information substation. Creating traceability between digital measuring value and analog measuring value and developing relevant calibration devices are key issues in smart substation construction. A new device for calibration digital energy meters using integral calibration technology is described in this paper. Based on the Standard Comparison Method, the device establishes a connection between the digital power and the traditional analog power with the integral calibration technology, and solves the problem that digital power was not able to be traced. The device is designed with final measuring voltage level: 220kV/110kV, measuring current range: 5A～5000A, and extended uncertainty less than 0.05%. Furthermore, the estimating process of the uncertainty of the device is discussed emphatically.

1 Introduction

Smart grid construction is an important field of growing concern in the world, and the smart substation is one of the important nodes in building smart grid, which bears electricity settlement gateway function and is an important guarantee for power efficient transmission and reliable operation of grid.

Different from traditional substation, the communication platform of smart substation is established on the basis of IEC61850 international unified network communication standard [1-2]. The electric energy metering device in smart substation contains Electronic Transformer, Merging Unit and Digital Input Electricity Meter, these components are fundamentally different from traditional ones just like traditional electromagnetic Transformer and Electricity Meter on the methods of measuring principle, and data transmission. Their metrological characteristics are clearly different from traditional energy metering device too. Analog signals are transmitted by cable in the traditional power measurement mode, while digital signals are transmitted by fibre optic Ethernet in the digital power measurement mode, so the traditional calibration method and device cannot meet the requirement of digital electric energy metering system.

In addition, metering is featured by scientific and legality, and is the process to unify the unit and make the measuring value accurate and reliable. The electric energy measurement device used for trade account must be traced to the national standard indirectly by compulsory verification. Up till

[a] Corresponding author : zhaosha@epri.sgcc.com.cn

now, China does not yet possess the capability of digital electric energy metering traceability. Fortunately, many scholars have made lots of related researches on calibration methods and traceability, and have come to many important conclusions [3-7]. However, most of these researches consider Electronic Transformer and Merging Unit as a whole for calibrating, while the Digital Input Electricity Meters are calibrated separately, there are not many studies by considering all digital electric energy measurement device as a whole [8-10], and are still not very mature.

The tests of Electronic Instrument Transformer (ECT/EVT) are commonly carried out with the help of Merging Unit (MU), ECT/EVT and MU are indivisible in practice, such as the tests of clock synchronization and data modification, it is difficult to make separate calibration for them. As a power calculation and cumulative unit, the Digital Input Electricity Meter does not have any measuring function for physical parameters, so there should be no error in it [11]. The quantization error, synchronization error, even network traffic anomaly from Merging Unit will all have an impact on Digital Input Electricity Meter. In addition, as one of the smart substation equipments, there are many parameters to be set for the Digital Input Electricity Meter during the debugging process of smart substation, which will also affect the calibration.

Aiming at making the digital power be able to trace to the traditional power standard, this paper presents an integral calibration method by considering the digital electric energy measurement device as a whole, to avoid the problem that Electronic Transformer and Merging Unit can't be calibrated separately, to calibrate the composite error of the whole set more comprehensively, and to reflect the performance of the whole set, including but not limited to network status.

2 Principle Introduction

Based on the Standard Comparison Method, our digital electric energy calibration system can measure the global error of electric energy measurement device. It is a single phase digital input energy standard device, and its principle block diagram is as shown in Figure 1:

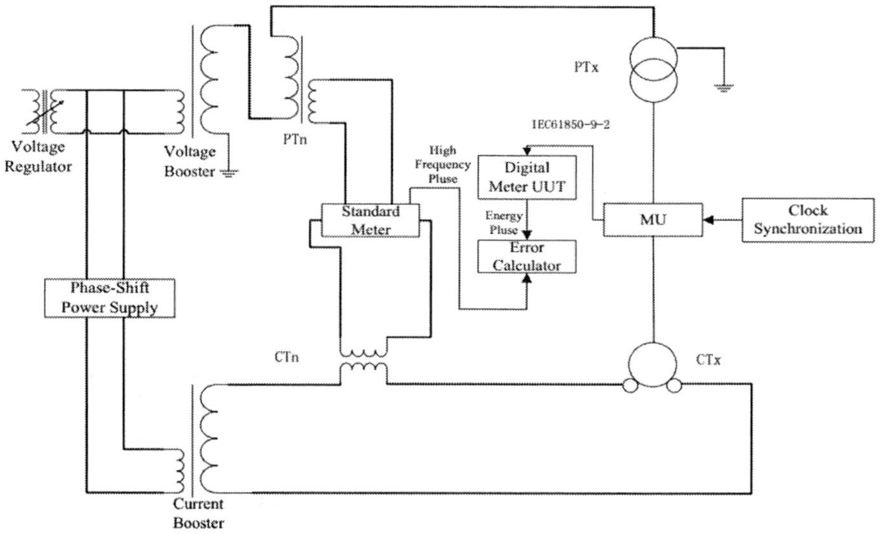

Figure 1. The principle block diagram of electric energy standard device

The phase-shift power supply shifts the voltage phase angle in voltage circuit and current circuit, and generates two different phases of AC voltages according to the phase of test point. Using a voltage regulator and a voltage booster, the input voltage in voltage circuit can be convert to high voltage and output to the Electronic Voltage Transformer PTx of the digital energy measurement device under test and the standard Voltage Transformer PTn of the standard equipment, the input voltage in current circuit through a voltage regulator and a current booster can be converted to high current and output to the Electronic Current Transformer CTx of the digital energy measurement device under test and the standard Current Transformer CTn.

The digital energy measurement device is made up by an Electronic Current Transformer CTx, an Electronic Voltage Transformer PTx, a Merging Unit and a Digital Input Electricity Meter under test. For digital output Electronic Transformer, ECT/EVT converts primary current (voltage) signal to digital signal and send to Merging Unit over optical fiber. For analog output Electronic Transformer, ECT/EVT converts primary current (voltage) signal to secondary small voltage signal and sends the signals to Merging Unit over shield cable. The A/D converter in Merging Unit transforms analog electronic signals into digital sampling values. These digital current and voltage sampling values can be assembled into network frames according to the IEC61850-9-2 standard by Merging Unit and sent to Digital Input Electricity Meter, which extracts current and voltage information from network frames to calculate electric energy, and sends out low—frequency electric energy pulses.

On the other hand, the electric energy standard device is made up by a traditional standard Current Transformer CTn, a standard Voltage Transformer PTn and a standard Electric Energy Meter. The traditional standard Energy Meter measures the standard electric energy directly using the output secondary signals from standard Current Transformer CTn and standard Voltage Transformer PTn, and then sends out high—frequency electric energy pulses. Low—frequency electric energy pulses from Digital Input Electricity Meter and high—frequency electric energy pulses from standard Energy Meter are putted into an error calculator simultaneously, so the global error of digital energy measurement device can be obtained.

According to the IEC61850-9-2 standard, the sampled values that the Digital Input Electricity Meter received are the primary current and voltage of the digital transformer, so the output pulse of the Digital Input Electricity Meter is proportional to the primary electric energy pulse. The following equation should apply to calculate the global error of the digital energy measurement device under test:

$$\gamma = \frac{E_{\mathrm{X}} - K_U K_I E_{\mathrm{N}}}{K_U K_I E_{\mathrm{N}}} \times 100\% \tag{1}$$

where

E_{N}—energy value of analog standard Energy Meter tested

E_{X}—energy value of Digital Input Electricity Meter tested

K_U—transformation ratio of standard Voltage Transformer

K_I—transformation ratio of standard Current Transformer

The standard Voltage Transformer has four taps: $\frac{220\mathrm{kV}}{100\mathrm{V}}$, $\frac{220\mathrm{kV}/\sqrt{3}}{100\mathrm{V}}$, $\frac{110\mathrm{kV}}{100\mathrm{V}}$ and $\frac{110\mathrm{kV}/\sqrt{3}}{100\mathrm{V}}$. The standard Current Transformer is of wide-range with the primary rated current 5A~2000A, and the secondary rated current 5A. These transformers meet the requirements of energy measurement device calibration of the substation which has voltage levels of 110kV, 220kV and various loads. With a synchronous clock device which has optoelectronic pulse conversion function, high precision clock synchronization is realized between the standard device and the Digital Input Electricity Meter under test. In such a way, the additional error caused by synchronization is avoided.

3 Uncertainty Analysis

3.1 Mathematical model

Using direct comparison method, the measurement error γ of the digital energy measurement device can be calculated by using the following formula:

$$\gamma = \frac{E_X - K_V K_I E_N}{K_V K_I E_N} \times 100\% \tag{2}$$

where

E_N—energy value of analog standard Energy Meter tested

E_X—energy value of Digital Input Electricity Meter tested

K_U—transformation ratio of standard Voltage Transformer

K_I—transformation ratio of standard Current Transformer

3.2 Sources of uncertainty

According to the principle and the mathematical model of the device, it can be made out that the sources of uncertainty include the following parts:

(1) Uncertainty of the digital energy measurement device under test, mainly refers to the uncertainty component introduced by random factors such as stability.

(2) Uncertainty component introduced by the ratio error and phase error of Voltage Transformer.

(3) Uncertainty component introduced by the ratio error and phase error of Current Transformer.

(4) Uncertainty component introduced by the standard Energy Meter.

This paper only analyses the uncertainty components introduced by the standard device, do not contain the uncertainty component introduced by the digital energy measurement device to be tested. Based on the Standard Comparison Method, the standard device and the device under test are connected with the same set of voltage and current circuit, therefore, the contribution of power supply to the uncertainty of the whole device can be ignored.

3.3 Uncertainty component introduced by ECT/EVT

For convenience, we will analyze the uncertainty components step by step. First, transform section of the transformer is discussed. The influence of transformer on electric energy of output terminal is presented here.

Instead of making modifications on the ratio error and phase error traceable value of ECT/EVT, uncertainty was evaluated by the accuracy class of the transformer itself. Input voltage and current are putted into the standard Energy Meter through ECT/EVT, the influence on electric energy should include ratio error and phase error.

The ratio error $f_V(f_I)$can be calculated by:

$$f_V = \frac{K_V V_2 - V_1}{V_1} \times 100\% \tag{3}$$

$$f_I = \frac{K_I I_2 - I_1}{I_1} \times 100\% \tag{4}$$

where

$V_2(I_2)$—the actual second voltage(current)

$K_V(K_I)$—rated transformation ratio

$V_1(I_1)$—the actual primary voltage (current)

Phase error$\delta_V(\delta_I)$ is also called angle error, which can be expressed as the phase displacement between the vectors that obtained by revolving the secondary voltage (current) vector $V_2(I_2)$for 180 degrees counterclockwise and the primary voltage (current) vector $V_1(I_1)$.

According to the calculation formula of single-phase electric energy $P = VI\cos\varphi$, the following is valid:

$$\frac{\partial P}{\partial V} = I\cos\varphi$$

$$\frac{\partial P}{\partial I} = V\cos\varphi$$

$$\frac{\partial P}{\partial \varphi} = -VI\sin\varphi \qquad (5)$$

V, I and φ are mutual independence, so according to the uncertainty conduction formula, ΔP can be calculated from:

$$\Delta P = \sqrt{\left(\frac{\partial P}{\partial V}\Delta V\right)^2 + \left(\frac{\partial P}{\partial I}\Delta I\right)^2 + \left(\frac{\partial P}{\partial \varphi}\Delta\varphi\right)^2}$$

$$= \sqrt{(I\cos\varphi\Delta V)^2 + (V\cos\varphi\Delta I)^2 + (VI\sin\varphi\Delta\varphi)^2} \qquad (6)$$

Divided by active power deviations:

$$\frac{\Delta P}{P} = \frac{\sqrt{(I\cos\varphi\Delta V)^2 + (V\cos\varphi\Delta I)^2 + (VI\sin\varphi\Delta\varphi)^2}}{VI\cos\varphi} \qquad (7)$$

Put the formula (3), formula (4) into formula (7):

$$= \sqrt{\left(\frac{\Delta V}{V}\right)^2 + \left(\frac{\Delta I}{I}\right)^2 + (\mathrm{tg}\varphi\cdot\Delta\varphi)^2}$$

$$= \sqrt{f_V{}^2 + f_I{}^2 + \mathrm{tg}^2\varphi\cdot\Delta\varphi^2} \qquad (8)$$

When $\Delta\varphi = \delta_V - \delta_I$,formula (8) is expressed as:

$$= \sqrt{f_V{}^2 + f_I{}^2 + \mathrm{tg}^2\varphi\cdot\delta_V{}^2 + \mathrm{tg}^2\varphi\cdot\delta_I{}^2 - 2\mathrm{tg}^2\varphi\cdot\delta_V\delta_I} \qquad (9)$$

When the variation of V、I、φ is very small, the unrelated higher order term could be neglected, meanwhile, $\delta_V \approx \sin\delta_V, \delta_I \approx \sin\delta_I$, then,

$$\frac{\Delta P}{P} = \sqrt{f_V{}^2 + f_I{}^2 + \mathrm{tg}^2\varphi\cdot\sin\delta_V{}^2 + \mathrm{tg}^2\varphi\cdot\sin\delta_V{}^2} \qquad (10)$$

where

$f_V = f_I = 1\times10^{-4}$

$\delta_V = \delta_I = 0.3$

$\sin\delta_V = \sin\delta_V = 87.3\times10^{-6}$,$k$=$\sqrt{3}$

3.4 uncertainty component introduced by standard energy meter

This device does not use the tracing error to correct the measurement results, but only through tracing to confirm the error limit of accuracy class that the standard Energy Meter met. Therefore, the contribution of the standard Energy Meter to the digital energy standard device is equal to the accuracy class 0.01%, which is processed by the average distribution, k=$\sqrt{3}$, the standard uncertainty of this component can be given as:

$$u_{\mathrm{rel}}(E_{\mathrm{N}}) = {}^{0.01\%}\!/_{\sqrt{3}} = 57.7 \times 10^{-6} \qquad (11)$$

3.5 Uncertainty component introduced by clock synchronization

The uncertainty introduced by clock synchronization device will have an effect on the clock of the Merging Unit and the measurement uncertainty of the whole set. The synchronization accuracy of the clock synchronization device is less than 100ns. By processing it with the average distribution, the relative standard uncertainty u_2 can be given as:

$$u_2 = \frac{100\mathrm{ns}}{1\mathrm{s}} \cdot \frac{1}{\sqrt{3}} = 5.8 \times 10^{-8} \qquad (12)$$

3.6 Combined standard uncertainty

The proportional change part of the voltage (current) and the electric energy measurement part of the standard Energy Meter are independent of each other, so the combined standard uncertainty of the whole set can be given as:

$$u_1 = \frac{0.001\%}{2\sqrt{3}} = 2.9 \times 10^{-6} \qquad (13)$$

Table 1. Uncertainty summary table.

NO	Uncertainty Factors		Sub-item Uncertaint y(×10-6)	Coverage Factor	Standard Uncertainty Component (×10⁻⁶) $u_i(x)$	Propagation Coefficient	Uncertainty(×10⁻⁶) $u_i(y)$		
							PF= 1.0	**PF= 0.5**	**PF= 0.8**
1	Uncertainty Component Introduced By Standard Energy Meter		100	$\sqrt{3}$	57.7	1	57.7	57.7	57.7
2	Uncertainty Component Introduced By Standard Transformer	f_V	100	$\sqrt{3}$	57.7	1	57.7	57.7	57.7
		f_I	100	$\sqrt{3}$	57.7	1	57.7	57.7	57.7
		δ_V	87.3	$\sqrt{3}$	50.4	tgφ	0	87.3	40.4
		δ_I	87.3	$\sqrt{3}$	50.4	tgφ	0	87.3	40.4
3	Uncertainty Component Introduced By Error Processor		10	$2\sqrt{3}$	2.9	1	2.9	2.9	2.9
4	Uncertainty Component Introduced By Synchronous Device		0.1	$\sqrt{3}$	0.06	1	0.06	0.06	0.06
Combined Standard Uncertainty							100	159	116

3.7 Expanded uncertainty

Expanded uncertainty $U_{\mathrm{rel}}(P) = ku_{\mathrm{rel}}(P)$,$k$=2,then the expanded uncertainty of the digital energy standard device can be expressed as:

$$U_{\mathrm{rel}}(P) = 2\sqrt{{f_V}^2 + {f_I}^2 + \mathrm{tg}^2\varphi \cdot {\sin\delta_V}^2 + \mathrm{tg}^2\varphi \cdot {\sin\delta_V}^2 + u_{\mathrm{rel}}^2(E_N) + u_1^2 + u_2^2} \qquad (14)$$

Expanded uncertainty is as shown in the following table:

Table 2. Expanded uncertainty

Power Factor	Expanded Uncertainty(%)
$\cos\varphi$=1.0	0.020
$\cos\varphi$=0.5L(0.5C)	0.032
$\cos\varphi$=0.8C(0.8L)	0.024

Considering the influence of the random error caused by noise and measurement stability, the overall measurement uncertainty is less than 0.05%.

4 Conclusion

Along with the expansion of smart substation construction, digital electric energy device has been widely applied, in order to ensure the accuracy of electric energy metering, it must trace to the electric energy metering standard of higher level. In this paper, a new digital energy standard device is designed, where integral calibration technology is used to complete the tracing of electric energy. The expanded uncertainty of the device can reach 0.05% (k=2), that can meet the requirements of the calibration and acceptance tests of all levels of digital electric energy measurement devices with voltage level: 110kV/ 220kV, current range: 5A~5000A, and integrated error less than 0.1%.

By taking the traditional electric energy metering system as standard, a connection between digital power and traditional analog power is established. And by applying integral calibration method, the additional error caused by tracing digital power to traditional analog power when make separate calibration for digital input electricity meter is avoided, the unified and accurate measuring value is effectively guaranteed. The device provides a simple and effective way to make the digital electric energy metering technology meet the requirements of law, and provides reference for China to establish a national digital energy metering traceability system.

Acknowledgement

Supported by the National High Technology Research and Development of China("863" Project)(Grant No. 2015AA050404)

References

1. Y.M. Ren, L.J. Qin, Q.X. Yang, Automat. Electr. Pow. Syst **24**, 8(2000)
2. W.S. Tan, Power. Syst. Technol **25**,9(2001)
3. L. Yao, S.T. Li, C.G. Lu, Electr. Meas. Instr **47**, 28—32(2010).
4. G.Y. Lin, S.L. Zhou, W.M. Sun, S.C. Wu, Proc. CSU-EPSA **23**, 3(2011).
5. Y. Xiao,B. Jiang,W. Zhao,B. Ai, S.L. Huang, Y. Tang, X.S. Chen, Electr. Meas. Instr **51**,1(2014)
6. R.M. Chen, X. Li, Z.M. Kong, Y.J. Fang, Electr. Meas. Instr **49**, 10(2012)
7. R.M. Chen, Z.Y. Wang, Z.M. Kong, Y.J. Fang, Electr. Meas. Instr **49**, 9(2012)
8. Z. Xu, S. Zhao, R.M. Chen, Y. Xiao, W.M. Sun, Control. Instrum. Chem. Ind 12(2012)
9. W.M. Sun, G.Y. Lin, S.L. Zhou, K.C. Li, Electr. Meas. Instr **47**, 8(2010)
10. Q. Li, S.H. Zhang, Y.B. Lu, D. Han, H.L. Hu, H. Li, D.Y. Li, Electr. Meas. Instr **47**, 10(2010)
11. H.J. Yu, YunNan. Electr. Pow **36**, 5(2008)

MATEC Web of Conferences **55**, 04001 (2016) DOI: 10.1051/matecconf/20165504001
ACPEE 2016

Modelling and Design of a 3 kW Permanent Magnet Synchronous Generator suitable for Variable Speed Small Wind Turbines

Acharya Parash[1,a], Papadakis Antonis[2], Shaikh Muhammad Naveed[3]

[1] *Lecturer, Department of Electrical and Electronics Engineering, Kathmandu University, Dhulikhel, Nepal*
[2] *Assistant Professor, Department of Electrical Engineering, Frederick University, Nicosia, Cyprus*
[3] *PHD student, Department of Electrical Engineering, Frederick University, Nicosia, Cyprus*

Abstract. This paper presents the modeling and design of a 3 kW Permanent Magnet Synchronous Generator (PMSG) used for a variable speed wind turbine. Initially, the PMSG is modeled in the d-q reference frame. Different optimized parameters of the generator are extracted from the design and used in simulation of the PMSG. The generator output power is matched with the power of the turbine such that the generator is not either over-sized or under-sized.

1 Introduction

Wind turbine is considered as a very promising renewable energy source. The performance of wind turbine depends firstly upon the optimum design of the system and secondly on the control strategies of different possible parameters that can operate efficiently under extreme variations in wind conditions. Generally, both induction and synchronous generators are used in wind power conversion systems. Within an induction generator, three different types of rotors can be used: (a) squirrel cage rotor, (b) wound rotor with slip control and (c) doubly fed induction rotors. From the above three aforementioned technologies, the doubly fed is normally used in wind speed generation as it provides a wide range of speed variation.

Variable speed multi-pole PMSG is normally chosen for small wind turbines as it offers better performance due to its higher efficiency and less maintenance because it does not have any rotor current (no field). PMSG can also be used without a gearbox, which implies a reduction in the weight of the nacelle and cost. Optimum wind energy extraction is achieved by operating the wind turbine generator at variable speed because of the higher energy gain and reduced stresses. The recent advancements in power electronics and control strategies have made it possible to regulate the output voltage of the PMSG, thereby optimizing the generator's power.

2 Model of PMSG

The mathematical model of a PMSG is generally based on the following assumptions [1].

[a] Corresponding author : parash.acharya@ku.edu.np

1. The stator windings are positioned sinusoidally along the air-gap as far as the mutual effect with the rotor is concerned.
2. The stator slots cause no significant variations of the rotor inductance with rotor position.
3. Magnetic hysteresis and saturation effects are negligible.
4. The stator windings are considered symmetrical.
5. Damper windings are not considered.
6. Capacitance of all the windings is neglected.
7. Resistances of the coils are assumed to be constant.

The dynamic model of a PMSG is derived from a two-phase synchronous reference, direct (d) and quadrature (q) axis frame in which the q-axis is 90° ahead of the d-axis with respect to the direction of rotation. In the case of a balanced three phase system, the dq reference frame transformation reduces three AC quantities to two DC quantities, allowing simplified calculations to be performed. The dq transformation applied to a three phase system is as shown by equation (1) and its inverse transform is given by equation (2) [1], where F can represent voltages, currents or inductances whose values depend upon the rotor position.

$$\begin{bmatrix} F_d \\ F_q \end{bmatrix} = \frac{2}{3} \begin{bmatrix} \sin\omega t & \sin\left(\omega t - \frac{2\pi}{3}\right) & \sin\left(\omega t + \frac{2\pi}{3}\right) \\ \cos\omega t & \cos\left(\omega t - \frac{2\pi}{3}\right) & \cos\left(\omega t + \frac{2\pi}{3}\right) \end{bmatrix} \begin{bmatrix} F_a \\ F_b \\ F_c \end{bmatrix} \tag{1}$$

$$\begin{bmatrix} F_a \\ F_b \\ F_c \end{bmatrix} = \begin{bmatrix} \sin\omega t & \cos\omega t \\ \sin\left(\omega t - \frac{2\pi}{3}\right) & \cos\left(\omega t - \frac{2\pi}{3}\right) \\ \sin\left(\omega t + \frac{2\pi}{3}\right) & \cos\left(\omega t + \frac{2\pi}{3}\right) \end{bmatrix} \begin{bmatrix} F_d \\ F_q \end{bmatrix} \tag{2}$$

The d and q-axis currents in the frequency (s) domain can be represented as in equations (3) and (4) respectively which can be obtained through equivalent circuit model of PMSG as shown in Figure 1 [1 ,2].

$$I_{ds} = \frac{(-V_{ds} - R_s I_{ds} - \omega_r)}{sL_{ds}} \tag{3}$$

$$I_{qs} = \frac{[-V_{qs} - R_s i_{qs} - \omega_r (L_{ds} + L_{ls}) I_{ds} + \omega_r \varphi_r]}{sL_{qs}} \tag{4}$$

where, V_{ds} and V_{qs} are the d and q-axis stator voltages, I_{ds} and I_{qs} are the d and q-axis stator currents, R_s is the stator resistance and ω_r is the angular speed of the generator, L_{ds} and L_{qs} are the stator d and q-axis self inductances and ϕ_r is the rotor flux. L_{ds} and L_{qs} are given by equations (5) and (6) respectively [2].

$$L_{ds} = L_{ls} + L_{dm} \tag{5}$$

$$L_{qs} = L_{ls} + L_{qm} \tag{6}$$

where, L_{dm} and L_{qm} are the magnetizing inductances in d and q-axis, and L_{ls} is the leakage inductance. For a non-salient pole PMSG, d and q-axis magnetizing inductances are equal (i.e. $L_{dm} = L_{qm}$), whereas for a salient pole PMSG, d-axis magnetizing inductance is normally lower than the q-axis magnetizing inductance (i.e. $L_{dm} < L_{qm}$) [2]. The simplified dq-axis model in the rotor-field synchronous frame is as shown in Figure 1.

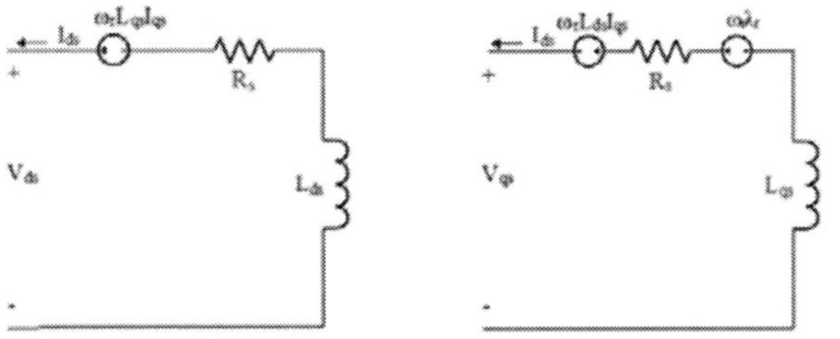

(a) d-axis circuit (b) q-axis circuit

Figure 1. Simplified dq-axis model of PMSG in the rotor field synchronous reference frame

The electromagnetic torque (T_e) and the rotor speed (ω_r) of the PMSG are calculated as shown in equations (7) and (8) respectively [2].

$$T_e = \frac{3N_{pp}}{2}[\varphi_r i_{qs} - (L_{ds} - L_{qs}) i_{ds} i_{qs}] \quad (7)$$

$$\omega_r = \frac{N_{pp}}{JS}(T_e - T_m) \quad (8)$$

where, N_{pp} is the number of pole pairs of the rotor, J is the rotational inertia of the generator and T_m is the mechanical torque for the generator (in the case of the PMSG connected to a wind turbine, T_m is the torque from the turbine).

Using equations from (1) to (7), a PMSG model is developed in Matlab/Simulink. Inputs for the model are d-q axis voltages and mechanical torque. Rotor speed (ω_r), is calculated through equation (8) and is used as feedback to the system. Number of pole pairs, d and q-axis inductances, leakage inductance, magnetic flux of the rotor magnets, stator resistance and moment of inertia of rotor and load can be initially set for a particular size of the generator. The d-axis current (I_{ds}), q-axis current (I_{qs}) and electromagnetic torque (T_e) are the outputs from the system as calculated by equations (3), (4) and (7), respectively.

3 Design Considerations for PMSG

Axial flux coreless PMSG is regarded as the suitable generator for small wind turbine because of its simplicity of design and manufacture. The absence of cogging torque for coreless design eliminates any magnetic pull between stationary and rotating parts [3]. Most of the axial flux machines have two sided rotor configuration rather than the single sided rotor to balance the axial forces [4]. Coreless design reduces the mass and increases the efficiency as compared to the conventional design [5].

Matching the generator size with power captured by the blade from wind is most important. A small rotor blade with big generator will not generate the full power while small generator with big rotor blades will generate full power below rated speed. Hence the generator is designed such that it extracts maximum power from the wind. Tip Speed Ratio (TSR) is the vital factor in matching generator power with the wind power. The optimum TSR depends upon number of rotor blades i.e. fewer the number of blades, faster should be the turbine speed to extract the available wind power. For small value of TSR (i.e. for improper design of rotor blades), the rotor will turn too slowly or even stall. For high value of TSR, rotor will spin very fast through turbulent air, optimum power cannot be extracted, and the turbine will be at great risk to have structural failure.

The value of TSR for a well designed three bladed wind turbine system will be around 6 to 7 [6]. The system considered here is geared three bladed systems and hence the generator rotors will have speed higher than that of the blades. Considering TSR value as 6 for wind speed of 7 m/s and the blade with radius 4 m (for 3 kW wind turbine), the blade speed of around 100 rpm is achieved. With gear ratio 1:4, the generator parameters are optimized such that it produces power output of around 3 kW at around 400 rpm. The general optimized parameters of the generator are type and size of the magnet, number of rotor poles, air gap flux density, number of coil turns, air gap distance and stator coil diameter. The overall system parameters for the generator design are as shown in Table 1.

3.1 Design Parameters Calculations

The turbine will start generating useful power above the cut in speed. The cut in turbine speed can be calculated using equation (9).

$$TSR = \frac{\omega_{cutin} R}{V_{cutin}} \quad (9)$$

where, ω_{cutin} is the turbine cut in speed in rad/s, R is radius of the blade in metre, V_{cutin} is the cut in speed for the wind in m/s .Assuming that the turbine starts at 3 m/s wind speed with Tip speed Ratio (TSR) value of 6 and blade radius of 4 m, turbine cut in speed can be calculated as 4.5 rad/s (or 43 rpm).The cut in frequency can be calculated from equation (10) [7].

$$f_{cutin} = \frac{N_{cutin} P}{120} \quad (10)$$

where, N_{cutin} is the generator cut in speed; P is the total number of rotor poles. Considering gear box ratio of 1:4, the generator cut in speed is approximately 130 rpm. With 20 number of rotor poles (i.e. total number of magnets), cut in frequency is obtained as 20 Hz. The total number of turns per phase ($N_{turns/phase}$) for the stator winding can be obtained from equation (11) [7].

$$N_{turns/phase} = \frac{V_p}{4.44 f_{cutin} \varphi_g K_w} \quad (11)$$

where, V_p is the phase voltage at generator cut in speed, Φ_g is the magnetic flux in air gap and K_w is the winding factor which is normally taken as 0.85 to 0.95.

The inductance of the stator coil can be obtained considering size of the coil and distance between the two coils. The most efficient shape is to keep the ratio of the winding depth to length close to unity so that the winding have square cross section. The square shape will keep minimum distance between the turns. There will be a high level of magnetic coupling when the turns are kept together and hence inductance of a coil increases with the square of the number of turns. Similarly, resistance of the coil can be determined after determining total length and cross sectional area of the coil. The length of the coil is calculated taking average diameter of inner and outer dimension of the coil. The inner diameter is chosen such that it has same dimension as that of magnetizing face of the magnet.

Table 1. Overall system parameters

Parameters	Values
Rated wind turbine power (P)	3kW
Rated wind speed (V_w)	7 m/s
Density of wind (ρ)	1.2 kg/m^3
Power Coefficient (C_p)	0.3
Number of blades	2
Tip Speed Ratio (TSR)	6
Area of the blade (A)	48.6 m^2
Diameter of the blade (D)	8 m
Rated turbine speed (N_t)	102 rpm
Cut in wind speed	3 m/s
Cut in turbine speed	43 rpm
Gear ratio	1:4
Cut in generator speed (N_{cutin})	130 rpm

Table 2. Optimized generator parameters

Parameters	Values
Number of rotor pole pairs (N_{pp})	10
Number of stator coils	9
Number of turns per phase ($N_{turns/phase}$)	442
Flux density in the air gap (φ_g)	4.84X10^{-4} Tesla
Magnet Dimensions	55X55X20 mm^3
Air Gap	2.3 mm
Generator outer diameter	500 mm
Stator coil diameter	1.628mm
Inductance of the coil/Phase ($L_{coil/phase}$)	0.405 mH
Resistance of the coil/phase ($R_{coil/phase}$)	0.9067 Ω
Moment of inertia	0.00062J Kg.m^2
Viscous damping	0.0003035 N.m.s

3.2 Optimization of Generator Parameters

The choice of phase number in the stator is directly related to the pole pair number which determines the output frequency as well as the total number of coils. The nominal frequency at 400 RPM should be around 50-60 Hz and each phase should have equal number of coils. The thickness of the stator support, that is able to provide enough mechanical strength, must not exceed that of the coil thickness. The diameter of the wire used in the stator should be as small as possible leading to high current density at the nominal point. Short circuit current would be the maximum current that might flow in the stator coil. Assuming that the system is designed to be stalled after 500 RPM, a coil size of 1.628 mm is recommended which can carry maximum enclosed 17 A current with current density of 8.2 A/mm^2 [8]. A large number of coils lead to large number of stator supports, thus increasing magnet and winding thickness. To compromise all these parameters, a phase number of 9 and pole pair of 10 are chosen.

In order to reduce eddy current losses, the air gap flux density should be kept low. Hence, magnet is chosen with flux density of around 0.2-0.4 T. It is better to increase the wire thickness and decrease the air gap flux density in order to reduce copper and eddy current losses. The generator outer diameter should not exceed 10 to 15 % of the blade diameter [9]. There are different parameters which affects output of the generator at the particular speed. Parameters such as flux density of the magnet; magnet size; wire diameter and phase number were varied in order to obtain generator output power of 3 kW when rotating at 400 RPM. A Mat lab program was developed and the output voltage, Current and Power was plotted varying different parameters until the required power output was obtained at the desirable speed. Output powers at various turbine speed and generator speed is plotted as shown in Figure 2 which shows that wind turbine power of 3 kW is achieved at turbine speed of nearly 100 rpm. Similarly, generator output power of 3 kW is achieved at the speed of nearly 400 rpm. Different optimized generator parameters, in order to obtain 3 kW output at 400 RPM, are summarized in Table 2.

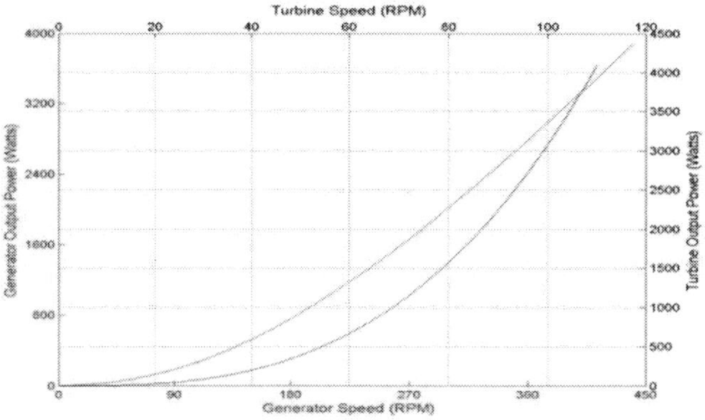

Figure 2. Matching wind turbine and generator power outputs

4 Simulation Results

The dq axis stator currents, i_{ds} and i_{qs}, in the synchronous frame rotating at the synchronous speed (ω_r), is calculated within the PMSG model. They are then transformed into the abc-axis stator currents, i_{as}, i_{bs}, and i_{cs}, in stationary frame through dq/abc transformation. The calculated load voltages v_{as}, v_{bs}, and v_{cs}, which are also the stator voltages, are transformed to the dq- axis voltages v_{ds} and v_{qs} in the synchronous frame, which are then fed back to the PMSG model.

The simulation was performed using optimized generator parameters as shown in Table 2. The generator output power, three phase voltages and currents are as shown in Figure 3 for different values of mechanical torque which represents the wind turbine torque. As shown in Figure 3, high torque represents high wind speed and low torque represents low wind speed. The output power of the generator also varies according to the input torque. The generator is designed to produce nearly 3000

W at the rated wind speed i.e. 7 m/s. The generator speed at 7 m/s is calculated to be 400 rpm with TSR value of 6 and hence the rated torque is calculated to be around 70 Nm. As shown in Figure 3 (b), the output of the generator is around 3000 W at this value of mechanical torque.

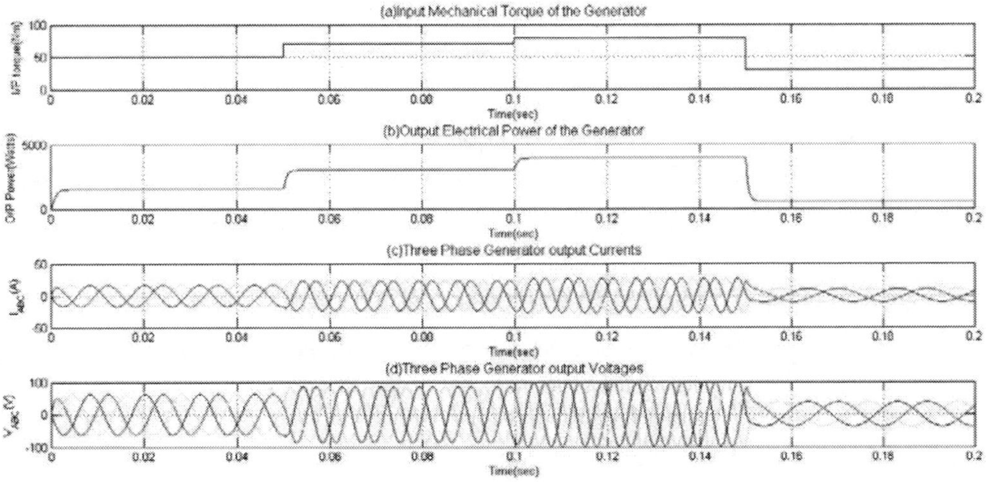

Figure 3. Electrical output power, voltages and currents at different generator mechanical input torque

5 Conclusions

The 3 kW PMSG is designed for variable speed wind turbine. Different Generator parameters were optimized in order to obtain 3 kW power output at wind speed of 7 m/s. The power output from optimized generator parameters were initially matched with the turbine output power in order to avoid under-sizing or over-sizing of the generator. These generator parameters were used within PMSG model in order to estimate performance of the generator. Result shows that the generator almost matches with the turbine output power at the rated wind speed. The simulation is performed at different input torques, which resembles different wind speed out of wind turbine. Result shows that the generator output voltages, currents and power varies with the wind speed and output power of the generator is around 3 kW at rotor speed of 400 rpm (i.e. turbine speed of around 100 rpm).

References

1. Krause, C. Analysis of Elecric Machinery, Wiley, second ed. (2004)
2. Wu, B., Lang, Y., Zargari, N. and Kouro, S., Power Conversion and Control of Wind Energy systems, Wiley. (2011)
3. T. F. Chan and L. L. Lai. An axial flux permanent magnet synchronous generator for a direct-coupled wind-turbine system. In IEEE Transactions on Energy Conversion, volume 22 of 01, pages 86-94. IEEE
4. Min-Fu Hseih, D. G. Dorrell, Yu-Han Yeh, and Samsul Ekram. Cogging torque reduction in axial flux machines for small wind turbines. IEEE. (2009)
5. Jacek F. Gieras, Rong Jie Wang, and Maarten J. Kamper. Axial Flux Permanent Magnet Brushless Machine. Kluwer Academic Publishers. (2004)
6. Tip Speed Ratio. http://www.reuk.co.uk/Wind-Turbine-Tip-Speed-Ratio.htm
7. D.P.Kothari and I.J. Nagrath, Electrical Machines, Third Ed.
8. http://wiki.xtronics.com/index.php/Wire-Gauge_Ampacity
9. L.Soderlund, A.Koski, H. Vihriala, J.T. Eriksson, and R.Perala. Design of an Axial Flux Permanent Magnet wind power generaion. pages 224-228, September. (2009)

MATEC Web of Conferences **55**, 04002 (2016) DOI: 10.1051/matecconf/20165504002
ACPEE 2016

Modeling of Self-Excited Isolated Permanent Magnet Induction Generator Using Iterative Numerical Method

Mostafa R. Mohamed, Karam M. Abdel-Latif [a] and Mahmoud R. Ahmed

Department of Electrical Machine and Power Engineering, Faculty of Engineering-Helwan University at Helwan - Egypt

Abstract. Self-Excited Permanent Magnet Induction Generator (PMIG) is commonly used in wind energy generation systems. The difficulty of Self-Excited Permanent Magnet Induction Generator (SEPMIG) modeling is the circuit parameters of the generator vary at each load conditions due to the a change in the frequency and stator voltage. The paper introduces a new modeling for SEPMIG using Gauss-sidle relaxation method. The SEPMIG characteristics using the proposed method are studied at different load conditions according to the wind speed variation, load impedance changes and different shunted capacitor values. The system modeling is investigated due to the magnetizing current variation, the efficiency variation, the power variation and power factor variation. The proposed modeling system satisfies high degree of simplicity and accuracy.

1 Introduction

Permanent Magnet Induction Generator (PMIG) is a new type of induction generator that has a stator equipped with a three phase-winding and two rotors mounted on the machine rotor shaft. The outer rotor has squirrel-cage-windings while the inner free rotor is a Permanent Magnet (PM). The outer squirrel- cage rotor rotates at an angular speed faster than that of the rotating field generated by the generator stator current while the inner PM rotor is free to rotate on the same angular speed of the stator magnetic field. The induced stator voltage due to the PM is compensating the reactive stator current and improves the generator power factor [1]-[5]. The PMIG equivalent circuit is similar to the conventional induction generator equivalent circuit. As many papers mentioned [1], [3] and [5] the flux produced by the internal PM rotor can be modeled as internal voltage source (E_{pm}) in series with the magnetizing reactance as shown in Figure 1. This paper introduces the PMIG model as an isolated unit in wind energy system so this machine is called Self- exited Permanent Magnet Induction Generator (SEPMIG). The PMIG obtains the necessary reactive power (Q) for initiating the magnetic flux in the air-gap from its connection to the grid. In case of SEPMIG there is no connection to the grid so the machine received the necessary reactive power by connecting its terminals with shunt capacitor as shown in Figure 2.

[a] Corresponding author : Karam106@yahoo.com

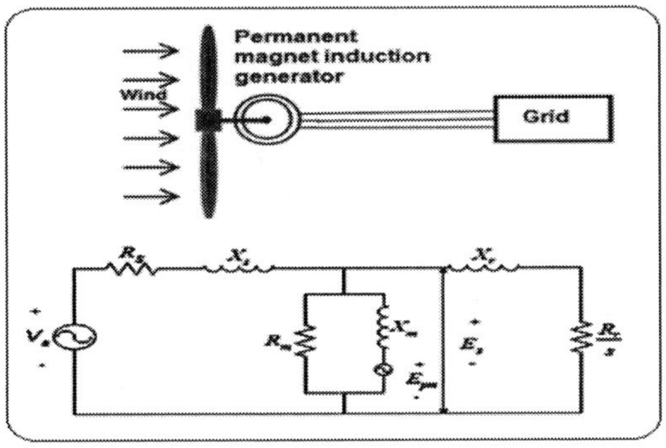

Figure 1. PMIG Equivalent Circuit Connected to Grid [5].

The rest of the paper is organized as follows; The SEPMIG proposed model is given in Section 2, the computational procedure of the proposed model is given in Section 3, case study and the performance results are illustrated in Section 4, and the conclusion is presented in Section 5.

2 Proposed Mathematical Model of SEPMIG

In SEPMIG modeling methods, the terminal machine voltage and generated stator frequency are varying with the operating load conditions. The difficulty of PMSEIG modeling is that all the circuit parameters shown in Figure2 of the generator vary according to the load condition changes due to the changes in frequency and stator voltage. The proposed model will produce a solution to overcome these difficulties through the following modifications.

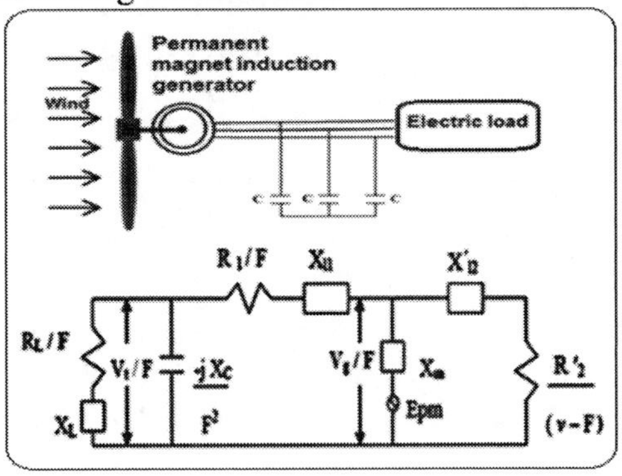

Figure 2. SEPMIG in Wind Energy System [6].

Step:1

The parameters of the equivalent circuit of the SEPMIG are presented in per-unit value based on stator frequency (F).

Step: 2

Appling a Gauss- Sidle iterative solution on SEPMIG to determine the machine parameters such as the magnetizing reactance, stator frequency and induced air gap voltage under specified machine load conditions.

Step: 3

Determine the SEPMIG performance using the machine parameters given in step 2.

The per-unit single phase equivalent circuit of SEPMIG is shown in Figure 3. The parameters are expressed in per unit by taking the line Frequency (F) asbase value.

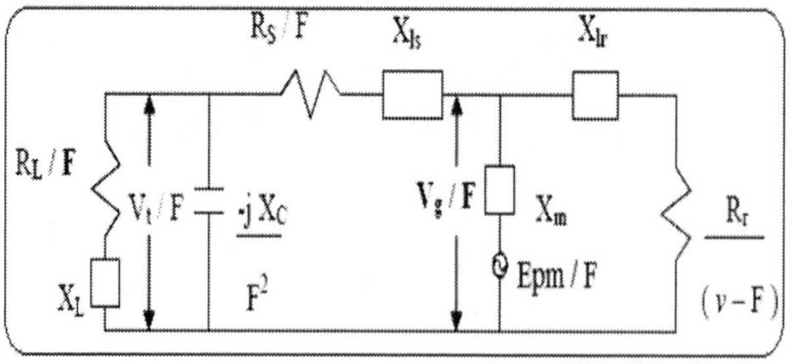

Figure 3. Per Phase SEPMIG Equivalent Circuit.

In this circuit the loading type of SEPMIG are considered as static resistive-inductive load connected in parallel with the shunt capacitor C sh. In this model the core resistance was neglected with respect to shunt magnetizing reactance. The stator resistance and leakage reactance per phase are $R_1 and X_{l1}$ respectively while the rotor resistance and leakage reactance per phase are $R_2' and X_{l2}'$ respectively.

2.1 Proposed Model Equations

Based on the concept of assuming the linearity of the machine parameters except its magnetizing inductance [7] and [8], the nodal current equation for the single phase equivalent circuit shown in Figure3under the steady state operation, will lead to the following equation.

$$I_s + I_r + I_m = 0 \tag{1}$$

where $\mathbf{I_s}$, $\mathbf{I_r}$ and $\mathbf{I_m}$are defined as stator current, rotor current and magnetizing current respectively. From the equivalent circuit given in Figure3, the different currents are obtained as follows:

$$I_s = \frac{V_g/F}{(R_1/F + jX_1) + ((R_L/F + jX_L)//(jX_c/F^2))} \tag{2}$$

$$I_r = \frac{V_g/F}{R'_2/(V-F) + jX'_2} \tag{3}$$

$$I_m = I_s + I_r \tag{4}$$

where V_g , R_L , X_L , F and $X_C$are defined as the induced air gap voltage,load resistance, load leakage reactance, Per unit frequency with respect to the line frequency and shunt capacitance respectively.

The induced air gap voltage across the magnetizing branch (Vg) and the terminal voltage (V_t) are calculated using equations (5& 6)as follows.

$$\frac{V_g}{F} = E_{pm} + jI_m X_m \tag{5}$$

$$V_t = V_g - I_s\left(\frac{R_s}{F} + jX_s\right) \tag{6}$$

where E_{pm} is the internal armature induced voltage. As the currents of the generator under the steady state condition must be not equal zero, the magnetizing current can be obtained from equation (7) as.

$$I_m = -\frac{E_{pm}}{Z_{eq} + jX_m} \tag{7}$$

where $\mathbf{Z_{eq}}$is the equivalent impedance of SEPMIG that obtained using equation (8)

$$Z_{eq} = \frac{Z_s Z_r}{Z_s + Z_r} \tag{8}$$

$$Z_r = \frac{r'_2 + jX'_{l2}(v-F)}{(v-F)} \tag{9}$$

$$Z_s = \frac{\frac{XcXl}{F^2} + \frac{r_1 R_L}{F^2} + \frac{X_{l1} X_c}{F^2} - X_{l1}X_L + j(\frac{X_{l1} R_L}{F} + \frac{r_1 X_L}{F} - \frac{r_1 X_C}{F^3} - \frac{X_C R_L}{F^3}}{\frac{R_L}{F} + j(X_L - \frac{X_C}{F^2})} \tag{10}$$

where $\mathbf{Z_s}$ and $\mathbf{Z_r}$ are the stator and rotor equivalent impedances respectively.
The solution of equation (5) leads to develop a 5th order polynomial expressed in terms of frequency (F) as expressed in equation (11).

$$A_6F^5 + A_5F^4 + A_4F^3 + A_3F^{25} + A_2F + A_1 = 0 \qquad (11)$$

where (A1- A6) represent the Polynomial Coefficients, The function of F, V_g, E_{pm}, X_{m12}, S (the machine slip), and the machine parameters. The polynomial equation obtained from equation (11) willbe solved using the iterative solution based on Gauss-Sidle relaxation Method.

3 The Computational Procedure

The iterative solution is based on the Gauss-sidle relaxation technique to solve the polynomial equation for the unknown F by assuming the value X_{m12}.The initial values of X_{m12}areassigned carefully to reduce the number of iteration required to converge the solution. The induced armature voltage E_{pm} is considered as a constant value equal to the system reference voltage. The solution is used to predict a convergent value for the $V_g$and to be matched with the values "F" and "X_{m12}". The full performance of SEPMIGis obtained by solving the equivalent circuit taking into consideration the predicted matched values of X_{m12}, F, and V_g. The computational procedure of the proposed model could be summarized in the flowchart shown in Figure4.

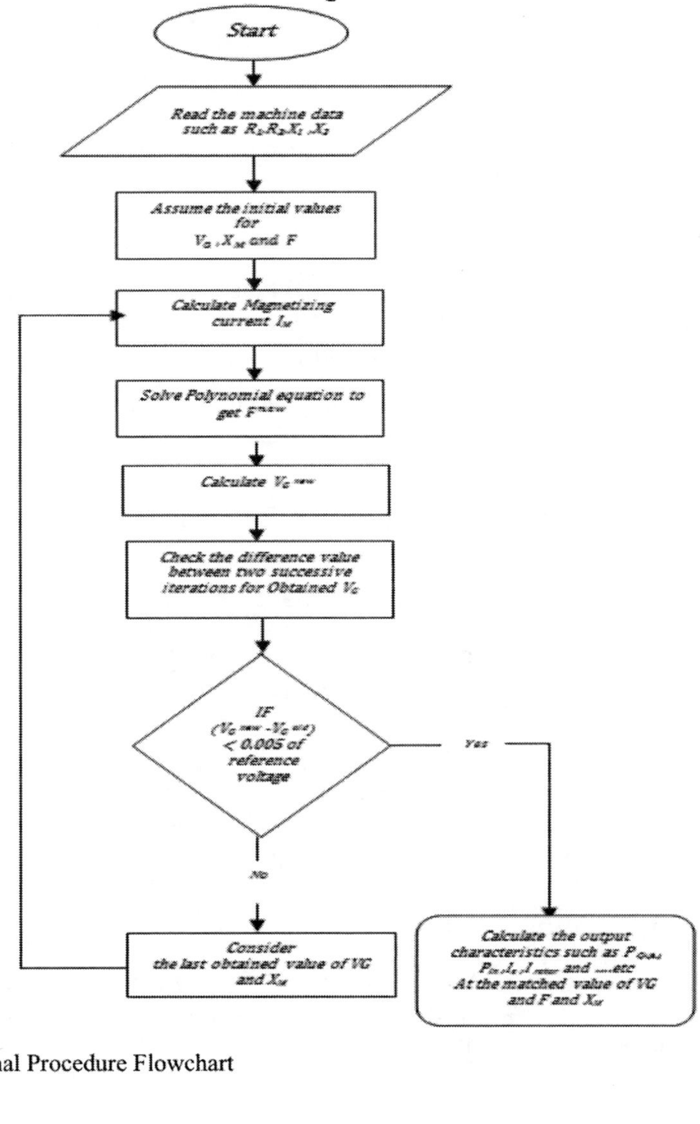

Figure 4. Computational Procedure Flowchart

 DOI: 10.1051/matecconf/20165504002

4 The Estimated Initial Value of the Magnetizing Reactance

The initial values of X_{m12} would be assigned carefully to reduce the number of iterations required to converge the solution. The linear value of X_{m12} is 167 Ω and the full parameters of the IG used are given in Table1. The proposed PMSEIG would be needed to modify the machine rotor [1] by considering a PM of approximately is 6 mm thickness placed on the surface of the inner rotor. Both outer and inner rotors are separated by air-gap of 1 mm thickness. Consequently, the linear value of X_{m12} is reduced to 20% approximately 32 Ω. However, the difficulty of assigning initial values for X_{m12} is due to the effect of the machine core saturation in addition to the effect of the wind-speed variation as shown in Figure5.

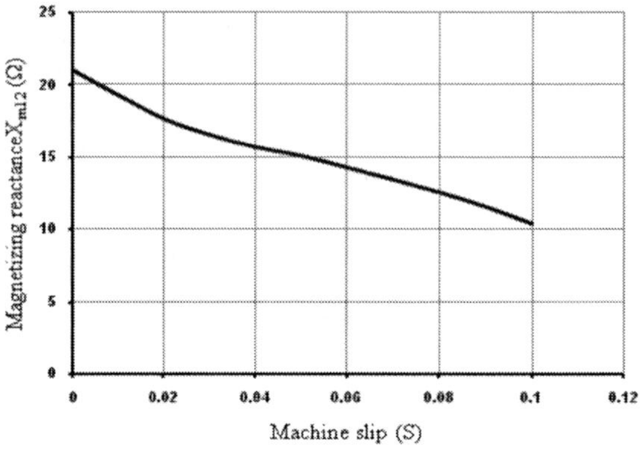

Figure 5. The Linear Magnetizing Reactance Variation against the Machine Slip

5 Case Study and Performance

The proposed model is tested at different load conditions. The load conditions are considered as operation at different wind speed, different load impedances and different values of shunt capacitor. The Induction machine parameters are shown in Table 1.

Table 1. Induction Machine Parameters.

Rated power	175 W
Voltage	206 V
Input current	1.2 A
Frequency	60 Hz
Pair pole	2
Rated speed	1670 r.p.m
Stator resistance	12.675 Ω
Rotor resistance	12.675 Ω
Stator reactance	12 Ω
Rotor reactance	8.4 Ω
Mutual reactance	184.5 Ω

The following cases show the model performance due to many cases of effects; the magnetizing current variation, efficiency variation, power variation and power factor variation.

5.1 The Magnetizing Current Variation

Figure6(a,b&c) shows the variation of the magnetizing current against the induced air gap voltage at shunt capacitors equal to 25µf, 45 µf and 65 µf respectively and wind speed equals to 1p.u and 1.05p.u. The magnetizing current equals to zero when the induced air-gap voltage equals the internal armature induced voltageE_{pm}. The results given by the model take into consideration the effects of the wind-speed and the shunt capacitance values. The result shows that increasing the value of the shunt capacitance leads to increase in the magnitude of the magnetizing current; however, the increase in the wind-speed lead to decrease in the rate of changing of the magnetizing current with respect to the induced air-gap voltage. Figure6 shows these results.

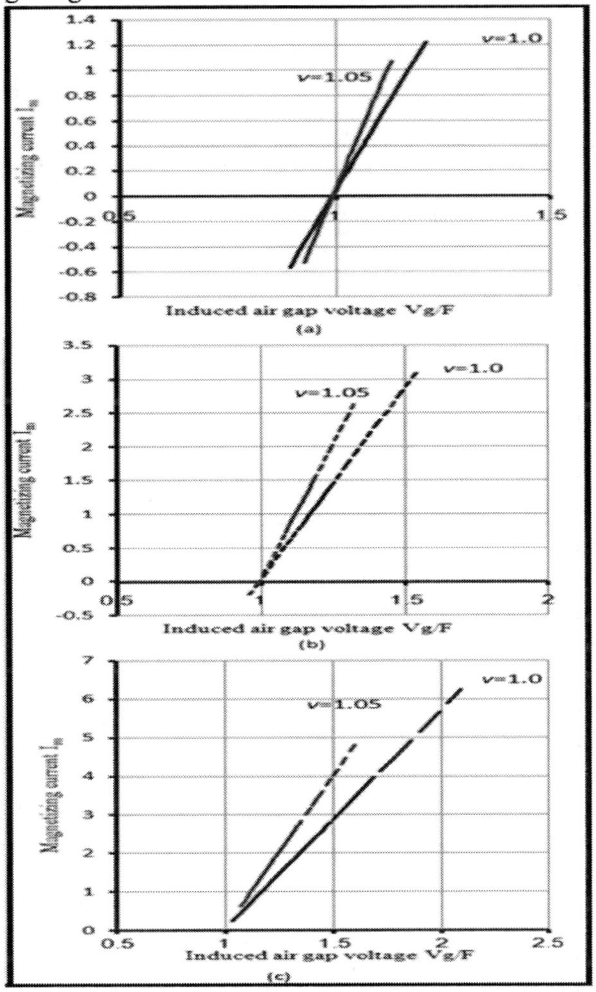

Figure 6. Magnetizing Current Variation against the Induced Air-Gap Voltage at Different Speed And Capacitor Values.

5.2 The Variation of the Nonlinear Magnetizing Reactance

Figure7 (a) shows the relation between the magnetizing reactance against the internal induced air gap voltage (Eg12) given by reference [6], Figure7(b) shows the relation between the magnetizing reactance against the internal induced air gap voltage (Eg12) that obtained from the proposed model of PMSEIG. By comparing the experimental and curve fitting values of Figure7 (a) [6] with the calculated results obtained by the proposed model plotted in Figure7 (b), it's clear that there is a fairly agreement with a satisfied accuracy above 91% as shown in Table 2. However, this error 9% is attributed to many reasons such as the error of measuring instruments used in experiment, the error due to the way of computing the results (the proposed iterative solution) and the error due to neglecting the core losses in the proposed model.

 DOI: 10.1051/matecconf/20165504002

Table 2. Magnetizing Reactance Values.

E_{g12} (V)	10	30	40	50
$X_{m12\,(Experimental)}$ (Ω)	8.9	11.8	13.8	14.9
$X_{m12\,(Predicted)}$ (Ω)	8.1	11.8	13.5	14.9
$\%\ error = \frac{Xm12\,(Experimental) - Xm12\,(Predicted)}{Xm12\,(Experimental)} \times 100$	8.9	0	2.1	0

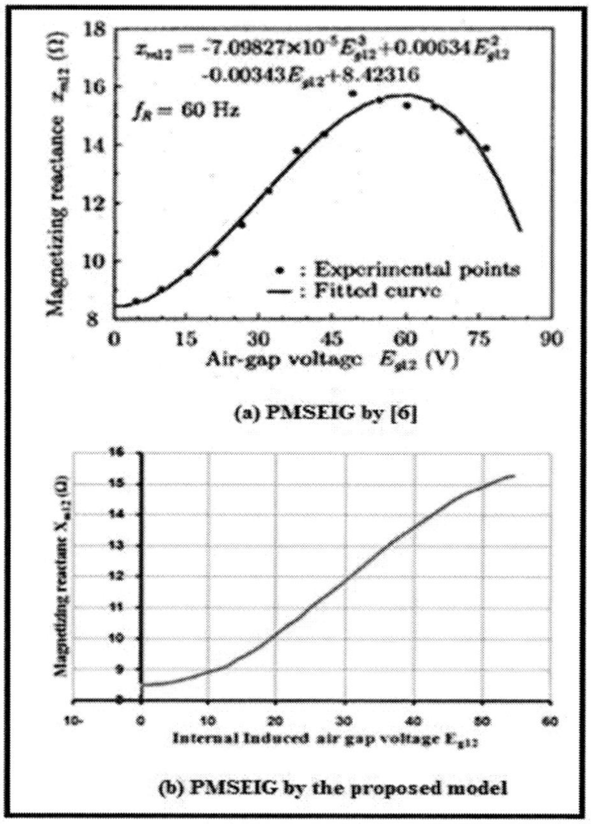

Figure 7. Magnetizing Reactance as A Function of Air-Gap Voltage.

5.3 The Terminal Voltage Variation

The terminal voltage variation against the shunt capacitance values at wind speed v =1.05 p.u. is given by reference [6] in Figure8 (a) while the corresponding results obtained by the proposed model of PMSEIG are shown in Figure8 (b). Both trends show that the terminal voltage is increasing as the shunt capacitance value is increased. Comparing both trends shows that the accuracy of the results obtained by the proposed model is above 94%. Table3 shows the comparison of different shunt capacitance values. The error is limited to 5.5% taking into consideration all the reasons of the error that discussed previously.

DOI: 10.1051/matecconf/20165504002

Table 3. Terminal Voltage Values at Different Shunt Capacitance.

C (μF)	20	40	60
$V_{t(Experimental)}$ (V)	64	68	77
$V_{t(Predicted)}$ (V)	65.9	71.8	81.2
$\% \text{ error} = \frac{Vt(Experimental) - Vt(Predicted)}{Vt\,(Experimental)} \times 100$	2.9	5.5	5.5

5.4 Terminal Voltage and Output Frequency Variations Against the Load Current

The corresponding results obtained by the model are given in Figure9 for (R-L) load with 0.7 PF. The figurer shows different variation for both the output frequency and the terminal voltage with respect to the load current. The output frequency and terminal voltage decreased with the load current increasing.

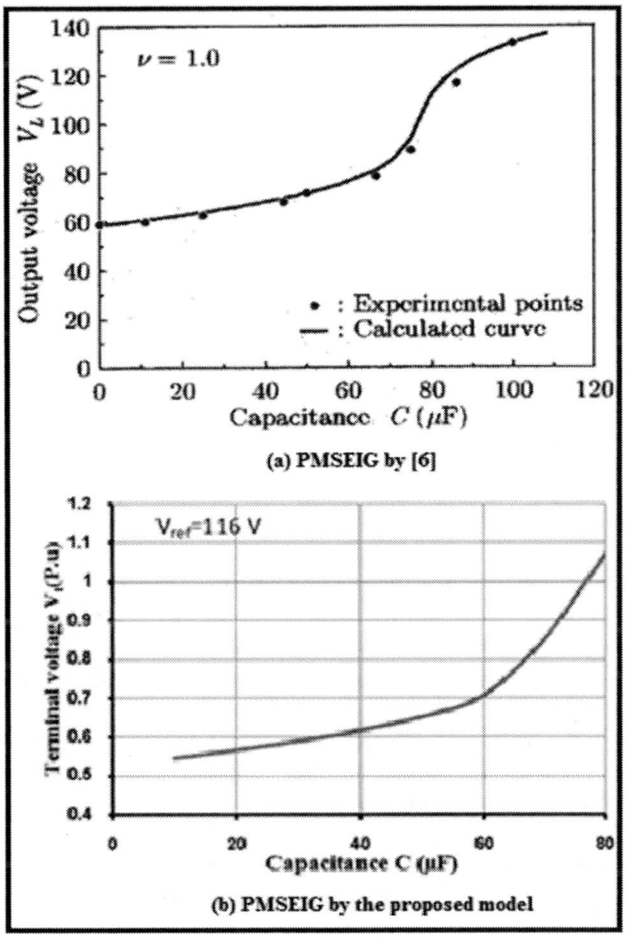

Figure 8. Terminal Voltage Variation Against The Shunt Capacitance

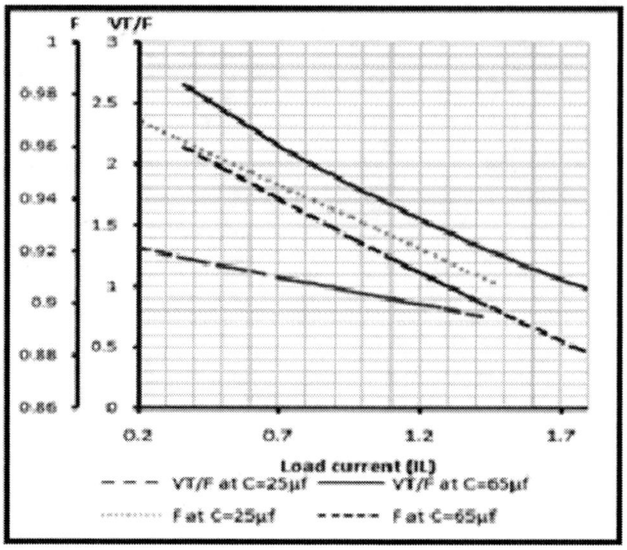

Figure 9. Output Frequency And Terminal Voltage Variation Against The Machine Slip

6 Conclusions

The paper is aimed to present the PMIG as a standalone unit in a wind energy system (PMSEIG) and to develop a simple technique based on iterative solution. From the analysis of the obtained results in this paper, the contributions are summarized as follows:

1. The difficulty of assessing varied value of the magnetizing reactance and variation of both the terminal machine voltage and generated stator frequency with the operating load conditions such as wind-speed, shunt capacitance value, and load impedance has been overcame through the following steps:
   - ➢ The equivalent circuit of the PMSEIG its developed in the per-unit stator frequency of the machine
   - ➢ The predicted values obtained from iterative solution with prior unknown machine parameters such as the magnetizing reactance, the per-unit stator frequency, and the induced air-gap voltage under a well specified machine load condition are predicted.
   - ➢ The full performance of the PMSEIG is solved using the equivalent circuit taking into consideration the obtained predicted values of the circuit parameters.
2. The solution convergence is tested for either resistive loading or resistive-inductive loading condition of 0.7 lagging PF.
3. The accuracy of the proposed iterative solution is satisfied as the results are tested and compared with the trends and results of the corresponding experimental and numerical results published in three different references (where the maximum error of the solution obtained is less than 9% compared to [6]).
4. The discussions of PMSEIG results are concluded and introduced the full performance of the machine taking into the consideration the effects of the machine parameters and loading types.

References

1. Potgieter, J.H.J., Lombard, A.N., Wang, R.J., and Kamper, M.J. ,*"Evaluation of permanent-magnet excited induction generator for renewable energy applications"*, Proceedings of the 18th Southern African Universities Power Engineering Conference, Stellenbosch, South Africa, 299-304, (2009).

2. Sung-Hooii Kiln and Song-yop Hahn" *Analysis and design of a induction generator with a super conducting bulk magnet rotor"*, IEEE Transactions on Applied Super conductivity, **Vol. 1**, pp.931-934, March (2000).
3. Fukami, Nakagawa, K., Kanamaru, Y., and Miyamoto, T. *"A technique for the steady-state analysis of a grid-Connected permanent-magnet induction generator"*, IEEE Transactions on Energy Conversion,**Vol.19**, pp. 318-324, June (2004).
4. Troster, E.; Sperling, M.; and, Hartkopf, T.H, *"Finite element analysis of a permanent magnet induction machine"*, Proceeding of International Symposium on Power Electronics, Electrical Drives, Automation and Motion (SPEEDAM), (2006).
5. Tsuada, T., Fukami, T., Kanamaru, Y., and Miyamoto, *"Effects of the built-in permanent magnet rotor on the equivalent circuit parameters of a permanent magnet induction generator*", Proceeding of IEEE Transaction on Energy Conversion, **Vol. 22**, pp. 798-799 , September (2007).
6. Fukami, Tadashi; Shimizu, Bungo; Hanaoka ,Ryoichi; Takata, Shinzo; Miyamoto, Toshio" *Performance prediction of a stand-alone permanent-magnet induction generator"* IEEE Transactions on Industry Applications, Volume 123, Issue 7, pp. 817-823 (2004). [7] S.C.Tripathy ,M.Kalantar and N.D.Rao "Wind turbine druven self-excited induction generator" proceeding of Energy Convers.Mgl , **Vol. 34** ,vol8,pp. 641-648,(1993).
7. S.C.Tripathy, M.Kalantar and N.D.Rao *"Wind turbine driven self-excited induction generator"* Proceeding of Energy Convers.Mgl, **Vol. 34**, pp. 641-648, August (1993).
8. A. Abbou,M. Barara, A.Ouchatti,M. Akherraz and H. Mahmoud, *"Capacitance required analysis for self excited induction generator"* Journal of Theoretical and Applied Information Technology, **Vol. 55** No.3, pp. 382-389, September (2013).

MATEC Web of Conferences **55**, 04003 (2016) DOI: 10.1051/matecconf/20165504003
ACPEE 2016

Influence of Load Modes on Voltage Stability of Receiving Network at DC/AC System

Chizu Mao[a], Shenghong Lin

School of Electric Power, South China University of Technology, Guangzhou, China

Abstract. This paper analyses influence of load modes on DC/AC system. Because of widespread use of HVDC, DC/AC system become more complex than before and the present modes used in dispatch and planning departments are not fit in simulation anymore. So it is necessary to find load modes accurately reflecting characteristics of the system. For the sake of the voltage stability, commutation failure, etc. the practical example of the receiving network in a large DC/AC system in China is simulated with BPA, and the influence of Classical Load Mode (CLM) and Synthesis load model (SLM) on simulation results is studies. Furthermore, some important parameters of SLM are varied respectively among an interval to analyse how they affect the system. According to this practical examples, the result is closely related to load modes and their parameters, and SLM is more conservative but more reasonable than the present modes. The consequences indicate that at critical states, micro variation in parameters may give rise to change in simulation results radically. Thus, correct mode and parameters are important to enhance simulation accuracy of DC/AC system and researches on how they affect the system make senses.

1 Introduction

Owning to enormous advances in large transmission capacity, low cost and strong regulating ability, HVDC technique is used widely. China Southern Power Grid (CSG) has constructed "Eight AC and Eight DC" west-to-east power transmission channel, whose large capacity and long distance make it become one of the most complex DC/AC system in the world. In the east is its receiving network, where lack of reactive power supply and closed electrical distance among inverter stations bring a severe test to the area. Even a fault from a long distance may give rise to commutation failure, which is a common error causing large impact to the system[1,2]. There are a great deal of studies about load modes, which play an important role in simulation of power system and are the key factors at voltage stability[3]. Incorrect mode gives rise to separation or even opposite conclusions between simulation and reality[4]. For example, difference modes result in difference results on voltage fall during fault period and make opposite judgement in commutation failure. As HVDC is developing quickly and used widely, the DC/AC system becomes complex, so static mode or impedance-motor mode is not able to meet requirement of simulation and evaluating system stability.

This paper analyses the influence of different kinds of load mode and parameters on DC/AC system by BPA, which is introduced by China Electric Power Research Institute (China EPRI) from

[a] Corresponding author : 249559676@qq.com

Bonneville Power Administration. After continuous development, it turns out to be Chinese style BPA suitable to Chinese reality and is used widely in various power sectors including CSG.

2 Synthesis Load Modes (SLM)

2.1 Introduction of the Modes

The existing modes are directly connected to the 110kV bus. In fact, it is impossible[5]. Paper [6] calls these kinds of mode Classical Load Mode (CLM). Usually CLM is a parallel combination of impedance and motors. Its defects are following:

(1)The impact of distribution network on static loads is ignored. Impedance of distribution, influenced by non-linear loads, cannot be separated by mean of present mathematical technique.

(2)If motors consume more reactive power than the bus supplies, consumption at constant current and power loads become negative. In other words, they are equivalent to reactive sources. Thus, voltage stability is improved significantly. In fact, there are no such sources in actual network.

China EPRI propose Synthesis Load Mode (SLM)[5,6] taking distribution network into account, who is more comfort to structure of power system and gradually researched in power grid. Its structure is shown at Figure 1. Below equivalent impedance is the virtual bus, which is connected directly to reactive power compensation, static loads, motors and equivalent generators.

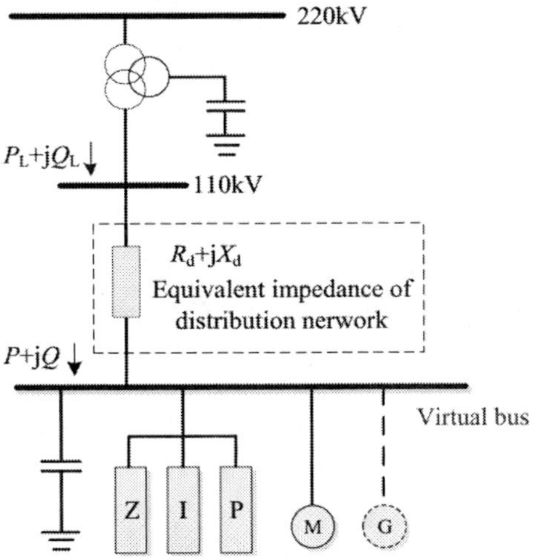

Figure 1. Structure diagram of SLM.

Before simulation, initiation is needed. Voltage (V_{vir0}) and apparent power (P_0+jQ_0) on virtual bus can be worked out through flow calculation. Other running parameters can be easily calculated. Due to limited space, this section only states how to calculate apparent power on static loads (P_{sta0}+jQ_{sta0}) and Capacity of reactive power compensation (Q_{C0}). These are the most different points in algorithm compared with CLM. Subscript '0' means initial value prepared for transient calculating and it is eliminated in the following text for convenience.

Power from the equivalent generator (P_{G0}+jQ_{G0}), active power consumed on motors (P_M) are known.

Power of static loads is shown as below:

$$\begin{cases} P_{sta} = P + P_G - P_M \\ Q_{sta} = P \cdot \tan \alpha_{sta} \end{cases} \tag{1}$$

where tanα_{sta} is power factor of static load; P_M is active power on motors.

Capacity of reactive power compensation is as follow:

DOI: 10.1051/matecconf/20165504003

$$-Q_C = Q + Q_G - Q_M - Q_{sta} \tag{2}$$

where Q_M is reactive power on motors, and it can be calculated according to its operating parameters.

The front minus sign means capacitive reactive power. Formula (1) guarantees positive value of reactive consumption on static loads so that there are no unreasonable reactive sources occurring in the process of calculation. SLM will be more conservative than CLM in voltage stability.

2.2 Comparisons on Fault Response CLM and SLM

Some short circuit accidents are set up. A fault occurs at 0.2s and the breakers on both sides work at 0.3s. It tempts to make comparison on fault response of eight HVDC systems, so as to study the distinction between SLM and CLM. Parameters of CLM, recommended by EPRI, are show in Table 1, while that of SLM are shown in Table 2.

Table 1. Parameters of CLM.

X_S	R_S	X_m	X_r	R_r	A	B	T_j	Z_p/%
0.02	0.18	3.5	0.12	0.02	0.85	0.15	2.0	100

Table 2. Parameters of SLM.

X_S	R_S	X_m	X_r	R_r	A	B	T_j	R_d	X_d	Z_p/%	Z_p/%	Z_p/%
0.03	0.115	3.2	0.15	0.012	0.85	0.15	2.3	0.025	0.05	26	57	17

Results of simulation are shown in Tables 3-5. Two systems of N-C DC share the same AC bus in an inverter station and their fault response are similar, so only one of them is recorded. Numbers on the both side of '/' respectively means results of SLM (left) and CLM (right), and '*' represent fault side of a line. Table 3 shows Instant fault voltage of AC bus of inverter stations, while Table 5 shows transient lowest voltage of them after breakers work in 2s. Table 4 shows instant extinction angle(Unit :°). If γ<8°, it is judged that commutation happens.

Table 3. Instant fault voltage of AC bus of inverter stations.

Fault line	T-G	G-Z	X-A	J-C	Y-G	P-Q	N-C
P-D*	0.94/0.97	0.96/0.98	0.89/0.90	0.72/0.74	0.85/0.87	0.97/0.98	0.85/0.87
Y-X*	0.81/0.87	0.44/0.47	0.93/0.94	0.89/0.92	0.85/0.88	0.52/0.54	0.81/0.85
J*-R	0.99/1.00	0.99/0.99	0.97/0.97	0.85/0.83	0.93/0.93	0.99/1.00	0.92/0.92
S-G*	0.88/0.86	0.90/0.91	0.74/0.77	0.82/0.85	0.55/0.57	0.93/0.94	0.83/0.86

Table 4. Instant fault extinction angle of inverter stations.

Fault line	T-G	G-Z	X-A	J-C	Y-G	P-Q	N-C
P-D*	14.5/15.5	16.3/16.7	14.2/15.0	0.0/0.0	8.7/10.8	16.6/16.9	18.0/18.0
Y-X*	0.0/4.8	0.0/0.0	15.7/16.4	11.0/12.4	3.5/7.9	0.0/0.0	18.0/18.0
J*-R	16.2/16.7	17.1/17.3	17.1/17.4	5.6/7.6	14.2/14.9	17.3/17.5	18.0/18.0
S-G*	6.5/8.9	13.4/14.2	0.0/5.4	0.0/0.0	0.0/0.0	14.3/15.0	18.0/18.0

Table 5. Transient lowest voltage of those buses in 2s after fault line being cut.

Fault line	T-G	G-Z	X-A	J-C	Y-G	P-Q	N-C
P-D*	0.81/0.87	0.88/0.91	0.78/0.82	0.58/0.62	0.71/0.76	0.91/0.93	0.71/0.76
Y-X*	0.41/0.6	0.58/0.81	0.63/0.90	0.55/0.78	0.50/0.87	0.61/0.85	0.51/0.82
J*-R	0.97/0.99	0.97/0.98	0.95/0.96	0.91/0.92	0.92/0.94	0.97/0.98	0.92/0.94
S-G*	0.57/0.97	0.77/0.97	0.73/0.95	0.70/0.91	0.64/0.92	0.83/0.97	0.64/0.92

DOI: 10.1051/matecconf/20165504003

According to Tables 3-5, commutation failure happens if voltage of a Converter bus is below about 0.85~0.87(p.u.). It is appropriately 1.01(p.u.) when the system is on steady-state operation. Although commutation failure is related to various factors, it still can draw a conclusion that low voltage easily brings out commutation failure. Table 3 and 5 indicate that voltage values of SLM is less than that of CLM as a whole. Stability of CLM system can be kept in four situations. Commutation failure happens inevitably, but voltage is recovered finally owning to regulation of system in HVDC.As for SLM, system is not stable except the fault happens at J-R line (J side), and the system is near a critical state in the first situation (P-D*).

Table 4 indicates that commutation failure happen in T-G, G-Z, Y-G and P-Q DC if a fault occurs at Y-X line (X side), either SLM or CLM is simulated. Their inverter stations are closer than that of the others. Only Y-G DC is discussed here. Figures 2-3 show the curves of extinction angle and voltage of AC bus.

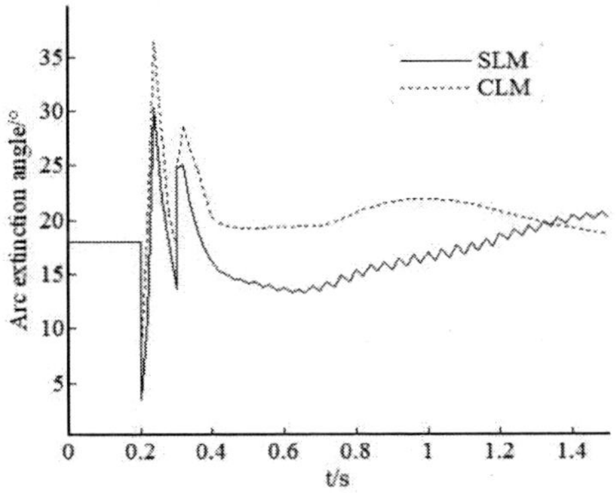

Figure 2. Extinction angle.

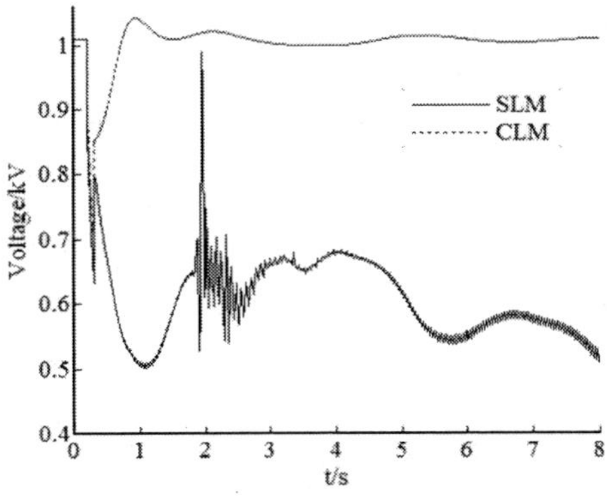

Figure 3. Voltage of Converter bus of Y-G DC.

Figure 2 and Table 4 show that commutation failure happens in either SLM or CLM system as soon as a fault occurs. Advance firing angle β is enlarged forcedly to magnify γ, because their relation can be express as $\gamma = \beta - \mu$ (μ is commutation angle). At 0.2s(+), when a fault occurs, voltage of CLM is 0.88 (p.u.), while that of SLM is 0.85 (p.u.), but the latter drop more quickly before the line is cut (see Figure 3). In 0.3s(-), the former is 0.70 (p.u.), while the latter is 0.63 (p.u.); in 0.3s(+), the former is 0.85(p.u.) and begins to increase, while the latter is 0.795 (p.u.) and begins to decrease. As a result, voltage of CLM system return a stable level and voltage collapse happens to SLM system. Reasons of their discrepancy are following:

(1) Static part of CLM only contains constant impedance loads, while that of SLM consist of three kinds of loads in a certain proportion[7]. Constant impedance and current loads command less reactive power if voltage decreases, so, to some extent, reactive shortage is compensated and voltage drop is avoided, while constant power loads do not. Aside, motors at SLM produce less short circuit current (reactive power) than CLM. Therefore, SLM curve is always below CLM.

(2) Motors demand a great deal of reactive power during recovery period. Algorithm of SLM ensures that constant current and power loads are impossible to become reactive source. So SLM is not as stable as CLM but more proper.

Simulations above indicate that SLM is conservative, but it is not sure that these parameters are the most reasonable. Furthermore, they vary among various regions and change as time goes by. Thus, it is necessary to research how they affect the stability of the system.

3 Impact of Parameters of SLM on Stability

SLM is more appropriate than others modes. But irrelevant parameters lead to large separation and even conflicting conclusion between simulation and reality as well as inappropriate modes[8]. It is necessary to research how they affect the system. Due to limited space, this chapter only discusses the motor ratio (M_P) and ratio of constant load (P_P) in static modes. They are likely to influences the stability dramatically. Situation (P-D*) in Chapter 2 is selected to be simulated and curve of voltage and extinction angle in the inverter of Y-G DC is studies. Because the system is actually near a critical state in this situation, contract among difference value of parameters may be very evident. Each simulation only changes one parameter.

3.1 Changes of Motor Ratio

Fault response is analysed when the value of M_P is 0%, 20%, 30%, 40%, 50% or 55%. Particularly, 50% is the initial value. Figures 4-5 show curves of voltage and extinction angle. Interval of time in Figure 5 is 0 to 1.5s, which is the same as below (for extinction angle).

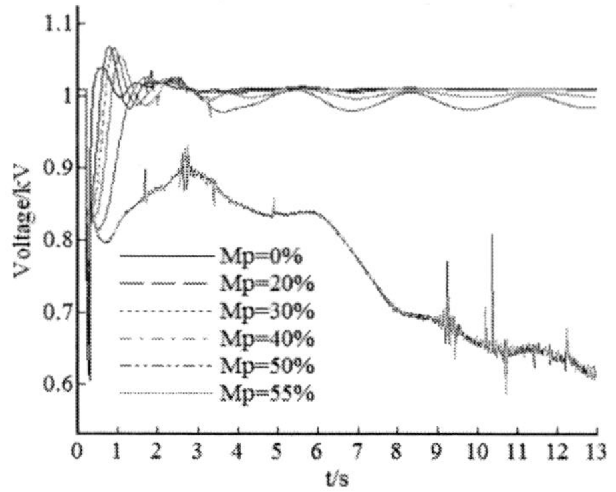

Figure 4. Voltage at different M_P.

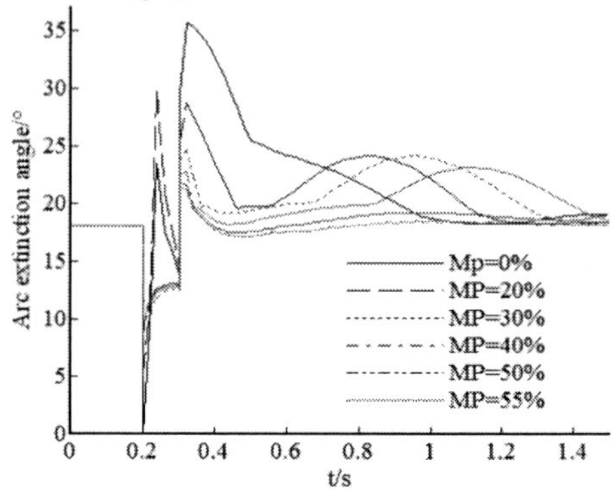

Figure 5. Extinction angle at different M_P.

Figure 5 indicates that the high motor rate reduces the possibility of commutation failure but increases the possibility of voltage collapse. If M_P>30%, γ>8° at 0.2s(+). But figure 4 shows that if M_P >55%, voltage is not able to be recovered because too much reactive power is demanded. To sum up, system is under the most stable condition if M_P is over 30% but less than 50% in this situation.

3.2 Changes of Ratio of Constant Power Load

Ratio of constant power loads in static part of SLM is not as high as common static mode. Initial value of P_P is 17%. The value of P_P is set to be 8%, 12%, 17%, 20% and 23%. Figures 6-7 show curves of voltage and extinction angle.

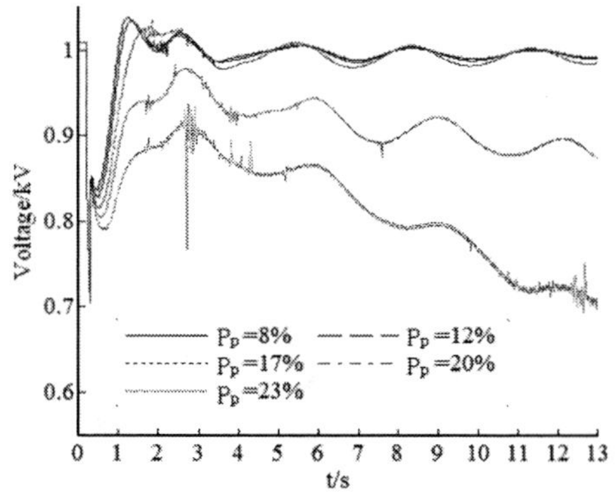

Figure 6. Voltage at different P_P.

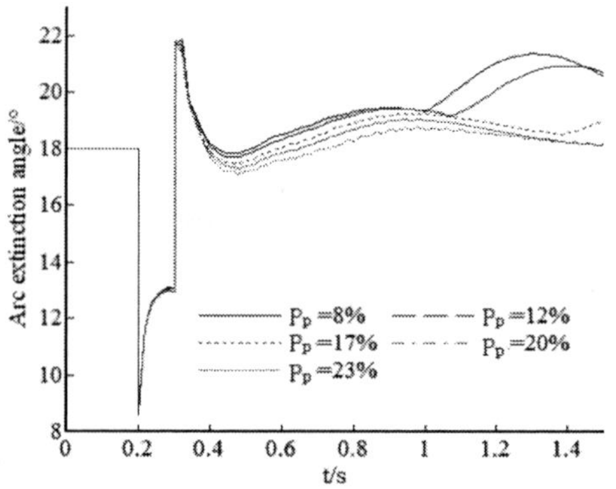

Figure 7. Extinction angle at different P_P.

Increasing P_P or M_P brings out remarkable impact on voltage stability, but the former is more significant. Figures 6-7 show that variety of P_P exerts little influence on voltage fall and commutation fault during fault period but makes a difference in transient voltage stability. Because static load produces no reactive power like motors, voltage fall is not changed as P_P varies. On the other hand, constant power loads demand constant power, so reactive shortage is enlarged if voltage decreases.

therefore, large value of P_P leads to severe power shortage and poor condition on voltage stability. Voltage collapse will happen if P_P is over 20%.

3.3 Analysis of the Simulations

This chapter studies how the two important parameters exert impact on voltage stability of DC/AC system. Only little changes on sensitive parameters may change the system evidently. Furthermore, simulation results may be changed qualitatively if the parameters are near a critical value. Voltage curve changes evidently as P_P vary from 12% to 23%, and it draws totally different conclusions about voltage changes between situation that M_P is 50% and that M_P is 55%. So, Parameters of SLM are important for simulation especially they are near critical states. In actually network, motor ratio is about 30%~60% and ratio of constant power in static loads varied between about 12%~25%.

4 Conclusions

This paper makes comparisons on various load modes and draw following conclusions.

(1) Actual loads are not directly connected to 110 kV or 220 kV buses. SLM takes into account effect from distribution network and the reactive power compensation, and its physics structure is more reasonable. Power factor of static load is defined to eliminate negative constant current and constant power loads. Voltage curves of SLM are below those of CLM as a whole. Thus, SLM is more conservative but rational.

(2) Ratio of motors and ratio of constant power loads play an important part in stability. But increasing the latter is likely to bring out more serious consequence. Parameters vary among various regions and vary as time goes by. Little changes in them may cause huge differences in simulation results if the DC/AC system is running at a critical state. Therefore, accurate parameters are important for simulation of power system and researching how they influence the system makes senses.

References

1. Wenfeng Zhang, Minghui Huang, Zhiyao Liu. Variation analysis of fault feature in receiving power network of DC/AC system. Beijing: China Electric Power Press.pp.14-15. (2015)
2. Hee-Jin Lee. Effect of a SFCL on commutation failure in a HVDC system. IEEE Transactions on Applied Superconductivity, Vol. 23, No.3, pp.67-73. (2013)
3. IEEE Task Force on Load Representation for Dynamic Performance. Standard load models for power flow and dynamic performance simulation. IEEE Transactions on Power Systems, Vol. 10, No.3, pp. 1302-1313.(1995)
4. Popovic D. Investigations of load tap changer interaction. International Journal of Electrical Power & Energy Systems, Vol. 18, No.2, pp. 81-97. (1996)
5. Yong Tang. A synthesis load model with distribution network for power system simulation and its validation. Power Engineering Society General Meeting, Calgary.(2009)
6. Qian Chen, Jianbo Sun, Min Cai. Parameters Determination of Synthesis Load Model. Proceedings of the CSEE, Vol. 28, No.16, pp. 45-50. (2008)
7. Liping Qiu, Bing Zhao, Wenchao Zhang. Effects of Synthesis Load modes on Dynamic Stabilities of Large-Scale Interconnected Grid. Power System Technology, Vol. 34, No.10, pp. 82-87.(2010)
8. EL-Hawary ME. Incorporation of load models in load-flow studies: form of model effects. IEEE Proceedings C, Vol. 134, No.1, pp. 27-30. (1987)

MATEC Web of Conferences **55**, 04004 (2016) DOI: 10.1051/matecconf/20165504004
ACPEE 2016

Chaos Cooperative Particle Swarm Optimization Based Water Level Control for Nuclear Steam Generator

Guimin Sheng[1], Yu Mu[2,a]

[1] *School of Electrical Engineering, Suihua University, Suihua, China*
[2] *College of Physics Science and Technology, Heilongjiang University, Harbin, China*

Abstract. The Stability of SG (Steam Generator) water level plays an important role in the safety of nuclear power plants, but it is difficult to tune the parameters of water level PID controller. A proposed novel algorithm, CCPSO (chaos cooperative particle swarm optimization), is used for tuning PID controller parameters. The (chaos particle swarm optimization)CPSO algorithm has the ability to avoid falling into local minimum and the (cooperative particle swarm optimization)CPSO-S_k has fast convergence in certain functions, so CCPSO algorithm is proposed to utilize the advantages of CPSO and CPSO-S_k. Therefore, half of the particles are updated in the CPSO-S_k, and the other half are updated in the CPSO. The information exchange of the optimal solutions obtained after the end of each iteration is the performance of CPSO-S_k and CPSO collaboration.The simulation results: compared with the PID controller whose parameters are tuned by ZN method, CCPSO show smaller overshoot, better stability, and shorter adjustment time. The simulation results show that the proposed method is effective for tuning PID parameters.

1 Introduction

The SG water level system is highly nonlinear, with time-varying characteristics [1]. The parameters of system change greatly with the power of the reactor. The system exhibits serious "contraction" and "expansion" phenomenon resulting in false level signals. So it is difficult to tune the parameters of PID controller. As a result, the system hard to meet the requirements of all conditions, namely, it is difficult to take into account the best dynamic characteristics and static characteristics.

Particle swarm optimization algorithm is a stochastic global optimization evolutionary algorithm based on swarm intelligence. According to the advantages of CPSO and CPSO-S_k, where CPSO algorithm has the ability to avoid falling into local minimum and CPSO-S_k has fast convergence in certain functions[2,3]. a new algorithm CCPSO is proposed for PID controller tuning[4]. Simulation test is conducted by matlab.

2 Particle Swarm Optimization Algorithm

A swarm is composed by *S* particle in *D* dimension space. The position of the *i*th particle is $x_i = (x_{i1}, x_{i2}, \ldots, x_{iD})$, $i = 1,2,\ldots s$, and its speed is $v_i = (v_{i1}, v_{i2}, \ldots, v_{iD})$. The optimal position of the

[a] Corresponding author: muyufffff@126.com

particle to be searched is $P_i = (P_{i1}, P_{i2}, \ldots, P_{iD})$. Therefore the optimal positions to be searched by particle swarm optimization are $P_g = (P_{g1}, P_{g2}, \ldots, P_{gD})$. Then we compute the adaptive value of x_i is computed by (2). The updating strategy of particle status is as follows:

$$v_{id} = wv_{id} + c_1 r_1 (P_{id} - x_{id}) + c_2 r_2 (P_{gd} - x_{id}) \tag{1}$$

$$x_{id} = x_{id} + v_{id} \tag{2}$$

3 Improved Cooperative Particle Swarm Optimization Algorithm

Chaos cooperative particle swarm optimization (CCPSO) algorithm is proposed by introducing chaos particle swarm optimization (CPSO) into cooperative particle swarm optimization (CPSO-S_k).

The CPSO algorithm has the ability to avoid falling into local minimum and CPSO-S_k has fast convergence in certain functions, so CCPSO algorithm is proposed to utilize the advantages of CPSO and CPSO-S_k. When the CPSO-S_k algorithm stagnation, changes over to uses the CPSO algorithm. Therefore, half of the particles are updated in the CPSO-S_k, and the other half are updated in the CPSO. The information exchange of the optimal solutions obtained after the end of each iteration is the performance of CPSO-S_k and CPSO collaboration.

3.1 The cooperate Particle Swarm Optimization

The components of D dimension vectors are divided into K groups according to the correlation by CPSO-S_k algorithm. The correlation components constitute a group. The sub vectors which are compose by each group, correspond to a particle vector, and the K sub population co-evolute. If the relationship of the k components can not be identified, then they are splited into $k_1 = D \bmod k$ groups, where each group contain $\lceil D/k \rceil$ components. Or they are splited into $k_2 = k - k_1$ groups, where each group contain $\lceil D/k \rceil$ components. Global optimal value is computed by

$b(j,z) = (P_1.\hat{y}, P_2.\hat{y}, \ldots, P_{j-1}.\hat{y}, z, P_{j+1}.\hat{y}, \ldots, P_k.\hat{y})$, where $P_j.\hat{y}$ is the global optimal value of the *j*th particle swarm. Global optimal value is computed by the vectors which are searched as global optimal.

3.2 Chaos optimization method

Chaos is a common phenomenon in nonlinear systems. We do an optimization search in the solution space by using the randomness, ergodic and regularity of chaos, so that the local optimal solution can be avoided. In this study the chaotic variables is generated by the following Logistic mapping:

$$z_{n+1} = \mu z_n (1 - z_n) \quad n = 0,1,2,\ldots \tag{3}$$

where μ is control variables, take $\mu = 4$, set $0 \le z_0 \le 1$, system (3) is in a state of chaos completely. Iterates definite time series $z_1, z_2, z_3, \ldots$, and with arbitrary initial value $z_0 \in [0,1]$.

3.3 Chaotic cooperative particle swarm optimization algorithm(CCPSO)

A chaotic sequence is generated based on the optimal position which the whole particle swarm searched. The position of a particle in the current particle swarm is replaced by the position of the optimal position particle in chaotic sequence. In the iterative procedure, the search algorithm of introducing the chaotic sequence can generate many neighberhood of the local optimal solution, which can help the inert particles to escape from the local extreme points, and quickly find the optimal solution.

CCPSO operation steps are as follows:

Step 1: k_1 population are initialized, which are $\lceil D/k \rceil$ dimensional chaotic PSO population, Each dimension has s particles: $P_j, j \in [1..k_1]$; k_2 population are initialized, which are $\lceil D/k \rceil$ dimensional chaotic PSO population Each dimension has s particles: $P_j, j \in [(k_1+1)..k]$

Step 2: A D dimensional chaotic PSO population is initialized: Q

Step 3: The group P of Chaotic PSO is updated by using the basic particle swarm algorithm formula.

The circulations are carried out as follow regarding each group $j \in [1..k]$:

The circulations are carried out as follow regarding each group $i \in [1..s]$:

If $f(b(j, P_j.x_i)) < f(b(j, P_j.y_i))$

Then $P_j.y_i = P_j.x_i$

If $f(b(j, P_j.y_i)) < f(b(j, P_j.\hat{y}))$

Then $P_j.\hat{y} = P_j.y_i$

Step4: Select random $k \sim U(1, s/2) | Q \cdot y_k \neq Q \cdot \hat{y}$,then $Q.x_k = b(1, P_1.\hat{y})$, The group Q of Chaotic PSO is updated by using the basic particle swarm algorithm formula.

The circulations are carried out as follows regarding each group $i \in [1...s]$:

If $f(Q.x_j) < f(Q.y_j)$

Then $Q.y_j = Q.x_j$

If $f(Q.y_j) < f(Q.\hat{y}_j)$

Then $Q.\hat{y}_j = Q.y_j$

End the cycle of particles.

Step5: The circulations are carried out as follows regarding each group $j \in [1..k]$

Select random $k \sim U(1, s/2) | Q \cdot y_k \neq Q \cdot \hat{y}$

$P_j \cdot x_k \neq Q \cdot \hat{y}_j$

Step6: We judge whether the end condition is satisfied. If it is not satisfied, repeat step four, if it is satisfied, then the procedure is completed.

4 Experimental Analyses

Analysis and simulation are based on the simplified mathematical model of the steam generator water level which was proposed by E.Irving[5]. Tuned PID controllers of feed water, and the steam generator water level are controlled on this basis. The structure of control system is shown in Figure 1.

In this system, the inner loop uses PI controller to control the flow of feed water, and there is a PID controller for water level in the outer loop. At varying load, the role of the water level signal is ignored. The steam flow rate and reach of the basic balance are to feed water flow by PI controller. The system is basically stable, so the false water level is avoided[6]. Then the PID controller for water level participates control again. The control process is shortened and the range of water level is reduced by the PID controller.

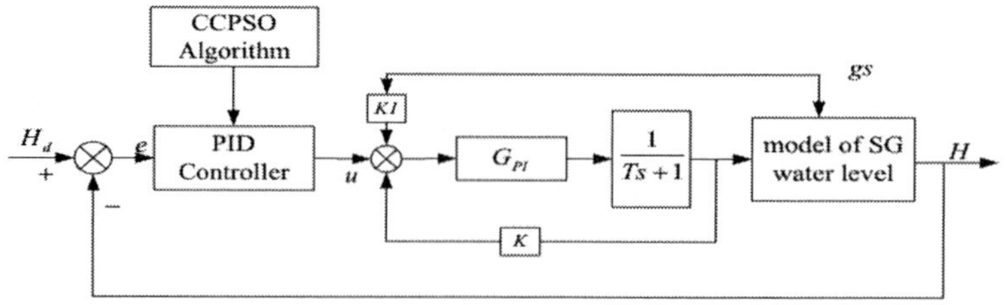

Figure 1. Structure of control system

Usually the parameters of SG level model change when adding perturbation. This is the difficulty of tuning parameters of the PID SG controller of water level. This situation is analysed and simulated as follow: Water level is at 0mm. Add perturbation of steam flow rate at 300s, namely, add steam flow, rate then the response of water level is simulated when steam flow rate step signal change.

1. Simulation of steam flow rate positive step signal change: At first the water level is stabilized at the level set value 0mm on the condition of 50% loading. The steam flow rate is increased from 660kg/s to 815kg/s at 300s, namely loading increase to 60%. As shown in Fig 2, the dashed line is the water level response curve when the parameters of PID controller are tuned by ZN method. Solid line is the water level response curve when the parameters of PID controller are tuned by CCPSO method. The results are shown in the graph; the PID controller whose parameters are tuned by CCPSO reduces overshoot, and raises its convergence speed, and enhances its stability, when adding perturbation of steam flow rate.

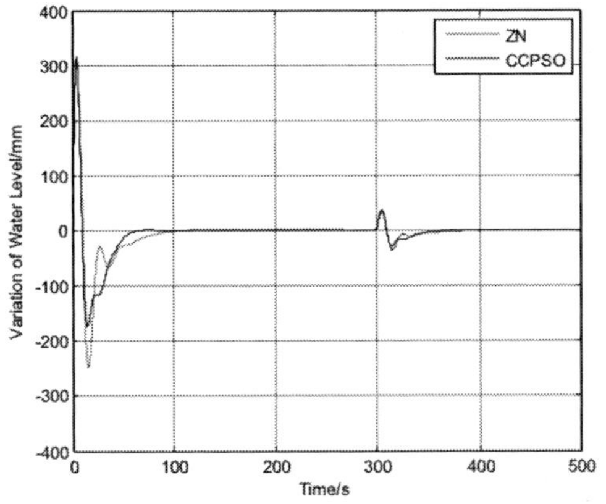

Figure 2. Water level response curve of positive step from 50% to 60%

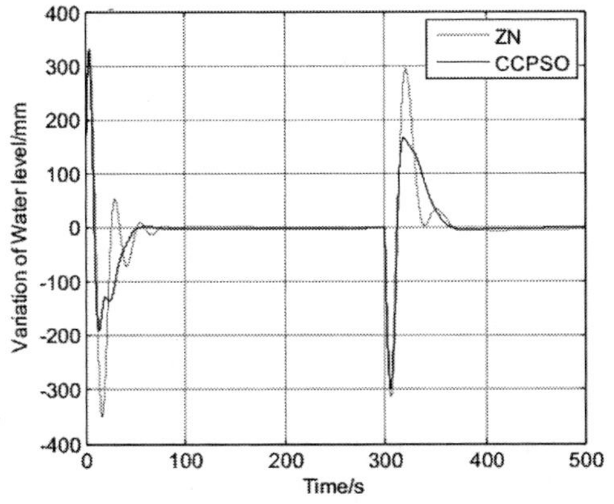

Figure 3. Water level response curve of negative step from 70% to 15%

2. Simulation of steam flow subjected to negative step signal change: At first the water level is stabilized at the level set value 0mm on the condition of 70% loading. The steam flow rate is decreased from 970kg/s to 180.8kg/s at 300s, namely loading decrease to 15%. As shown in Fig 3, the dashed line is the water level response curve when the parameters of PID controller are tuned by ZN method. Solid line is the water level response curve when the parameters of PID controller are tuned by CCPSO method. The results are shown in the graph; the PID controller whose parameters are tuned by CCPSO reduces overshoot, and raises its convergence speed and enhances stability, when adding perturbation of steam flow rate.

3. Now the PI controller is also tuned by CCPSO algorithm, the structure of the block diagram as shown in Figure 4. at first the water level is stabilized at the level set value 0mm on the condition of

DOI: 10.1051/matecconf/20165504004

50% loading, and then it decrease to 30%. As shown in figure5, The performance of tuning the PI controller by CCPSO algorithm is better than the conventional PI controller of given value.

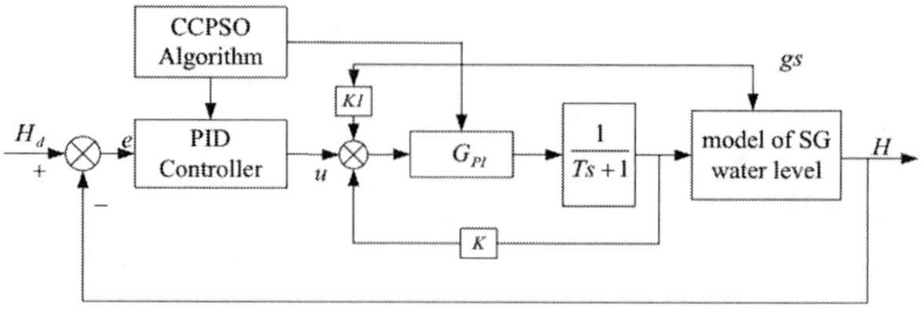

Figure 4. Structure of control system

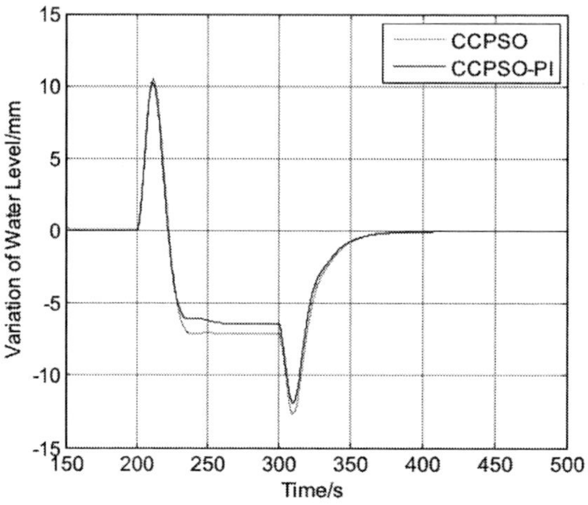

Figure 5. Water level response curve of negative step from 50% to 30%

5 Summary

According to the simplified mathematical model of the steam generator water level, we use a novel algorithm: CCPSO, which overcomes PSO algorithm in not guaranteeing the global convergence,with easiness to produce pseudo optimal value problems. CPSO-S_k algorithm improves the case by using the random and ergodic properties of chaos. We can update the information of each subgroup by chaotic motion, when the CPSO-S_k algorithm fall into local extreme value. And continue to search according to the new optimal values.

The simulation results shows: a comparison against the PID controller whose parameters are tuned by ZN method, where PID controller with tuned parameters by CCPSO has smaller overshoot, better stability, and shorter adjustment time. The results of simulation prove the effectiveness of CCPSO algorithm in tuning PID parameters.

Acknowledgement

The authors would like to thank the Science and Technology Projects of Heilongjiang Provincial Education Department, Natural Science Foundation of China (12531514).

References

1. Jizhou Zhu. The operation of PWR Nuclear Power Plant .M. Beijing: atomic energy press, 12(2000): p.138-139.
2. Van den Bergh F, Engelbrech A P. A cooperative apporach to particle swarm optimization. IEEE Trans on Evolution Computation, J. **8,** 3, (2004):225-239.
3. Liu Huailiang ,SU Ruijuan ,XU Ruoning ,GAO Ying. Cooperative particle swarm optimization. Journal of Computer Applications. J. **29,** 11, (2009): 3068-3073.
4. Man chuntao,shengguimin. An Improved Algorithm Based on Cooperate Particle Swarm Optimization, Journal of harbin university of science and technology. J. **15,** 6, (2010): 51-53,58
5. E.Irving. Toward Efficient Full Aromatic Operation of the PWR Steam Generator With Water Level Adaptive Control,. Proc 2nd Int Conf Boiler Dynamics and Control in Nuclear Power Stations2. Bournemouth, C, 1979: 309-329
6. SU Ying-bin, XIA Hong, SHEN Ji, CFNN based water level control for nuclear steam generator, Chinese Journal of Nuclear Science and Engineering.J. **28**, 2, (2008),: 158-162

MATEC Web of Conferences **55**, 05001 (2016) DOI: 10.1051/matecconf/20165505001
ACPEE 2016

Calculation of Onset Voltage of Sliding Discharge over a Dielectric Barrier

Mazen Abdel-Salam[1,a], Azza Hashem[2] and Hadeer El-Hawary[2]

[1] *Department of Electrical Engineering, 71516, Assiut University, Assiut, Egypt.*
[2]*Department of Physics, 71516, Assiut University, Assiut, Egypt.*

Abstract. This paper is aimed at calculating the onset voltage of a sliding discharge established between two electrodes at the upper and bottom surfaces of a dielectric barrier plate; named sliding dielectric barrier discharge (SDBD) driven by AC voltage during negative half cycle. The onset voltage is based on the condition of self-sustenance of avalanche growth in the vicinity of the stressed electrode. This calls at first for calculation of the electric field in the vicinity of stressed electrode. The dependency of onset voltage on the thickness and permittivity of the dielectric barrier as well as the thickness of the stressed electrode and the inter-electrode spacing between the stressed and ground electrodes is investigated. The obtained results are discussed in the light of gas discharge physics.

1 Introduction

The sliding discharge over a dielectric barrier (SDBD) has been studied for more than fifteen years to harness it for use in electro hydrodynamic actuators. Such discharge is generated on the surface of a dielectric plate sandwiched between upper and bottom electrodes. The upper electrode is stressed by DC and AC voltages with respect to the bottom one [1] with a focus on the discharge current pulses.

The velocity induced by the plasma actuator and dissipatedpower were recorded as a function of the applied voltage [2, 3]. A numerical modeling of a single micro discharge together with experimental finding made it possible to develop a simplified analytical model of the body force induced by sliding discharge driven by AC voltage [3, 4].All those studies [1–4] did not report on the assessment of the onset voltage of sliding (or dielectric barrier) discharge SDBD.

This paper is aimed at calculating the onset voltage of sliding discharge over a dielectric barrier based on the condition of self-sustenance of avalanche growth within the ionization zone around the AC stressed electrode during negative half-cycle. This calls at first for calculation the electric field in the vicinity of stressed electrode. The dependency of onset voltage on the thickness and permittivity of the dielectric barrier as well as the thickness of the stressed electrode and the inter-electrode spacing between the stressed and ground electrodes is investigated. The obtained results are discussed in the light of gas discharge physics.

2. Method of Analysis

[a] Corresponding author : mazen2000as@yahoo.com

2.1 Electric field calculation

The calculation of the onset voltage calls for electric field assessment at the barrier surface using the charge simulation method.

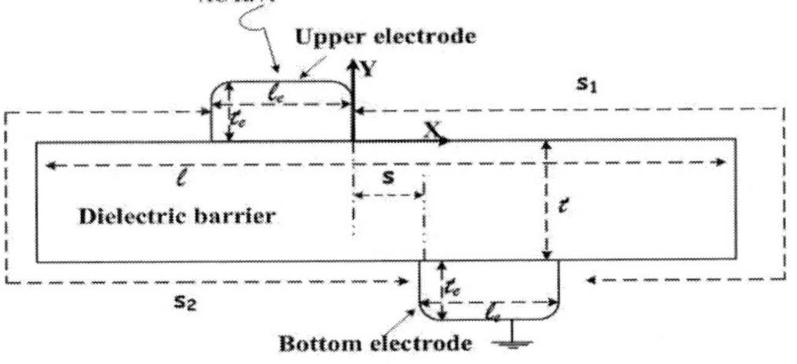

Figure 1. Upper and bottom electrodes sandwiching a dielectric barrier

2.1.1 Electrodes and dielectric arrangement for SDBD

Figure 1 shows a cross-section as viewed in the x-y plane of an electrode arrangement selected to assess the onset voltage of sliding discharge over a dielectric plate. The electrode arrangement is considered long along the z-direction. This electrode arrangement was selected before to investigate the surface dielectric barrier discharge (SDBD) [1]. The dielectric plate with width l, thickness t and permittivityε_r extends between upper and bottom electrodes as shown in Fig. 1. The width and thickness of both the upper and bottom electrodes are l_e andt_e, respectively. The spacing S between the front edge of the upper electrode and the back edge of the bottom one is variable by moving the bottom electrode with respect the upper one. The distance S_1 over the dielectric surface is longer than S_2 to ensure the occurrence of the sliding discharge overS_1.The upper electrode is stressed by AC voltage with respect to the bottom one. Therefore, the two electrodes are charged provided that the dielectric plate and the surrounding air are perfect dielectrics.

2.1.2 Charge simulation of electrodes and dielectric barrier

With the applied voltage, each charged electrode is simulated by a set N_e of line charges extending parallel to its axis. Fig. 2.The interface between two dielectrics (dielectric plate and air) is simulated by two sets of line charges, each of number $N(= N_a = N_d)$, one in the dielectric plate and the other in the air, extending parallel to the axis of the dielectric plate, Fig. 2. Therefore, the total number of the line charges simulating the upper and bottom electrode as well as the dielectric plate is $2N_e + 2N$.

A set of $2N_e + 2N$ simulations equations is to be formulated and solved to determine the unknown simulation charges. This set is to satisfy pertinent boundary conditions at boundary points selected over the boundaries ofFig. 1. Boundary points of number N_e are selected over the surface of each electrode inFig. 2, so each boundary point corresponds to a simulation charge in the electrode. The distance d_1 is chosen equal to the distance d_2 as shown in Fig 3 (a). Also, N boundary points are selected over the periphery of the interface between the dielectric plate and the surrounding air inFig. 2, so each boundary point corresponds to two simulation charges, one in the air and the other in the dielectric plate. The distance d_3 between two successive boundary points on the interface is equal to $d_4(= d_5)$ provided that d_4 and d_5 are the distances between the boundary point and the corresponding simulation charges in air and dielectric plate, respectively as shown in Fig. 3 (b).With the determination of the simulation charges, the electric potential and field can be calculated anywhere around the electrodes and the dielectric barrier.

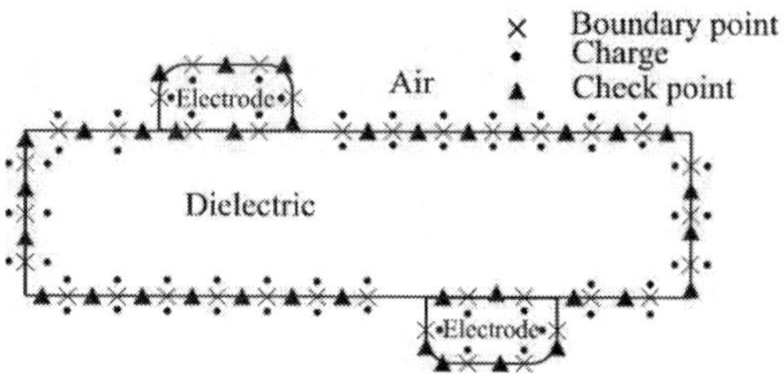

Figure 2. Simulation charge arrangement for the electrodes and dielectric barrier.

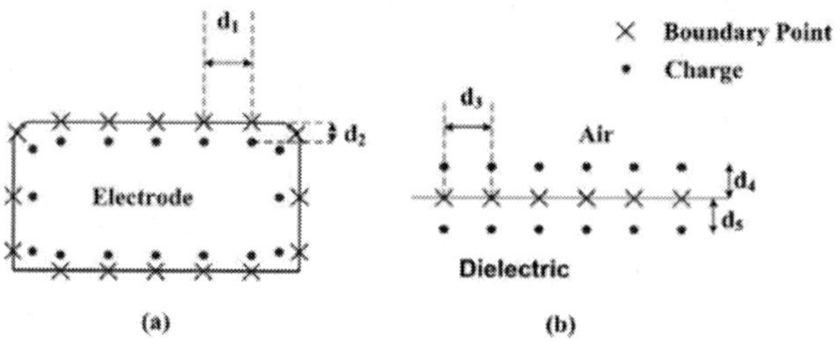

Figure 3. Boundary points arrangement at the electrode bound and air-dielectric interface.

2.1.3 Pertinent boundary conditions

The boundary conditions are Dirichlet condition at the electrodes' surface where the calculated potential is equal to the applied voltage as well as the equality of air-side and dielectric-side potentials and Neumann condition at the interface between the dielectric plate and surrounding air, where the calculated normal electric flux is continuous.

2.1.4 Describing equations

For the jth simulation charge q_j at(x_j, y_j),the potential at point P_i with (a_i, b_i) is expressed as

$$\emptyset(i,j) = P_{ij}\, q_j \tag{1}$$

whereP_{ij} is the potential coefficient due to the jth simulation charge at the ith point as expressed as

$$P(i,j) = \frac{1}{4\pi\varepsilon_\circ}\frac{1}{r_{ij}}, \qquad r_{ij} = [(x_j - a_i)^2 + (y_j - b_i)^2]^{0.5} \tag{2}$$

where $\varepsilon_\circ$ is the permittivity of the space.

The x and y components of the electric field are expressed as

$$E_x(i,j) = F_x(i,j)q_j, \qquad E_y(i,j) = F_y(i,j)q_j \tag{3}$$

where$F_x(i,j)$ and $F_y(i,j)$ are x- and y- field coefficient due to the jth simulation charge at the ith point along x- and y- directions,

$$F_x(i,j) = \frac{1}{4\pi\varepsilon_\circ}\frac{x-a_i}{{r_{ij}}^3}, \qquad F_y(i,j) = \frac{1}{4\pi\varepsilon_\circ}\frac{y-b_i}{{r_{ij}}^3} \tag{4}$$

Satisfaction of the above mentioned boundary conditions at the selected boundary points formulates a set of equations,whose simultaneous solution of this set of equations determines the unknown simulation charges.

2.1.5 Accuracy of charge simulation methodology

To check the accuracy of the chosen charge simulation, check points have been located midway between the boundary points. Satisfaction of the above-mentioned boundary conditions at the check points is a measure of the solution accuracy.

2.2 Calculation of onset voltage of SDBD.

When the instantaneous value of the electric field at the stressed electrode during the negative half cycle of the applied voltagereaches the threshold value for ionization by electron collision, an electron avalanche starts its growth. The avalanche extends along the maximum electric field stream line in the ionization zone surrounding the stressed electrode.The number of electrons at a distance x from the starting point of the avalanche growth, Fig. 4 is expressed as:

$$N_e(x) = \gamma_{ph} \int_0^x [\alpha(x) - \eta(x)]\, dx \tag{5}$$

where α is the Townsend's first coefficient of ionization, η is the electron attachment coefficients and γ_{ph} is the Townsend's second coefficient of ionization. α and η depend on the strength of electric field at the head of avalanche growing withinthe ionization-zone shown in Fig. 5. Expressions ofα and η as functions of electric field E and air pressure P were reported before [5].

The avalanche continues to grow until it reaches the ionization zone boundary, where $\alpha = \eta$. The number of the emitted photoelectrons N_{eph}, emitted from the stressed electrode during the avalanche growth up to distance x from the stressed electrode,

$$N_{eph} = \gamma_{ph} \int_{x_0}^{x} \alpha(x) N_e(x) g(x) e^{-\mu x}\, dx \tag{6}$$

where $g(x)$ is a geometry factor to account for the fact that some photons are not received by the stressed conductor as estimated before [5] andμ is the coefficient of photon absorption in air [5].

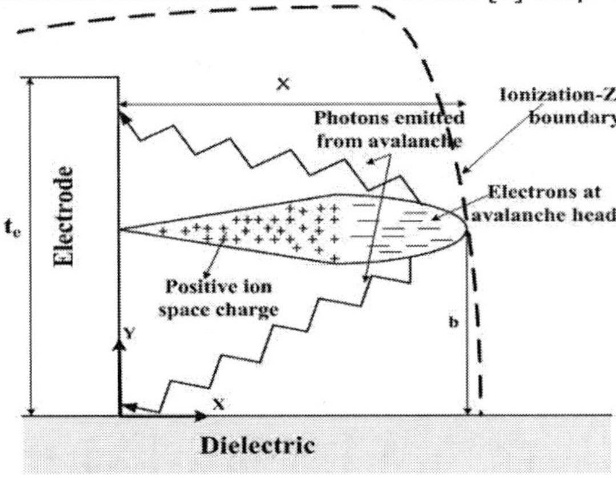

Figure 4. Growth electron avalanche within ionization zone around the stressed electrode edge.

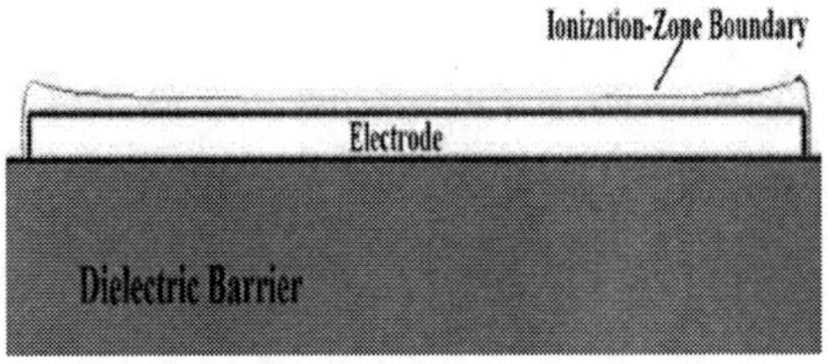

Figure 5. Ionization zone boundary around the stressed electrode in the plane $x - y$.

Therefore, the onset voltage condition is satisfiedwhen the total number of photoelectrons is equal to or more than unity in order to sustain the sliding discharge over the dielectric surface [5].

$$N_{eph} \geq 1 \tag{7}$$

The onset voltage does not appear explicitly in the equation. However, the applied voltage effects in the values of x, $\alpha(x)$ and $\eta(x)$ along the growth of the avalanche. The peak value of the applied voltage which satisfies the relation (7) is the onset voltage of sliding discharge (SDBD) over the dielectric barrier.

3 Results and Discussion

The width l_e and thickness t_e are equal $10mm$ and $0.1\ mm$ respectively. The thickness $t$lies in the range $2 - 6\ mm$. The spacing S is varying over the range $0 - 20\ mm$. The relative permittivity ε_r is chosenequal to 4 and 8.With $N_e = 540$ and $N = 2300$, Dirichlet and Neumann conditions are satisfied with acceptable accuracy.

Figure 6 shows the increase of the onset V_0 of the sliding discharge over the dielectric surface with the increase of the dielectric thickness t for different inter-electrode spacing S. For the same values of the spacing S and relative permittivityε_r, the inter electrode capacitance C decreases with the increase of the dielectric barrier thickness t with a subsequent decrease of the electrodes charge at a given applied voltage.This results in a decrease of the electric field in the vicinity of the stressed electrode, i.e. within the ionization-zone surrounding the stressed electrode as shown in Fig. 5. Within this zone, the electron avalanches grow to sustain the sliding discharge over the dielectric surface. To compensate for the decrease of the electric field within the ionization zone, the applied voltage has to increase. This is why the onset voltage V_0 of the sliding discharge increases with the increase of the thickness of the dielectric barrier as shown in Fig. 6. For a given thickness t value, the onset voltage V_0 increases with the increase of the spacing S as shown in Fig. 6. This may be attributed to the fact that the streamlines of the electric field which emanate from the stressed electrode extend over a large surface area of the dielectric barrier. Subsequently, the electric field over the dielectric surface decreases and the applied voltage to sustain the sliding discharge increases with the increase of the spacing S. This is why the onset voltage V_0 increases with the spacing S for the same value of dielectric thickness t as shown in Fig. 6.

Figure 7shows the increase of the onset voltage V_0 of sliding discharge over the dielectric surface with the increase of spacing S for different thickness t values of the dielectric barrier. For the same values of thickness t and relative permittivity ε_r, the increase of the inter electrode spacing S as stated above makes the streamlines of the electric field emanating from the stressed electrode to extend over a large surface area of the dielectric barrier with a subsequent decrease of the electric field over the dielectric surface. This calls for an increase of the applied voltage to sustain the sliding discharge. This is why the onset voltage V_0 increases with the increase of spacing as shown in Fig. 7 for the same thickness t and relative permittivity ε_r. For a given spacing S, the onset voltage V_0 increases with the dielectric thickness t in agreement withFig. 7.

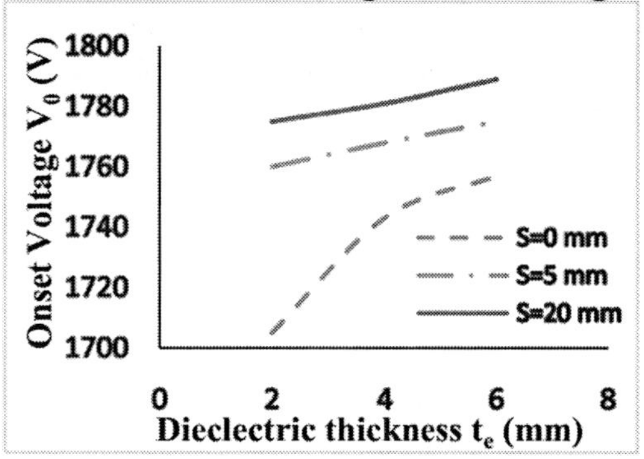

Figure 6. Onset voltage ($\boldsymbol{V_0}$) of sliding discharge as influenced by dielectric barrier thickness ($\boldsymbol{t}$) for different spacing (S) values between electrodes, (ε_r=4, $\boldsymbol{t_e}$=0.1mm).

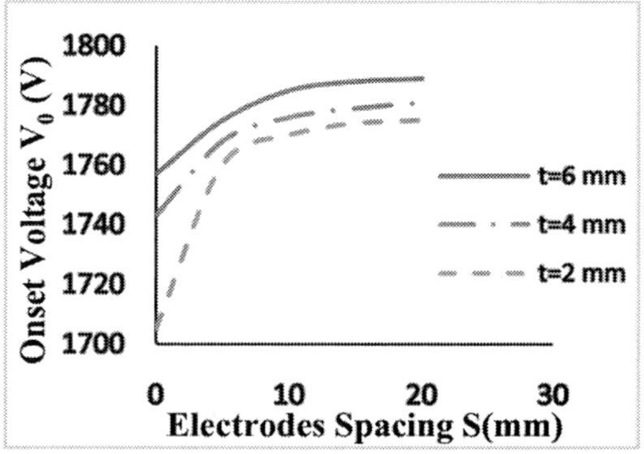

Figure 7. Onset voltage ($\boldsymbol{V_0}$) of sliding discharge as influenced by the spacing (S) between the electrodes for different thickness ($\boldsymbol{t}$) values of dielectric barrier, (ε_r=4, $\boldsymbol{t_e}$=0.1mm).

Figure 8 shows the increase of the onset voltage V_0 of the sliding discharge over the dielectric surface with the increase of spacing S for different relative permittivity ε_r values. The increase of V_0 with the increase of S is explained inFig. 7. For the same values of spacing S and dielectric thickness$t$, the onset voltage V_0 decreases with the increase of the relative permittivity ε_r of the dielectric barrier as shown inFig. 8. This is because the inter electrode capacitance C increases with the increase of dielectric relative permittivity ε_r for constant values of S and t with a subsequent increase of the electrodes chargeat a given applied voltage.This results in an increase of the electric field within the ionization-zone surrounding the stressed electrode. To counter balance the increase of the electric field within the ionization-zone, the applied voltage has to decrease. This is why the onset voltage V_0 of the sliding discharge decreases with the increase of the relative permittivity ε_r as shown in Fig. 8 for the same values of thickness t and spacing S.

Figure 9 shows the increase of the onset V_0 of the sliding discharge with the increase of the electrode thickness t_e for different spacing$S$ values. For the same values of the spacing S and relative permittivityε_r, the inter electrode capacitance C decreases with the increase of the thickness t_e with a subsequent decrease of the electrodes charge at a given applied voltage. This results in a decrease of the electric field within the ionization-zone surrounding the stressed electrode as shown inFig. 5. To compensate for the decrease of the electric field within the ionization zone, the applied voltage has to increase. This is why the onset voltage V_0 of the sliding discharge increases with the increase of the electrode thickness as shown inFig. 9.

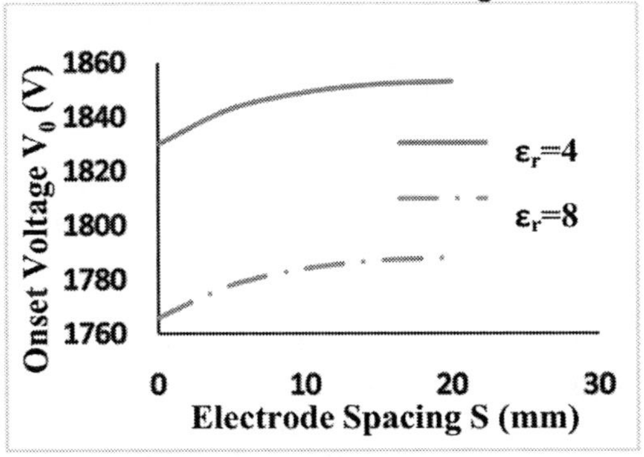

Figure 8. Onset voltage ($\boldsymbol{V_0}$) of sliding discharge as influenced by the spacing (S) between the electrodes for different permittivity (ε_r) values of dielectric barrier, ($\boldsymbol{t}$=6mm, $\boldsymbol{t_e}$=0.1mm).

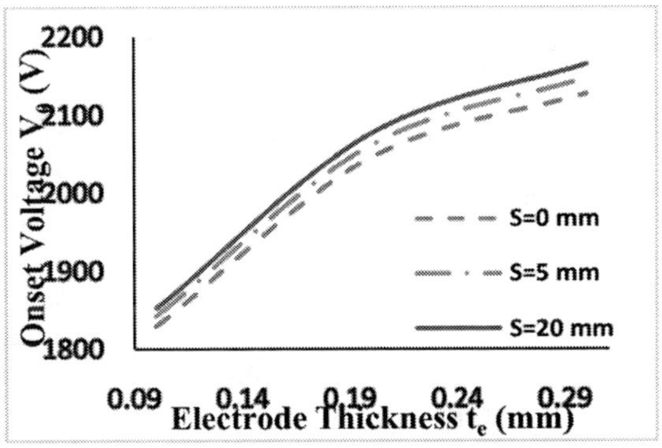

Figure 9. Onset voltage (V_0) of sliding discharge as influenced by electrode thickness (t_e) for different spacing (S) values between electrodes, (ε_r=4, t=6mm).

4 Conclusions

1. The onset voltage of sliding dielectric barrier discharge driven by AC voltage during negative half-cycle can be calculated by a theoretical method.
2. The calculation of the electric field values in the vicinity of the stressed electrode, which is a pre-requisite for onset voltage calculation, is made using the charge simulation method.
3. The calculated onset voltage decreases with the increase of the relative permittivity ε_r , as well as the decresce of t, S and t_e.

References

1. Louste, C., Artana, G., Moreau, E. & Touchard, G. Sliding discharge in air at atmospheric pressure: electrical properties. J. Electrostat.63, 615–620 (2005).
2. Enloe, C. L. *et al.* Mechanisms and Responses of a Dielectric Barrier Plasma Actuator: Geometric Effects. *AIAA J.***42,** 595–604 (2004).
3. Pons, J., Moreau, E. & Touchard, G. Asymmetric surface dielectric barrier discharge in air at atmospheric pressure: electrical properties and induced airflow characteristics. *J. Phys. D. Appl. Phys.***38,** 3635–3642 (2005).
4. Soloviev, V. R. Analytical estimation of the thrust generated by a surface dielectric barrier discharge. *J. Phys. D. Appl. Phys.***45,** 025205 (2012).
5. Khalifa, M., M. Abdel-Salam & M. Abou-Seada. Calculation of negative corona onset votage. *IEEE-PES conferance Pap. C 73 160-9, Winter Meet. New York, N.Y, USA* (1973).

MATEC Web of Conferences **55**, 05002 (2016) DOI: 10.1051/matecconf/20165505002
ACPEE 2016

Corona Performance in Wire-cylinder ESP without Particle Loading

Mohamed Th. El-Mohandes[1], Mazen Abdel-Salam[1, a] and Seham Kamal El-deen[1]

[1]*Electrical Engineering Department, Assiut University, Assiut, Egypt*

Abstract: This paper is aimed at investigating thoroughly the corona performance in the wire-cylinder electrostatic precipitor (ESP) without suspended particles. The onset voltage of negative corona on the discharge wire is calculated based on the criterion of self-sustained discharge. The ionized field in the ESP is mathematically modeled for calculating the spatial distribution of the space-charge density due to the corona ions as well as the components of the electric field including the applied field and the field due to the space charge. This is in addition to the calculation of the current-voltage characteristics of the ESP.

1 Introduction

The electrostatic precipitator is the most effective device to collect industrialparticulates before escaping into the atmosphere to pollute it. This is why the researchers over the years work on modeling the processes inside the electrostatic precipitors (ESPs)including corona discharge seeking improvement of their performance.

A model was described for computation of charge density and electric field structures without particle loading. The model employed an iterative model based on the finite element method for computing electric potential structure in wire-duct electrostatic precipitators for an assumed charge density distribution by solving Poisson's equation, and the method of characteristics for computing charge density structure for an assumed electric field distribution[1-3].

The charge simulation technique has been adapted [4] tomodel the electrostatic and the corona characteristics in clean air of a duct-type electrostatic precipitator where the corona onset voltage is predicted by Peek's formula [5].

In a wire-duct electrostatic precipitator, the current-voltage characteristics under clean air conditions were predicted [6]. The approach was based on the simultaneous solution for the governing Poisson's and current-continuity equations using a combined boundary element and finite difference method.

An iterative numerical methodology was developed [7] for the calculation of the electric field in wire-duct precipitation systems using finite differencing in orthogonal curvilinear coordinates to solve the potential equation. Charge density distribution was obtained by the method of characteristics or using a prediction-correction scheme. To predict the I-V characteristic of wire-duct ESP which agrees

[a] Corresponding author: mazen2000as@yahoo.com.

reasonably with the measured one, an empirically estimated thickness of the ionization zone was considered based on Peek's formula for assessing the corona onset voltage.

In wire-duct ESP, Poisson's equation and the continuity equation of corona current density were solved [8] simultaneously after being transformed into integral form. The calculated spatial distribution of the electric potential and charge density were compared with those obtained by White's formula in wire-cylinder ESP [9].

The current-voltage formula for wire-duct ESP was derived by solving Poisson's equation in one dimension for ion large current density values where the electrostatic field is large compared to the space charge field[10]. The current-voltage formula for wire-duct ESP was also derived for high current density values where the space charge is large[9].

There are several attempts to compute the electric field and I-V characteristics of different ESP geometries including wire-cylinder[8, 10, 11]and wire-ductarrangements[1-4,6-11]. The analysis of the ionized-field in these arrangements is based onboundary element method (BEM)[3], charge simulation method (CSM)[4],finite volume characterization method (FVM)[8],finite element method combined with the method of characteristics (MOC) [1, 2], FDM combined with MOC [7]and BEM combined with FDM [6].

Simplifying assumptions commonly adopted in the ESP modeling which are justified:

1. Kaptsov' assumption which states that the field at the coronating surface remains constant at the corona onset value as determined by Peek's formula[5]irrespective of the value of the applied voltage. The ions emanating from the corona electrode have polarity the same as that of the discharge wire. They tend to diminish the electric field at the electrode surface by a value approximately equal to that due to the excess of applied voltage above the onset value, so the field remains constant at its onset value.
2. The mobility of ions remains constant as their transit time from the discharge wire to the collecting electrode is so small for the mobility to change with the ions' life time.

This motivates the authors to investigate thoroughly the corona performance in the wire-cylinder ESP without loading with suspended particles.

2 Method of Analysis

The equations describing the electric field and flow of ions in the ESP are the Poisson's equation for the electric field, the equation for current density and the equation of current continuity[12].

Poisson's equation for the electric field is expressed as:

$$\nabla . E = \frac{\rho}{\epsilon_0} \qquad (1)$$

where E is the electric field with space charge including corona ions, ρ is the volume charge density of ions and ϵ_o is permittivity of free space.

The electric field E is expressed as:

$$E = E_f + E_s \qquad (2)$$

where$E_f$is the electric field without space charge i.e.; the field due to the applied voltage and E_s is the field due to space charge of ions.

In a wire-cylinder geometry of the precipitator, the space charge affects only the magnitude but not the direction of the electric field, i.e.;

$$E = \xi E_f \qquad (3)$$

where ξ is a scalar which changes over the space between the discharge wire and the collecting cylinder.

From equations (2) and (3), the field E_s due to space charge is expressed as:

$$E_s = (\xi - 1)E_f \qquad (4)$$

The equation of precipitator current density is expressed as:

$$J = k_i \rho_i E \qquad (5)$$

where k_i is the mobility of corona ions.

Combination of equations (1) and (3) results in the following differential equation, whichdefines the spatial distribution of the scalar ξ over the space between the discharge wire and collecting cylinder:

$$\frac{d\xi}{dr} = \frac{\rho_i}{\epsilon_0 E_f} \tag{6}$$

The continuity equation of the precipitator current density is expressed as:

$$\nabla . J = 0 \tag{7}$$

Combination of equations (3), (5) and (17) results in the following differential equation:

$$\frac{d\rho_i}{d\xi} = -(\frac{\rho_i}{\xi}) \tag{8}$$

Combining of equations (6) and (8) results in the following differential equationwhich defines the spatial distribution of the ion charge density ρ_i over the space between the discharge wire and collecting cylinder:

$$\frac{d\rho_i}{dr} = -\left(\frac{\rho_i}{\xi}\right)\left(\frac{\rho_i}{\epsilon_0 E_f}\right) \tag{9}$$

The differential equations (6) and (9) are integrated along the radial direction to determine the spatial distribution of ξ and ρ_i over the space between the discharge wire and collecting cylinder. The initial conditions for such integrals are (i) ξ at discharge wire is equal to V_o/V, where V_o is the corona onset voltage and V is the applied voltage. (ii) The electric field at the surface of the discharge electrode remains constant at the onset value, which is Kaptsov's assumption. The onset value of negative corona in the ESP is calculated based on the criterion of self-recurrence of avalanche growth in the ionization zone surrounding the discharge electrode[13].

3 Results and Discussions

3.1. Performance of ESP without particle loading:

Figure (1) shows how the electric field E changes along the radial direction from the ionization-zone boundary to the collecting cylinder. It is satisfactory that the calculated spatial distribution of the electric field agrees well with the analytical solutions reported in Appendix 1.

Figure (2) shows the spatial distribution of the electric field E and its components E_f and E_s. The integration value of E_f along the spacing between the wire and cylinder is 59.98 kV which is almost equal to the applied voltage V (= 60 kV). The integration of E_s along the spacing between the wire and cylinder is almost equal zero. This is why the integrationof the electric field E is equal to the applied voltage V.

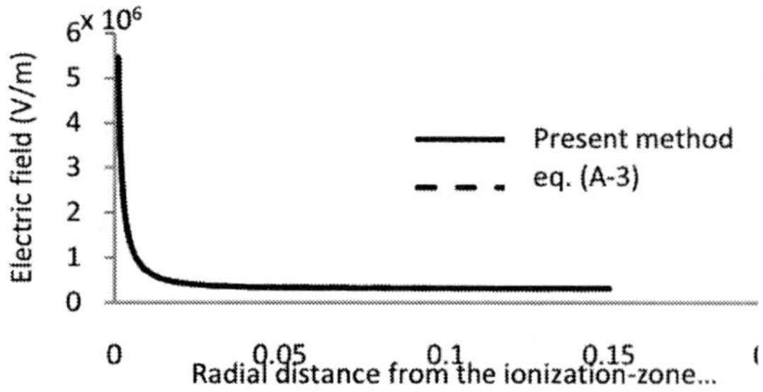

Figure 1: Calculated spatial distribution of the electric field E between the discharge wire and collecting cylinder compared with that obtained by the analytical equations (A-3) and (A-5). r_o = 1 mm, R = 15 cm, V_o = 30.659 kV and V = 60 kV.

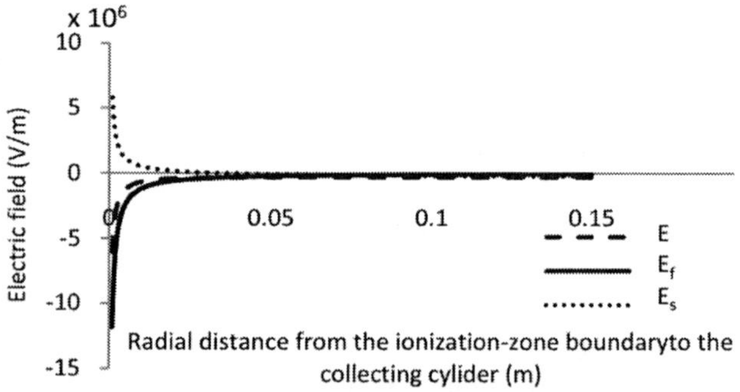

Figure 2: Calculated spatial distribution of the electric field E and its components E_f and E_s between the discharge wire and collecting cylinder. r_o = 1 mm, R = 15 cm, V = 60 kV and Vo = 30.659 kV

As the electron ionization coefficient increases with the electric field, the ion charge density ρ_i also increases with the increase of the electric field. Therefore, ρ_i assumes high values near the discharge wire and decreases gradually toward the cylinder as shown in Fig. (3). It is satisfactory that the calculated spatial distribution of ρ_i between the wire and cylinder agrees well with the analytical solution eq. (A-4) reported in Appendix 1. The deviation between the present calculated values and those predicted by the analytical solution does not exceed 0.01 %.

The calculated current-voltage characteristic of the ESP agrees satisfactory with that predicted by the analytical solution as shown in Fig. (4). The deviation between the calculated and predicted values of corona current at the same applied voltage does not exceed 2.2%.

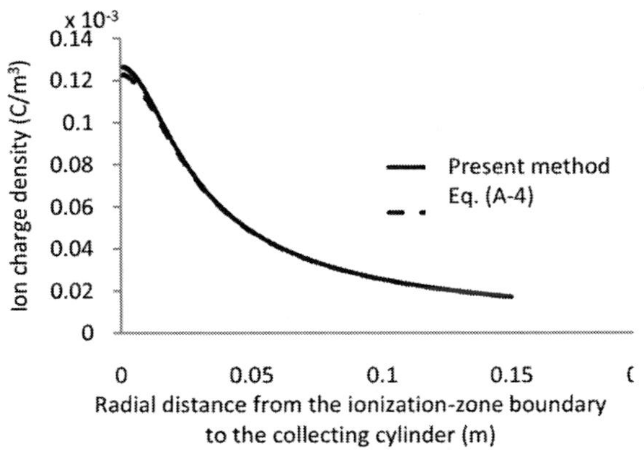

Figure 3: Calculated spatial distribution of the ion charge density between the discharge wire and collecting cylinder compared with that obtained by the analytical equation (A-4). r_o = 1 mm, R = 15 cm, V_o = 30.659 kV and V = 60 kV.

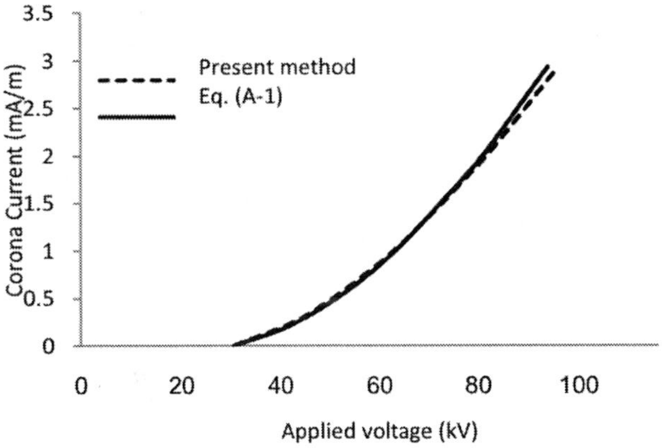

Figure 4: Calculated corona current-voltage characteristics of the ESP compared with that obtained by the analytical equation (A-1) for wide range of the corona current. r_o = 1 mm, R = 15 cm and V_o = 30.659 kV

Figure (5) shows the current-voltage (I-V) characteristics of the ESP at varying radius r_o of the discharge wire for the same radius R of the collecting cylinder. The onset voltage increases with the increase of the discharge wire r_o as given in Table (1) for the same radius R of the cylinder.

Figure (6) shows the I-V characteristics of the ESP at varying radius R of the collecting cylinder for the same radius r_o of the discharge wire. The onset voltage decreases with the increase of R for the same radius r_o as given in Table (2).

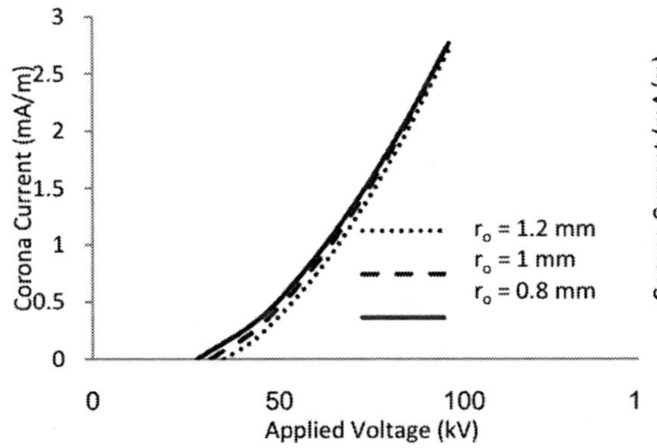

Figure 5: Calculated corona I-V characteristics of the ESP for different values of the discharge wire radius r_o at the same radius R (=15 cm) of the collecting cylinder.

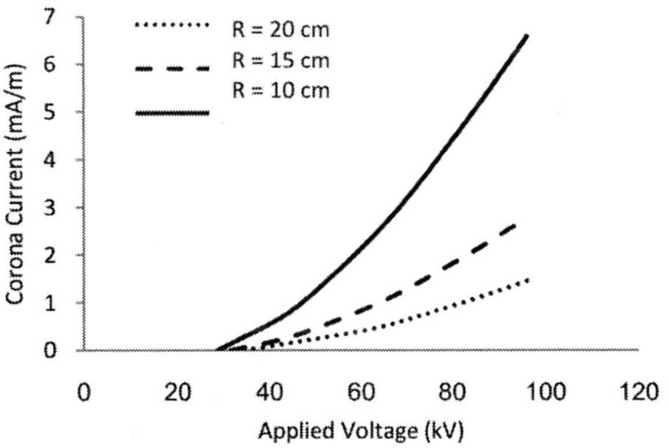

Figure 6: Calculated corona I-V characteristics of the ESP for different values of the collecting-cylinder radius R at the same radius of the discharge wire r_o (= 1 mm)

Table 1: Calculated corona onset voltage V_o as influenced by the radius of the discharge wire at the same radius R (= 15 cm) of the collecting cylinder.

r_o (mm)	0.8	1	1.2
V_o (kV)	28.05942	31.57891	35.21861

Table 2: Calculated corona onset voltage V_o as influenced by the radius of the collecting cylinder R at the same radius of the discharge wire r_o (= 1 mm).

R (cm)	10	15	20
V_o (kV)	28.74174	31.57891	34.36458

4 Conclusions

1. The onset voltage of negative corona on the dischargewire of the ESP is calculated based on the criterion ofself-sustained discharge instead of using the empirical equation of Peek.
2. The ionized field in the ESP is mathematically modeled for calculating the spatial distribution ofspace charge density due to corona ions as well as the components of electric field including the applied field and the field due to space charge.
3. The calculated spatial distribution of the electric field between the discharge wire and cylinder of the ESP without particle loading agreed very well with that obtained analytically.
4. The integration of the electric field with and without space charge is equal to the applied voltage with a deviation not exceeding 0.01 %. This reflects itself on the accuracy of the present method of calculation.
5. The calculated spatial distribution of the ion charge density over the space between the discharge wire and cylinder agreed reasonably with that obtained analytically.
6. The calculated current-voltage characteristics of the ESP without particle loading agreed reasonably with that obtained analytically.
7. The calculated onset voltage of corona on the discharge wire of the ESP without particle loading increases with the increase of either the wire radius or the cylinder radius.

Acknowledgment

This paper is extracted from a research project sponsored through joint innovative projects fund STDF(Egypt) – IRD (France), project STDF ID-5457. The authors acknowledge such sponsorship.

References

1. J. L. Davis and J. F. Hoburg, "Wire-duct precipitator field and charge computation using finite element and characteristics methods," J. Electrostat., **14**, 2(1983).
2. A. J. Butler, Z. J. Cendes, and J. F. Hoburg, "Interfacing the finite-element method with the method of characteristics in self-consistent electrostatic field models," IEEE Trans. Ind. Appl., **IA-25**, 3(1989).
3. K. Adamiak, "Simulation of Corona in Wire-Duct Electrostatic Precipitator by Means of the Boundary Element Method," IEEE Trans. Ind. Appl. vol. IA-30, no. 2, pp. 381–386, 1994.
4. A. A. ElMoursi and G. S. P. Castle, "Modeling Corona Characteristics in a Wire-Duct Precipitator Using the Charge Simulation," IEEE Trans. on Ind. Appl., **IA-23**, 1(1987).
5. F.W. Peek, "Dielectric Phenomena in High-Voltage Engineering," McGraw-Hill, New Work, USA, (1929).
6. B. S. Rajanikanth and N. Thirumaran, "Prediction of pre-breakdown V–I characteristics of an electrostatic precipitator using a combined boundary element-finite difference approach," Fuel Process. Technol., **76**, 3(2002).
7. J. Anagnostopoulos and G. Bergeles, "Corona discharge simulation in wire-duct electrostatic precipitator," J. Electrostat., **54**, 2(2002).
8. N. Neimarlija, I. Demirdžić, and S. Muzaferija, "Finite volume method for calculation of electrostatic fields in electrostatic precipitators," J. Electrostat., **67**, 1(2009).

9. H. J. White, Industrial electrostatic precipitation. Reading, MA: Addison-Wesley, (1963).
10. P. Cooperman, "A theory for space charge limited currents with application to electrical precipitation," AIEE Trans., **79**, 47–50(1960).
11. A. D. Moore, "Electrostatics and its Applications", J. Wiley & Sons, New York, USA, (1973).
12. M. Abdel-Salam and Z. Al-Hamouz, "A New Finite Element Solution of Ionized Field in Coaxial Cylindrical Geometry", J. Phys. D: Applied Physics, **25**, 1551-1555(1992).
13. M. Abdel-Salam and D. Wiitanen, "Calculation of corona onset voltage for duct-type precipitators," IEEE Trans. Ind. Appl., **IA-29**, 2(1993).
14. M. P. Sarma and W. Janischewskyj And, "Analysis of Corona Losses on DC Transmission Lines: I - Unipolar Lines," IEEE Trans. Power Appar. Syst., **PAS-88**, 718–731(1969).

7 Appendices

7.1. Appendix 1: Analytical expressions defining wire-cylinder ESP performance without particle loading:

7.1.1. Discharge current-voltage characteristics:

The characteristics were expressed in the literature by equation (A-1) [14].

$$\frac{V-V_o}{V_o}\ln\frac{R}{r_o} = \sqrt{(1+Y)} - 1 + \ln\frac{2}{1+\sqrt{1+Y}} \qquad \text{(A-1a)}$$

where:

$$Y = \frac{I}{2\pi\epsilon_o K}\left(\frac{R}{E_o r_o}\right) \qquad \text{(A-1b)}$$

I is the current per unit length.

7.1.2. Spatial distribution of electric potential ϕ, electric field E and ion charge density ρ_i:

The expressions (A-2) and (A-4) were reported before for the potential $\phi(r)$ and ion charge density ρ_i along the radial distance r from the discharge wire axis [9]:

$$\phi(r) = \sqrt{\left(\frac{Ir^2}{2\pi\epsilon_o} + A^2\right)} + \sqrt{\left(\frac{IR^2}{2\pi\epsilon_o} + A^2\right)} + A\ln\left(\frac{A+\sqrt{\frac{IR^2}{2\pi\epsilon_o}+A^2}}{A+\sqrt{\frac{Ir^2}{2\pi\epsilon_o}+A^2}}\right) + A\ln\frac{r}{R} \qquad \text{(A-2a)}$$

where:

$$A = r_o\sqrt{E_o^2 - \frac{I}{2\pi\epsilon_o K}} \qquad \text{(A-2b)}$$

The electric field E(r) is obtained as equal to $\frac{d(\phi(r))}{dr}$:

$$E(r) = \frac{Ir/\pi\epsilon_o k_i}{2\sqrt{\frac{Ir^2}{2\pi\epsilon_o}+A^2}} + A\left[\frac{-Ir/\pi\epsilon_o k}{2\left(A^2+\sqrt{\frac{Ir^2}{2\pi\epsilon_o}+A^2}\right)\left(\sqrt{\frac{Ir^2}{2\pi\epsilon_o}+A^2}\right)}\right] + \frac{A}{r} \qquad \text{(A-3)}$$

$$\rho_i(r) = \frac{I}{2\pi r k_i}\frac{1}{\sqrt{\frac{I}{2\pi\epsilon_o K}\left(1-\frac{r_o^2}{r^2}\right)+\frac{r_o^2}{r^2}E_i^2}} \qquad \text{(A-4)}$$

7.1.3. Spatial distribution of electric field E

The spatial distribution of electric field E (r) was expressed before [11]:

$$E(r) = \left(\frac{I}{2\pi\epsilon_o k_i} + \left(\frac{r_o}{r}\right)^2\left(E_o^2 - \frac{I}{2\pi\epsilon_o k_i}\right)\right)^{1/2} \qquad \text{(A-5)}$$

MATEC Web of Conferences **55**, 05003 (2016) DOI: 10.1051/matecconf/20165505003
ACPEE 2016

Corona Performance in Wire-cylinder ESP with Particle Loading

Mazen Abdel-Salam[1,a], Mohamed Th. El-Mohandes[1] and Seham Kamal El-deen[1]

[1]*Electrical Engineering Department, Assiut University, Assiut, Egypt*

Abstract: This paper is aimed at investigating thoroughly the corona performance in the wire-cylinder electrostatic precipitor (ESP) with loading by suspended particles in the exhaust of a diesel engine. The onset voltage of negative corona on the discharge wire is calculated based on the criterion of self-sustained discharge. The ionized field in the ESP is mathematically modeled for calculating the spatial distribution of the space-charge density due to both the ions and the charged particles as well as the components of the electric field including the applied field and the field due to the space charge. This is in addition to the calculation of the current-voltage characteristics of the ESP with particle loading.

1 Introduction

The electrostatic precipitator is the most effective device to collect industrial particulates before escaping into the atmosphere to pollute it. This is why the researchers over the years work on modeling the processes inside the electrostatic precipitors(ESPs) including corona discharge seeking improvement of their performance.

An iterative model employed the finite element method for computing electric potential structure for an assumed charge density distribution and the doner cell method was used to compute charge density structure for an assumed electric field distribution. The model can account for particle loading of the ESP and bipolar ionic species resulting from back-ionization at the collecting electrode [1].

The Poisson's equation was solved in wire-duct ESP with particle loading using the finite difference method for a given ion density distribution. The ion charge density at the boundary of the ionization-zone was evaluated empirically and used to determine the ion-density spatial distribution. The effect of assumed uniform particle charge density on the current-voltage characteristic of the ESP was investigated, where the corona onset voltage increases with the increase of the particle charge density. Subsequently, the ion current decreases with the increase of particle charge density at the same applied voltage [2]. In wire-duct ESP, the effect of particle loading and applied voltage on the mean radius of the ionization zone around the discharge wires was investigated[3].

In wire-duct ESP, the finite element method was used to determine the electric field and current density distributions based on Kaptzov's assumption which was considered valid for each corona electrode[4]. The electric field on the corona electrode surfaces is equal to the corona onset field, given by Peek's formula[5]. Non-justified simplifying assumptions have been additionally introduced

[a] Corresponding author: mazen2000as@yahoo.com.

in the mathematical model: 1) It is supposed that the electric field on the surface of each corona electrode is virtually uninfluenced by the presence of particles. Subsequently, the corona onset voltage is not influenced by the presence of particles. 2) Ionic charge density is also virtually uninfluenced by the presence of particles.

The current-voltage formula for wire-duct ESP was derived where the particle space charge was assumed uniform over the cross-section of the precipitator. This concluded that the onset voltage of corona increases with the particle loading [6].

Simplifying assumptions commonly adopted in the ESP modeling which are justified:

1. Kaptsov' assumption which states that the field at the coronating surface remains constant at the corona onset value as determined by Peek's formula [5] irrespective of the value of the applied voltage.
2. The mobility of ions remains constant as their transit time from the discharge wire to the collecting electrode is so small for the mobility to change with the ions' life time.

Other simplifying assumptions were adopted without justification for the ESP with particle loading:

1) Some investigators [2, 3] evaluated empirically the ion charge density at the boundary of the ionization-zone surrounding the discharge wire in order to determine the ion density distribution.
2) Some investigators [4] assumed the electric field on the surface of the discharge wire is virtually uninfluenced by the presence of particles.
3) Some investigators [4] assumed the ion charge density is virtually uninfluenced by the presence of particles.
4) Some investigators [6] assumed uniform particle space charge density over the ESP cross section.
5) With particle loading, some investigators [7, 8] assumed that the charge acquired by particles is related to the field component due to the ion space charge with no justification.

This motivates the authors to investigate thoroughly the corona performance in the wire-cylinder ESP when loaded with suspended particles in the exhaust of diesel engine.

2 Method of Analysis

The equations describing the electric field and flow of charges carriers (ions and charged particles) in the ESP are the Poisson's equation for the electric field, the equation for current density and the equation of current continuity[9].

Poisson's equation for the electric field is expressed as:

$$\nabla . E = \frac{\rho}{\epsilon_0} \tag{1}$$

where E is the electric field with space charge including corona ions and charged particles, ρ is the total volume charge density of ions and particles and ϵ_o is permittivity of free space.

The total volume charge density is expressed as:

$$\rho = \rho_i + \rho_p \tag{2}$$

where ρ_i is the ion charge density and ρ_p is the particle charge density.

The electric field E is expressed as:

$$E = E_{f+}E_s \tag{3}$$

where E_f is the electric field without space charge i.e.; the field due to the applied voltage and E_s is the field due to space charge of ions and charged particles.

In a wire-cylinder geometry of the precipitator, the space charge affects only the magnitude but not the direction of the electric field, i.e.;

$$E = \xi E_f \tag{4}$$

where ξ is a scalar which changes over the space between the discharge wire and the collecting cylinder.

From equations (3) and (4), the field E_s due to space charge is expressed as:

 DOI: 10.1051/matecconf/20165505003

$$E_s = (\xi - 1)E_f \quad (5)$$

The equation of precipitator current density is expressed as:

$$J = k_i \rho_i E + k_p \rho_p E \quad (6)$$

where k_i and k_p are the mobility of corona ions and charged particles, respectively.

Equation (6) of the current density is rewritten as:

$$J = k_i E(\rho_i + k\rho_p) \quad (7)$$

where the mobility ratio k is expressed as:

$$k = \frac{k_p}{k_i} \quad (8)$$

The particles are charged by ion-bombardment process, which is the dominant charging mechanism for particles of radii greater than 1 μm [10]. The particle charge density is expressed as:

$$\rho_p = \epsilon_o \lambda E S \quad (9)$$

where:

$$\lambda = \frac{2\epsilon_r}{\epsilon_r + 3} \quad (10)$$

where ϵ_r is the relative permittivity of the particles and S is the particle specific surface.

Assuming the particles are spherical in shape with radius r_p, so S is expressed as:

$$S = 4\pi r_p^2 N_p \quad (11)$$

where N_p is the particles concentration being assumed constant over the cross-section of the particle.

The particle concentration N_p is as expressed as follows:

$$N_p = \frac{Z}{\gamma \frac{4}{3}\pi r_p^3} \quad (12)$$

where Z is the particles weight per unit volume and γ is the specific gravity of the particles.

With the use of equation (4), the particle charge ρ_p density of eq. (9) is expressed as:

$$\rho_p = A\xi E_f \quad (13)$$

where $A = \epsilon_o \lambda$ S is a constant for particles of the same radius and permittivity.

The mobility k_p of the particle with charge q_p is determined by balancing the electric force (=q_p.E) with Stokes drag force (= $6\pi\eta r_p k_p E$) where η is the viscosity of the gas inside the precipitator.

Therefore, the particle mobility k_p is expressed as:

$$k_p = \frac{q_p}{6\pi\eta r_p} \quad (14)$$

The particle charge q_p is expressed as:

$$q_p = \frac{\rho_p}{N_p} \quad (15)$$

Combination of equations (1), (2), (4) and (13) results in the following differential equation, which defines the spatial distribution of the scalar ξ over the space between the discharge wire and collecting cylinder:

$$\frac{d\xi}{dr} = \frac{\rho_i + A\xi E_f}{\epsilon_0 E_f} \quad (16)$$

The continuity equation of the precipitator current density is expressed as:

$$\nabla . J = 0 \quad (17)$$

Combination of equations (4), (7) and (17) results in the following differential equation:

$$\frac{d\rho_i}{d\xi} = -\left(\frac{\rho_i + 2kA\xi E_f}{\xi}\right) \quad (18)$$

Combining of equations (16) and (18) results in the following differential equation which defines the spatial distribution of the ion charge density ρ_i over the space between the discharge wire and collecting cylinder:

$$\frac{d\rho_i}{dr} = -\left(\frac{\rho_i + 2KA\xi E_f}{\xi}\right)\left(\frac{\rho_i + A\xi E_f}{\epsilon_o E_f}\right) \quad (19)$$

The differential equations (16) and (19) are integrated along the radial direction to determine the spatial distribution of ξ and ρ_i over the space between the discharge wire and collecting cylinder. The initial conditions for such integrals are (i) ξ at discharge wire is equal to V_o/V, where V_o is the corona onset voltage determined as explained elsewhere [11] and V is the applied voltage. (ii) The electric field at the surface of the discharge electrode remains constant at the onset value, which is Kaptsov's assumption.

3 Results and Discussions

Figure (1) shows the slight increase of E_s with particle loading when compared with that for the case without particle loading [11]. The integration of E_s along the spacing between the discharge wire and collecting cylinder remains very close to zero. Subsequently, the integration of the electric field E over the spacing between the wire and cylinder is almost equal to the applied voltage which satisfies the Dirichlet condition. The onset voltage V_o' with particle loading exceeds that V_o without particle loading in agreement with eq. (A-2) of Appendix 1.

Figure (2) shows the spatial distribution of the electric field E over the space from the ionization-zone boundary to the cylinder as predicted by eq. (A-1) of Appendix 1 in comparison with that obtained using the present method of calculation. The deviation in Fig. (2) from the present calculated values is attributed to the fact that eq. (A-1) is based on assuming the same particle charge density over the cross section of the ESP.

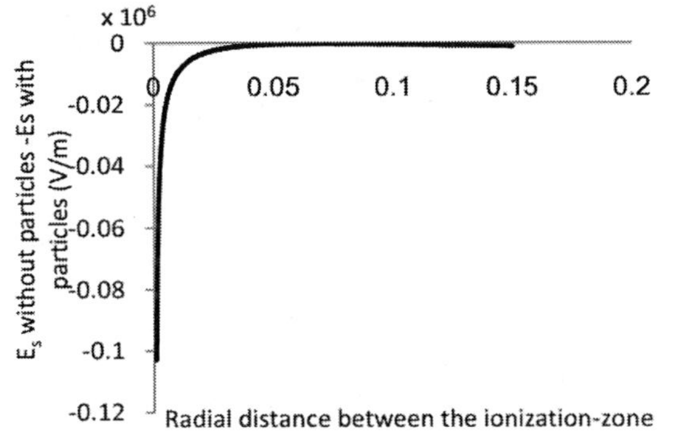

Figure 1: Excess of Es with particle loading over that without particle loading between the ionization-zone boundary and collecting cylinder. $r_o = 1$ mm, R = 15 cm, $V_o' = 30.659$ kV and V = 60 kV.

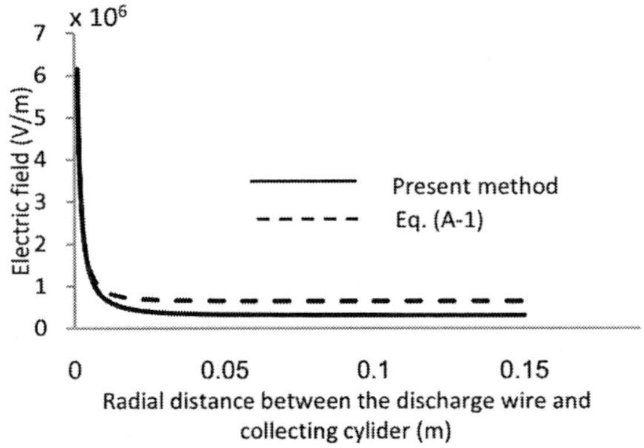

Figure 2: Calculated spatial distribution of the electric field E between the discharge wire and collecting cylinder compared with that obtained by eq. (A-1) for the ESP with particle loading. $r_o = 1$ mm, R = 15 cm, $V_o' = 30.659$ kV and V = 60 kV.

Figure (3) shows the spatial distribution of the ion charge density over the spacing between the wire and cylinder with and without particle loading [11]. The spatial distribution of the ρ_p in Fig. (3) follows that of the electric field which charges the particle. This does not support one of the assumptions proposed before [4] which consider no effect of particle loading on the ion charge density. With particle loading, it is worthy to notice in Fig. (3) that (i) ρ_p assumes values almost one order smaller in magnitude than those of ρ_i with particle loading in agreement with previous findings [1] and (ii) the sum $\rho_p + \rho_i$ with particle loading is almost equal to ρ_i without particle loading over the spacing between wire and cylinder. This is because the particles are charged at the expense of ions.

Figure (4) shows how the corona current increases with the applied voltage with and without particle loading. The figure shows a slight increase of the corona current value without particle loading over that with particle loading at the same voltage.

The decrease of the electric field within the ionization zone due to charged-particles field-shielding results in an increase of the onset voltage V_o' with the increase of particle radius r_p as given in Table (1). The corona I-V characteristics of the ESP at varying r_p for the same particle density N_p showed an increase of the corona current with the decrease of r_p due to decrease of V_o'. Such increase of the corona current reaches up to 10 % at applied voltage of about three times the corona onset voltage V_o'.

The decrease of the electric field within the ionization zone due to increase of N_p because of the above-mentioned field-shielding results in increase of the corona onset voltage V_o' as given in Table (2). The corona I-V characteristic of the ESP at varying particle concentration N_p for the same particle radius r_p showed a decrease of the corona current with the increase of the N_p because of the increase of V_o'.

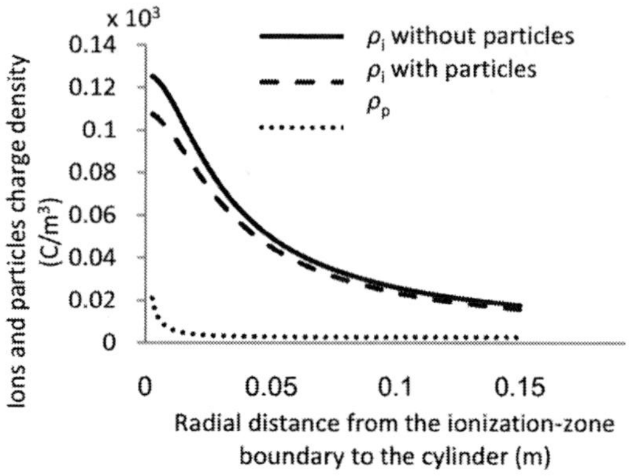

Figure 3: Calculated spatial distribution of the ion and particle charge density values between the discharge wire and collecting cylinder. r_o = 1 mm, R = 15 cm, V_o = 30.659 kV and V_o' = 31.0271 kV and V = 60 kV.

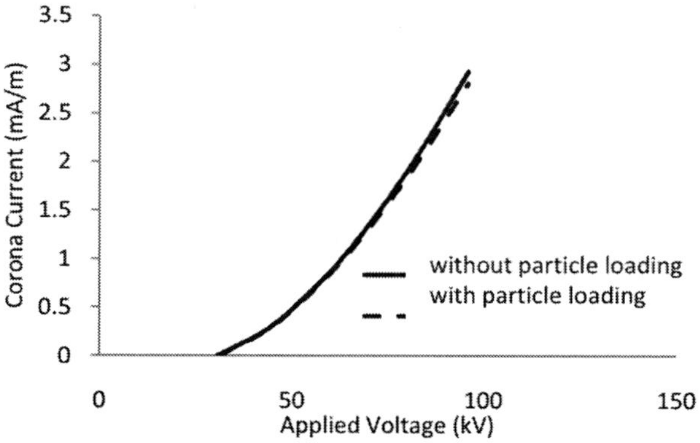

Figure 4: Calculated corona I-V characteristics of the ESP with and without particle loading. r_o = 1 mm, R = 15 cm, r_p = 30 µm, N_p = 0.094*10^9, V_o = 30.659 kV and V_o' = 31.0271 kV

Table 1: Calculated increase of the corona onset voltage V_o' with the increase of particle radius r_p at the same particle concentration N_p (=0.2*0.094*10^9).

r_p (µm)	10	30	50
V_o (kV)	30.6898	31.57891	33.72505

Table 2: Calculated increase of the corona onset voltage V_o' with the increase of particle concentration N_p at the same particle radius r_p (=30 µm).

r_p (µm)	10	30	50
V_o (kV)	30.6898	31.57891	33.72505

4 Conclusions

1. The ionized field in the ESP is mathematically modeled for calculating the spatial distribution of space charge density due to both the ions and charged particles as well as the components of electric field including the applied field and the field due to space charge.
2. The calculated spatial distribution of space charge density due to charged particles between the discharge wire and cylinder of the ESP with particle loading is one order higher than that due to ions.
3. The current-voltage characteristics are calculated for the ESP with particle loading at varying particle radius for the same particle concentration. The smaller the particle radius, the higher is the corona current at the same applied voltage.
4. The current-voltage characteristics are calculated for the ESP with particle loading at varying particle concentration for the same particle radius. The higher the particle concentration, the smaller is the corona current at the same applied voltage.
5. The calculated onset voltage of corona on the discharge wire of the ESP with particle loading increases with the increase of either the particle radius or the particle concentration.

Acknowledgment

This paper is extracted from a research project sponsored through joint innovative projects fund STDF (Egypt) – IRD (France), project STDF ID-5457. The authors acknowledge such sponsorship.

References

1. P. L. Levin and J. F. Hoburg, "Donor cell-finite element descriptions of wire-duct precipitator fields, charges, and efficiencies," IEEE Trans. Ind. Appl., **IA-26**, 4,(1990).
2. M. R. Talaie, M. Taheri, and J. Fathikaljahi, "A new method to evaluate the voltage–current characteristics applicable for a single-stage electrostatic precipitator," J. Electrostat., **53**, 3(2001).
3. M. R. Talaie, "Mathematical modeling of wire-duct single-stage electrostatic precipitators.," J. Hazard. Mater., **124**, 1–3, (2005).
4. S. Cristina and M. Feliziani, "Calculation of Ionized Fields in DC Electrostatic Precipitators in the Presence of Dust and Electric Wind", IEEE Transactions on Industry Applications, **IA-77**, 1446-1451(1995).
5. F.W. Peek, "Dielectric Phenomena in High-Voltage Engineering", New Work: McGraw-Hill, (1929).
6. G. Cooperman, "A New Current-Voltage Relation for Duct Precipitators Valid for Low and High Current Densities," IEEE Trans. Ind. Appl., **IA-17**, 2(1981).
7. A. D. Moore, "Electrostatics and its Applications", J. Wiley & Sons, New York, USA, 1973.
8. S. Abdel-Satar and H. Singer, "Electrical conditionsin -wire-duct electrostatic," Journal of Electrostatics, **26**, 1–20(1991).
9. M. Abdel-Salam, L. Fouad, S. El-Hazek and M. Abu-Hussin, "Wire-Tube Electrostatic Precipitator Cement Factories", Conference record - Industry Applications Socity Annual meeting, San Fransisco, USA, 1129-1135(1982).
10. M. Abdel-Salam and Z. Al-Hamouz, "A New Finite Element Solution of Ionized Field in Coaxial Cylindrical Geometry", J. Phys. D: Applied Physics, **25**, 1551-1555(1992).
11. M. Abdel-Salam, M. El-Mohandes and S, Kamal El-deen, "Corona Performance in Wire-Cylinder ESP withot Particle loading", Asia Conference on Power and Electrical Engineering (ACPEE 2016), Bangkok, Thailand, March 20-22, 2016.

Appendices

Appendix 1: Analytical expressions defining wire-cylinder ESP performance with particle loading:

1. Spatial distribution of electric field E:

It has been reported [7]that the spatial distribution of the electric field with space charge E (r) in wire-cylinder ESP is expressed as:

$$E(r) = \sqrt{\left[\left(\frac{r_o}{R}\right)^2\left(E_o^2 - \frac{I}{2\pi\epsilon_o K_i}\right) + \frac{I}{4\pi\epsilon_o K_i(\lambda Sr)^2}\right]e^{2\lambda Sr} - \frac{I}{4\pi\epsilon_o K_i}\left[\frac{2}{\lambda Sr} + \frac{1}{(\lambda Sr)^2}\right]} \quad \text{(A-1)}$$

2. Onset voltage of corona:

It has been reported [6, 7] that the corona onset voltage in wire-cylinder ESP increases with the presence of dust as expressed by:

$$V_o' = V_o + \frac{\rho_p}{4\epsilon_o}R^2 \quad \text{(A-2)}$$

where V_o and V_o' are the corona onset voltage values with and without particle loading.

The expressions (A-1) and (A-2) are based on assuming both particle ρ_p and ion ρ_i space charge density values are independent of position over the ESP cross-section which is not justified. This makes the formulation of these expressions are questionable [6, 7].

MATEC Web of Conferences **55**, 05004 (2016) DOI: 10.1051/matecconf/20165505004
ACPEE 2016

Research on Line Patrol Strategy of 110kV Transmission Line after Lightning Strike

LI Mingjun[1], SONG Kunyu[2,a], OU Yuqiang[1], GAN Tuanjie[1], FAN Yazhou[1] and CHEN Jianping[1]

[1]*Guangdong Power Grid Limited Liability Company, Jiangmen Power Supply Bureau, Jiangmen, 529000, China*
[2]*Wuhan Sunshine Power Science & Technology co., LTD, Wuhan, 430074, China*

Abstract. Lightning faults occupy in the majority of instantaneous fault and reclosing can usually be successful, so power supply can be restored without immediate patrol in many cases. Firstly, this paper introduces the lightning fault positioning and identifying method. Then test electrical performance of insulators after lightning strike from 110kV lines. Data shows that lightning strike has little effect on the electric performance of insulator. Finally, illustrating disposal process of the 110 kV transmission line after lightning fault, certifying that the power supply reliability be ensured without line patrol.

1 Introduction

Statistics show that lightning faults occupy more than 50% of the power system faults in general[1]-[4]. Frequent lightning outages bring operation and maintenance personnel heavy task of inspection[5]-[7]. According to operating experience, lightning failure is usually instantaneous fault and reclosing can generally be successful, so patrol is not required when some lightning fault occurs and power supply can be restored[8]-[9]. Therefore, precise positioning and identifying of lightning failure as well as assessing the electrical performance of insulators after strike by lightning can help to make more efficient transmission line strategy, which is important for the safe operation of transmission lines.

Paper[10]-[12] describes principle of fault positioning method basing on travelling wave. However, research on identification of fault is seldom reported.

In terms of insulators' electrical performance evaluation, paper[13] describes electrical performance of composite insulators in operation after several years. While research which focuses on performance of insulators after lightning is not very usual.

This paper firstly introduces basic principles of positioning and identification, analyzing the performance of insulator after lightning fault. Secondly, assesses the performance of insulators froma110kV transmission line after lightning strike, at last develop guideline patrol strategy for the operation and maintenance personnel.

2 Principle of Fault Positioning

[a]Corresponding author : 418668255@qq.com

When fault occurs, travelling wave will spread from lightning point to both ends of transmission, and the position of fault can be obtained through double terminal method which is as shown in Figure.1.

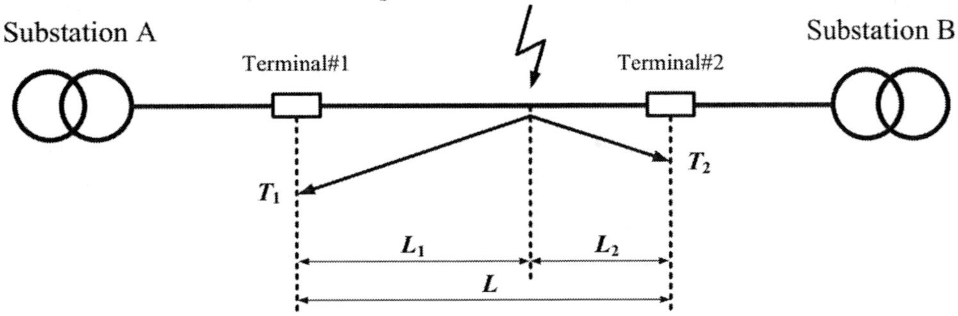

Figure 1. Double Terminal Method Schematic

In Figure.1, #1 and #2 are monitoring terminal devices, which are used to record travelling waveforms[14]. $T_1$and T_2refer to time that travelling wave arrives in #1 and #2 respectively, and L is the distance between #1 and #2 terminals, v is velocity of travelling wave. The distance of L_1 which is between fault point and #1 terminal can be obtained by formula (1), and L_2can be obtained by (2):

$$L_1 = (L + (T_1 - T_2)v)/2 \tag{1}$$

$$L_2 = (L - (T_1 - T_2)v)/2 \tag{2}$$

If there is only one monitoring terminal device on transmission line, we can also get the distance between lightning point and terminal through single terminal method[15], the principle of which is as shown in Figure.2.

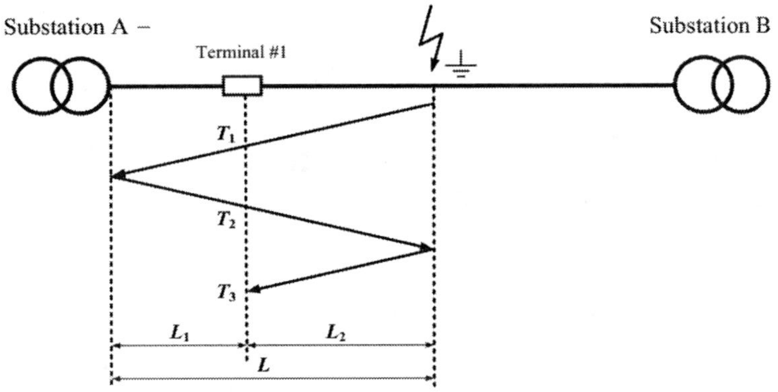

Figure 2. Single Terminal Method Schematic

When travelling wave reaches substation, it will be reflected by substation and then spreads to lightning point. When waveform reaches lightning point, then second reflection occurs and waveform spreads to both end of transmission line. In Figure.2, L_2can be given out by (3):

$$L_2 = \frac{(T_3 - T_2)v}{2} \tag{3}$$

3 Principle of Fault Recognition

In order to find out the difference between lightning and non-lightning travelling wave, distributed fault diagnosis devices installed on transmission lines are used to monitor kinds of fault for a long time. As a result, we have recorded and obtained many lightning and non-lightning travelling waveforms. Such waveforms are as shown as Figure.3.

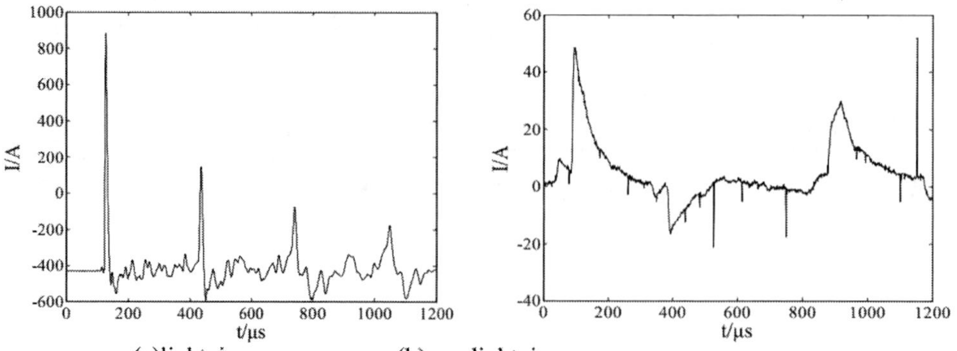

Figure 3. Lightning And Non-Lightning Travelling Wave

From Figure.3 we can see that lightning travelling wave is quite different from non-lightning travelling wave, such as large waveform amplitude, narrow pulse width, large steepness and so on. The tail time of lightning waveform tends to be less than 40μs, while the tail time of non-lightning wave is mostly more than 40μs, which can help to identify the lightning and non-lightning waveform. In addition, the fast Fourier transform (FFT)is adopted to convert the time domain waveform to the frequency domain[16]-[17]. And lightning and non-lightning travelling wave spectrum analysis results are as follows:

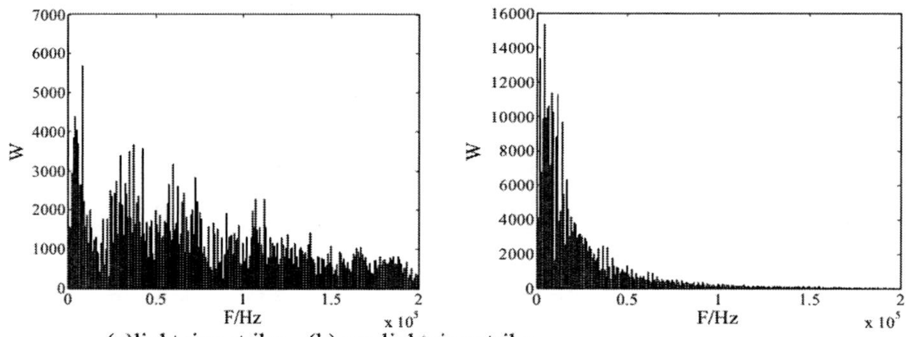

Figure 4. Frequency Graph Of Lightning And Non-Lighting Wave

The output of FFT results are discrete data points and each point corresponds to a frequency of the AC component. The energy of each point can be obtained by(4):

$$W(i) = |Y(i)|^2 \ (i = 0, \Delta f, 2\Delta f, 3\Delta f......) \tag{4}$$

$Y(i)$ is the modulus of point. $\Delta f = f_M / NFFT$, f_M is sampling rate.

The FFT results are fully symmetric, so the proportion of total energy between point 1 to i can be expressed by the following equation (5):

$$R(i) = \sum_{j=0}^{i} W(j) / \sum_{j=0}^{f_M/2} W(j) (i = 0, \Delta f, 2\Delta f, 3\Delta f ... f_M/2;) \tag{5}$$

Figure.4 shows that lightning travelling wave contains much high frequency component, which component of F>30kHzaccounts for a significant proportion, while component of F>30kHz for non-lightning travelling wave occupies very small.

In order to further analyze the numerical difference between the two values, the present paper analyzed more than 20 lightning travelling waves and 30 non-lightning travelling waves. The results show that R(0-30kHz) of lightning travelling wave is between 0.1 and 0.9, while R(0-30kHz) exceeds 0.93 for non-lightning travelling wave.

Overall, we can use these two methods to identify lightning and non-lightning travelling wave. Experience shows that recognizing accuracy ratio is over 95%.

4 Experimental Details

4.1 Experimental Equipment

4 glass insulators and 4 composite insulators are applied in the experiment, which are all under operation about 4 years. Before experiment, we number all these insulators, then 4 composite insulators are marked as 1, 2, 3, 4 and glass insulators marked as 5, 6, 7, 8. Insulator marked 1and 5 are normal, while other insulators have been struck by lightning.

A 1000kVA power frequency test transformer is adopted for the AC flashover experiment, which rated voltage and current is 1000kV and 1A respectively. While in the impulse flashover experiment we use a 2400kV/210kJ impulse generator, which waveform generated meet the requirements of IEC60060[18].

4.2 Test Results and Analysis

4.2.1 Lightning Impulse Experimental

Lightning flashover experiments are carried out under the prescribed procedures according to GB/T 775.2-2003[19], and each test has been repeated for 26~35 times. The lightning impulse $U_{50\%}$ is determined by formula (6)

$$U_{50\%} = \frac{\sum n_i U_i}{\sum n_i} \tag{6}$$

The test results are as shown in table 1:

Table 1. Insulator Lightning Impulse$U_{50\%}$

Type	Number	$U_{50\%}$/kV
Composite Insulator	1	798.6
	2	806.1
	3	792.7
	4	779.2
Glass Insulator	5	765.2
	6	766.6
	7	761.9
	8	771.0

Table 1shows that lightning impulse$U_{50\%}$of composite insulators change from 779.2 to 806.1 kV, while $U_{50\%}$of glass insulators are 765.2 to 771kV.

There are certain differences in $U_{50\%}$of different composite insulators and the deviation is within 2.4%compared with the #1 insulator, while $U_{50\%}$of #2 is higher than the normal insulator #1, so lightning fault has little effect on lightning impulse insulation level for composite insulator, and almost no effect on glass insulators.

4.2.2 Switching Impulse experimental

Test methods and data processing is similar to 3.2.1, the results are as shown in Table 2:

Table 2. Switching Impulse $U_{50\%}$

Type	Number	$U_{50\%}$/kV
Composite insulator	1	725.6
	2	727.9
	3	721.2
	4	712.5
Glass insulator	5	692.6
	6	689.7
	7	693.8
	8	696.5

Composite insulators switching impulse$U_{50\%}$are between 712.5 and 727.9kV,while $U_{50\%}$of glass insulators are 689.7 to 696.5kV. Compared with #1 insulator, deviation of $U_{50\%}$of #2~#4 is between -1.8%~0.3%, while -0.4%~0.4% for glass insulators.

4.2.3 Dry-Band Arcing Discharge Characteristic Experiment under Power Frequency

According to [18]-[19], dry-band arcing discharge voltage $U_d$of eight insulators are as shown in table 3.

Table 3. Dry-Band Arcing Discharge Voltage $U_d$under Power Frequency

Type	Number	U_d/kV
Composite insulator	1	417.5
	2	414.2
	3	396.1
	4	401.3
Glass insulator	5	382.4
	6	386.0
	7	371.5
	8	361.7

We can see from Table 3 that lightning fault has little effect on dry-band arcing discharge voltage of composite insulators and glass insulators. Compared with the normal case, the maximum deviation of U_d of composite insulators is -3.8%, while it is-5.5% for glass insulators. The value of U_d of insulators is all above 300kV, which is much higher than actual operating voltage of the 110kV transmission line.

4.2.4 Wet-Band Arcing Discharge Characteristic Experiment under Power Frequency

Vertical component of artificial rain is 1.4mm/min, the horizontal component is 1.3mm / min, rainfall temperature is 15℃, rain resistance is 114Ω·m, the experimental results are as shown in Table 4[18]-[19].

Table 4. Wet-Band Arcing Discharge Voltage U_w Under Power Frequency

Type	Number	U_w/kV
Composite Insulator	1	344.1
	2	352.2

	3	329.9
	4	333.4
Glass Insulator	5	317.5
	6	321.4
	7	306.5
	8	311.7

As it can be seen from Table 4, wet-band arcing discharge voltage of each insulator in power frequency is lower than its dry-band arcing discharge voltage. U_w of Composite Insulators is between 329.9 and 352.2kV, while glass insulators is 306.5~321.4 kV. All are above300kV.Therefore, wet-band arcing discharge will not happen on 110kV transmission line even though lightning strike occurs.

Synthesizing table 1-4 data and analysis, we can conclude that insulation performance of 110kV glass insulators and composite insulators changes little after lightning strike. Even though lightning fault occurs, such insulators still remains a high degree of insulation margin, which may still continue operating for a while longer.

5 Lightning Fault Line Patrol Strategy and Practical Case

A diagnosis case of 110kV lightning fault is as follows. Fault happened on a 110kV transmission line, and Figure.5(a) is the fault travelling wave, which FFT frequency graph is shown as Figure.5(b).

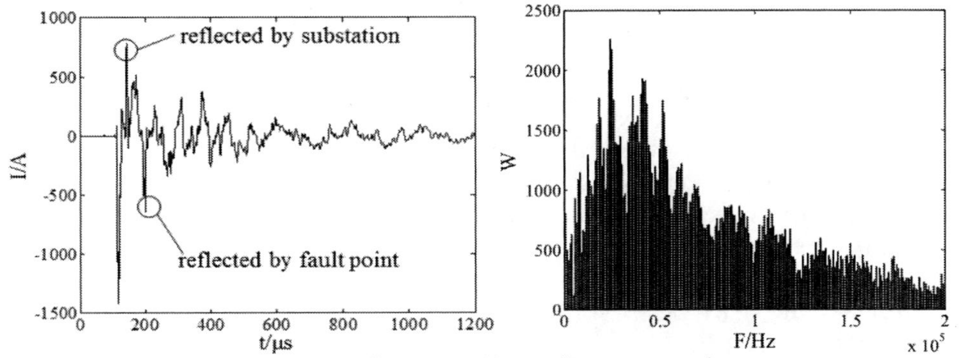

(a) travelling wave(b) FFT frequency graph

Figure 5. Travelling Wave and Its FFT Frequency Graph

The tail time of the fault travelling wave of Figure.5 (a) is 17μs, and R(0-30kHz)=0.45, so it can be concluded that the fault is lightning fault.

From Figure.5(a), T_3-T_2=71.2μs. Take the velocity of travelling wave as v=290m/μs. Single terminal method is used to determine the distance between terminal and fault point, $L_2 = 71.2 \times 290 / 2 = 10324m$, and it may be easily found out that the fault point is located in #27 tower A phase, which insulator is the same as #1~#4 in the experiment. From the experiment results mentioned above, lightning strike has little influence on electric performance of 110kV composite insulator, and it reclosed successfully, so it doesn't need line patrol. The following running experience shows that this insulator runs well, which has proved that the line patrol strategy is correct.

6 Conclusion

(1)Compared with non-lightning fault travelling wave, lightning fault travelling wave has characteristics such as a large component of high frequency components, tail time <40μs, and R(0-30kHz) between 0.1 and 0.9, which can be used to distinguish lightning and non-lightning fault.

(2)Experimental data shows that lightning strike has little effect on insulation performance of composite insulator and glass insulator. The actual operation result proves the patrol strategy is correct. Therefore, it is not necessary to immediately patrol for most transmission line after lightning failures.
(3) To make the results more convincing, more insulators after lightning fault need to be collected and tested. Research needs to be put forward further. The present study can provide some certain guidance for the operation and maintenance of transmission lines.

References

1. A. T. Johns and S. K. Salman, *Digital Protection for Power Systems*,1st ed. London, U.K.: Peregrinus, 1997.
2. H. Suad, G. Ferdinand, and F. G. Andrej, "Prediction of power systemsecurity levels," *IEEE Trans Power Syst*, vol. 24, no. 1, pp. 368–377,Feb. 200
3. *IEEE Standard Dictionary of Electrical and Electronics Terms*, ANSI/IEEE Standard 100, 1996.
4. A. A. M. Zin and S. P. A. Karim, "Protection system analysis usingfault signatures in Malaysia," *Int J Elect. Power Energy Syst.*, vol. 45,pp. 194–205, 2013.
5. G. Diendorfer and W. Shulz, "Ground flash density and lightning exposureof power transmission lines," presented at the IEEE Power TechConf., Bologna, Italy, 2003.
6. G. Andersson, P. Donalek, R. Farmer, N. Hatziargyriou, I. Kamwa,P. Kundur, N. Martins, J. Paserba, P. Pourbeik, J. Sanchez-Gasca, R.Schulz, and A. Stankovic, "Causes of the 2003 major grid blackoutsin North America and Europe andrecommended means to improvesystem dynamic performance," *IEEE Trans Power Syst*, vol. 20, no. 4,pp. 1922–1928, Nov.2005.
7. EventsAnalysis: System Disturbance Reports NorthAmerican ElectricReliability Corporation (NERC), USA, 1992–2010.
8. M. Evan, "Extreme grid disruptions and extreme weather," presentedat the U.S. Disaster Reanalysis Workshop, Asheville, NC, USA, May3–4, 2012.
9. J. O. Adepitan and E. O. Oladiran, "Analysis of the dependence ofpower outages on lightning events within the ijebu province, Nigeria,"*J. Environ. Earth Sci.*, pp. 850–856, 2012.
10. Magnago F H, Abur A. Fault location using wavelets[J]. IEEE Transactions on Power Delivery, 1998, 13(4): 1475-1480.
11. Y. F. Li and P. Chen, "The Principle of the Modem Travelling Wave Based on Fault Location and the Applications to Actual Fault Analysis," in Proc. 2009 Asia-Pacific Power and Energy Engineering Conference.
12. F. H. Magnago and A. Abur, " Fault location using wavelets," IEEE Trans. Power Delivery, vo13, pp. 1475-1480, Oct. 1998.
13. de T ourrei l C , Roberge R, Bourdon P .Long-Term Mechanical Properties of High Voltage Composite Insulators [J] .IEEE Transactions on Power Apparatus and Systems , 1985 , 104 (10):2918-2921 .
14. M. Vitins, "A Correlation Method for Transmission Line Protection," IEEE Trans. Power Apparatus and Systems, vol. PAS-97, pp. 1607-1617, Sep. 1978.
15. L. Jie, S. Elangovan and 1. B. X. Devotta, "Adaptive travelling wave protection algorithm using two correlation functions," IEEE Trans Power Delivery, vol. 14, pp. 126-131, Jan. 1999.
16. Y. Suzuki, Toshio Sone, and K. Kido, "A new FFT algorithm of radix 3,6, and 12," Acoustics, Speech and Signal Processing, IEEE Transactions on, vol. 34, no.2, pp. 380–383, Apr 1986.
17. T. Patyk, D. Guevorkian, T. Pitkanen, P. Jaaskelainen, and J. Takala, "Low-power application-specific FFT processor for LTE applications," in Embedded Computer Systems: Architectures, Modeling, and Simulation(SAMOS XIII), 2013 International Conference on, July 2013, pp. 28–32.
18. IEC 60060-1-1989, High-voltage test techniques Part 1 : general definitions and test requirements[S].
19. GB/T 775.2-2003, Test method for insulators--Part 2: Electrical test methods

Modeling and Simulation of Electrical Resonance in EHV Transmission Line Case Study in West Java Region 500 kV System

Putu Agus Aditya Pramana[1,a], Buyung Sofiarto Munir[1]

[1]*PLN Indonesia, Transmission and Distribution Research Group, Jakarta, Indonesia*

Abstract. Electrical transmission line can be approached by the second order system with the model of capacitive, inductive, and resistive elements. In the system, the capacitive element is represented by the capacitance value between phases and phase to ground. In the other side, inductance of reactor and phase conductor represent the inductance element, whereas resistance of the phase conductor represents the resistance value of the system. As a second order system, the electrical transmission line is potentially experiencing the resonance phenomena if the natural frequency reaches the 50 Hz operation frequency. This paper explains the study of resonance phenomena in Extra High Voltage (EHV) 500 kV transmission line between two 500 kV Substation in West Java Region by the modeling and simulation. The results show that the transmission line is very vulnerable to the resonance phenomenon which is indicated by the existence of voltage up to 800 kV when this transmission line is not yet energized and causing one of the line arrester broken. The results also show that under the resonance phenomena, single pole auto reclose process is potentially failed. This study is expected to increase the concern for the resonance phenomena in the transmission lines.

1 Introduction

Substation B is a new 500 kV extra high voltage (EHV) substation which is located between substation A and the substation C. Substation A is separated about 304 km to the Substation C, whereas Substation B intersects the line about 105 km from Substation A, and the first transpose point is located about 101 km from Substation A. EHV Transmission lines from Substation A to Substation C consist of two lines, line I and line II as shown in figure 1(a). Most of tower design which is used in this transmission lines are shown in figure 1(b). From the figure 1(b) we can find the dimension of the tower, the distance between the phase conductors, and the distance of phase conductors to the ground. Substation B is an EHV Substation with a 1½ breaker system. Currently, Substation B intersects transmission line I from Substation A to Substation C only, while the transmission line II from Substation A to Substation C just pass through Substation B.

Prior to the energize process in substation B, Earthing Switch (ES) in Substation B still on the closed position, a moment later the ES was opened and then the disconnecting switch (DS) in both side Substation B and Substation A were closed. In this condition, the reactors in line I Substation A were still opened and the induction voltages on line I Substation B arose which have value about 28

[a] Corresponding author : putu.agus.tf@gmail.com

kV for phase R-N, 19 kV for phase S-N, and 12 kV for phase T-N. On the further step, the reactors for line I in Substation A were closed for each phase and unexpectedly the voltages for each phase in this line increased significantly to the 261 kV for phase R-N, 44 kV for phase S-N, and 235 kV for phase T-N, moreover for a moment the voltage phase value reached 800 kV which is suspected due the electrical resonance in EHV transmission line I Substation A to Substation B.

Related to this study, there are some researchers who have been published the paper about electrical transmission line, electromagnetic and resonance phenomena. [1] explained about the transient process on reactors with controlled and uncontrolled switching process. They did the calculation process and propose the mitigation for measures and operating switching strategy of variable shunt reactor. [2] described about the ferroresonance modelling on transformer. Based on EMTP simulation and field tests data, they explained that unload transformer possible to affect the ferroresonance. [3] described the methods to find the capacitance and inductance value for transmission line by the interactions of each phase conductor to the others. They found that both results from finite element analysis and analytical methods are within reasonable agreement to each other. [4] explained about the electrical resonance due to Power Factor correction (PFC)/VAR compensation which affects the harmonic value of the electrical wave. [5] studied the harmonics and resonances within the wind power plants which included an introduction to series and parallel resonances, frequency scan analysis, and the harmonic source characteristics of wind turbine generator and of utility interconnections. [6] described about the ferroresonance in distribution system. [7] studied the transients of reactor when energized and de- energized process and found that reactor imposed heavy duty on circuit breaker. [8] described the process of electrical resonance in the transmission line and how to solve the problem. [9] studied the ferroresonance in voltage transformer and found that instant switching and capacitance value affect the ferroresonance.

This study focused on electrical resonance phenomena for the EHV Transmission line from Substation A to Substation B which is include the modelling and simulation, and then describe the electrical resonance effect on the auto reclose process.

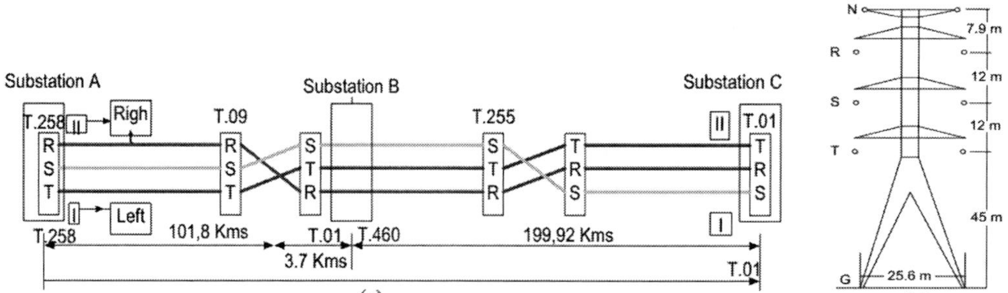

Figure 1. (a) Description of EHV Transmission Line from Substation A to Substation C, (b) Tower model for EHV Transmission line from Substation A to Substation B

2 Modeling

EHV Transmission line from Substation A to Substation B can be modelled using the capacitive, inductive and resistive elements as second order system that has natural frequency characteristic. The primary capacitance element of the transmission line model is the capacitance between the phases conductors and the capacitance between the phases to the ground. The primary inductance element for the model is the reactors for each phase which is located on Substation A and the inductance for the phase conductor, while the primary resistance element is the resistance of the phase conductor.

2.1 Model Assumptions

The modelling process has some assumptions to study the electrical resonance effect in the EHV transmission line. The assumptions are the EHV transmission lines from Substation A to Substation B assumed has the horizontal contour therefore the line tower can be modelled as figure 1(b), the effect conductor sagging is neglected, modelling process limited to the EHV transmission line Substation A to Substation B with type pi model, and the last one the transpose effect is included in the model.

2.2 Model Properties

The model described the two lines from Substation A to Substation B which each line consists of 3 phase and 4 wire transmission system. Interaction of each conductor phase and the conductor phase to the ground were modelled as capacitance values which given by the Equation 1 [3]. Equation 1 explained that the capacitance value of transmission system is affected by the dimension of the conductor phase, and the separated distance of the conductors. In the equations (1) to (3), C is capacitance per meter (F/m), d is the distance between conductors, r is the radius of conductor, ε is the air permittivity, L is inductance per meter (H/m), D is distance between conductor, *R* is resistance (Ohm), R is first phase, S is second phase, T is third phase, A is conductor area, ρ is resistivity, 1 is line I, 2 is line II, l is conductor length, and r is conductor radius.

$$C = \frac{\pi\varepsilon}{\ln\left(\frac{d}{r}\right)} \quad (1)$$

$$L = 2\,x\,10^{-7}\ln\left[\sqrt[3]{D_{RS'}D_{RT'}D_{ST'}}\Big/\left(\sqrt{re^{-\frac{1}{4}}}.\sqrt[6]{D_{R1R2}D_{S1S2}D_{T1T2}}\right)\right] \quad (2)$$

$$D_{RS'} = \sqrt[4]{D_{R1S1}D_{R1S2}D_{R2S1}D_{R2S2}},\ D_{RT'} = \sqrt[4]{D_{S1T1}D_{S1T2}D_{S2T1}D_{S2T2}},\ D_{ST'} = \sqrt[4]{D_{R1T1}D_{R1T2}D_{R2T1}D_{R2T2}} \quad (3)$$

$$R = \rho\frac{l}{A} \quad (4)$$

Table 1. Capacitance Model Value

Line	Capacitance values from Substation A to first transpose (uF)	Capacitance values from first transpose to Substation B (uF)	Line	Capacitance values from Substation A to first transpose	Capacitance values from first transpose to Substation B (uF)
Phase T1-S1	0.1513	0.0047	Phase R1-T2	0.1289	0.0040
Phase S1-R1	0.1513	0.0047	Phase R1-S2	0.1329	0.0041
Phase R1-T1	0.1361	0.0042	Phase R1-R2	0.1898	0.0049
Phase R1-G	0.1084	0.0033	N1- Phase-R1	0.1449	0.0045
Phase S1-G	0.1510	0.0034	N1- Phase-S1	0.1189	0.0037
Phase T1-G	0.1814	0.0035	N1- Phase-T1	0.1154	0.0035
Phase T2-S2	0.1513	0.0047	N1- Phase-G	0.1541	0.0047
Phase S2-R2	0.1513	0.0047	N1- Phase-N2	0.1310	0.0062
Phase R2-T2	0.1361	0.0042	N1- Phase-R2	0.1290	0.0040
Phase R2-G	0.1084	0.0033	N1- Phase-S2	0.1173	0.0036
Phase S2-G	0.1510	0.0034	N1- Phase-T2	0.1143	0.0035
Phase T2-G	0.1814	0.0035	N2- Phase-R2	0.1449	0.0045
Phase T1-T2	0.1898	0.0049	N2- Phase-S2	0.1189	0.0037
Phase T1-S2	0.1329	0.0041	N2- Phase-T2	0.1154	0.0035
Phase T1-R2	0.1289	0.0040	N2- Phase-G	0.1541	0.0047
Phase S1-T2	0.1329	0.0041	N2- Phase-R1	0.1290	0.0040
Phase S1-S2	0.1898	0.0049	N2- Phase-S1	0.1173	0.0036
Phase S1-R2	0.1329	0.0041	N2- Phase-T1	0.1143	0.0035

By the the results of Equation 1, we can find the values of each capacitance element as given in Table 1. "R,S,T,N" indicated the conductor phase, whereas "1,2" indicated the line number. In example, phase T1-S1 indicated the capacitance value between phase T in line I and phase S in line I, Phase T2-S2 indicated the capacitance value between phase T in line II and phase S in line II, and so on. By the Table 1, the total values for capacitance in phase R1, S1, T1 respectively about 1.2213 μF, 1.1454 μF, 1.1501 μF. Besides of the capacitance values, the inductance value for each phase in the model is defined by the Equation 2 and 3. By these equations we can find that the inductance value of each phase is about 73 mH. On the other hand, the resistance value of each phase in the model is about 0.35 ohm which is defined by the Equation 4. Definition of symbols are given as follows

2.3 Model Validation

The model validation based on the voltage response from the simulation results to the real voltage response for each phase in line I and was performed when the DS were disconnected and the reactors in Substation A were opened. Figure 2(a) shows the validation result. The entire graphical image of this paper refers to system responses of voltage (kV or MV) versus time (s).

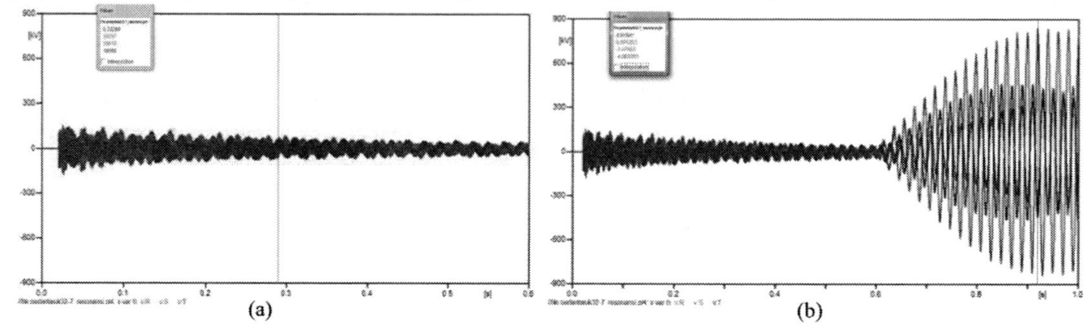

(a) (b)

Figure 2. (a)Voltage response of each phase in line I Substation B based on model simulation when the reactors on Substation A is opened, (b) Simulation result of voltage value on Line 1 Substation B when the reactor is closed

On this validation, the voltage phase values of the simulation are 36.2 kV, 29.5 kV, 18.9 kV for phases R1-N, S1-N, T1-N respectively. On the other hand the voltage values based on the real condition are 39.5 kV, 26.8 kV, 16.9 kV respectively for phases R1-N, S1-N, T1-N. By these results we can find that the voltage response of model reaches the voltage value on the real condition, and then we can use the model for the further simulation.

3 Simulations

3.1 Simulation when ES is Open, DS Closed, and Reactor in Substation A is Closed

The simulation for voltage response when the reactor is closed is given by the figure 2(b). Reactive power (MVAR) value of the reactor is 100 MVAR for all the three phase therefore each phase has about 33.3 MVAR. By the MVAR value, the inductance per phase of the reactor can be determined, which each value is about 7910 mH and 8320 mH for neutral to ground. On this simulation, the voltage response for each phase is captured when the reactors in Substation A was closed. The voltage phase values of the simulation are 809 kV, 347 kV, 363 kV for phases R1-N, S1-N, T1-N respectively. On the other hand the voltage values based on the real condition are 369 kV, 62 kV, 332 kV respectively for phases R1-N, S1-N, T1-N. The results show that there are significant difference between the simulation and the real condition. Simulation results tend to higher then the value of the real condition because the real value is captured only on a moment data which can change if the captured time changed. However, even if the simulation value is greater, the model can shows that the reactor can affect the voltage response of each phase. The entry of reactor make the natural frequency of model reaches the operation frequency which is 50Hz. The significant increase of voltage values especially for phase R can be indicated as the cause of the broken line arrester.

3.2 Simulation of Voltage Responses Related To the Single Pole Auto Reclose

These simulations described the voltage value of the line if the transmission line facing the disturbance and try to performing reclose procedure. These simulations show the study of the impact of the reactor in Substation A. Figure 3(a) shows the voltage response when the reactor is opened and figure 3(b) shows the voltage response when the reactor closed.

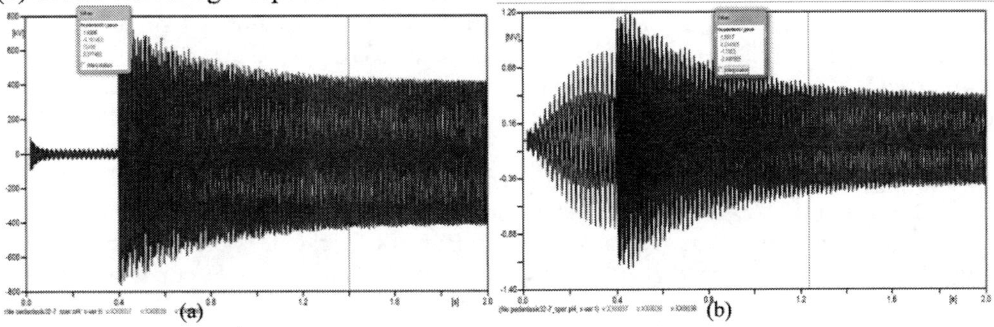

Figure 3. (a) Voltage Response on Line I Substation B without Reactor on Substation A, (b) Voltage Response on Line I Substation B with Reactor on Substation A

By the figure 3(a), the maximum voltage values for the R-N, S-N, T-N on the transients process respectively about 717 kV, 664 kV, 658 kV. But, for the steady state, the maximum voltage values for phase R-N, S-N, T-N is about 443 kV, 443 kV, and 443 kV. Figure 3(b) shows that the maximum voltage value for the R-N, S-N, T-N on the transients process respectively about 1126 kV, 684 kV, 1180 kV. But, for the steady state, the maximum voltage value for phase R-N, S-N, T-N is about 423 kV, 454 kV, and 477 kV.

4 Results

4.1 Resonance Phenomenon

The entry of reactors in the Substation A can be the cause of the electrical resonance phenomenon which is related to the values of the reactors in the Substation A. The existing reactors in Substation A have been designed to compensate the capacitive value along the EHV transmission line from Substation A to Substation C which is about 304 km. But, when the Substation B as a new substation intercepts the transmission line I of Substation A to Substation C in 105 km from Substation A, the inductive reactance value for each phase 7910 mH reactor is about 2484 Ω, where as the total reactive capacitance for phase R1 (C=1.2213 μF) is about 2608 Ω, for phase S1 (C=1.1454 μF) is about 2781 Ω, and for phase T1 (C=1.1501 μF) is about 2769 Ω. These values show that the capacitive reactance value of each phase in the transmission line from Substation A to Substation B reaches the inductive reactance values of this system which can cause the electrical resonance phenomenon.

This electrical resonance increases the voltage value for each phase of the transmission line especially for phase R which is the voltage maximum value increased until the 809 kV. The voltage exceeds the voltage value of the line arrester in Substation B which has the nominal voltage about 444 kV. Since this voltage needs to be neutralized to the ground through the line arrester, the phase R line arrester in Substation B experienced the excessive heating process until it broken.

4.2 Single Pole Auto Reclose (SPAR)

The entry of reactor with inductance 7910 mH per phase and 8340 mH neutral to ground in Substation A affect the voltage values in line I at Substation B. The maximum voltage value for each phase increase significantly especially for phase T-N increased until 1180 kV.

Maximum voltage value for steady state condition when the reactor in Substation A is opened is about 443 kV for phase R-N, 443 kV for phase S-N, 443 kV for phase T-N. The steady state value is defined as the time about 0.9 s after the disturbance. However, if the reactor is closed in Substation A, the maximum voltage value in the steady state condition is about 423 kV for phase R-N, 454 kV for phase S-N, 477 kV for phase T-N.

If we compare to the operation voltage value at 500 kV line to line (or maximum 408 kV line to neutral), the difference voltage value (at Substation A and Substation B) for each phase without reactor is stay in the normal operation range for the auto reclose procedure. But in the other hand, the entry of reactor makes the maximum difference voltage value exceed the operation range for auto reclose. Based on the data setting of the protection relay, the maximum acceptable voltage different value in both side Substation A and Substation B is about 10% from the nominal value. Therefore for the 500 kV operation voltages, the maximum acceptable for voltage difference value is about 50 kV. Based on these results, the EHV transmission line from Substation A to Substation B is potential to success in reclose process if the reactor is opened. But, if the reactor is closed, the reclose process is potentially failed. Besides on that, the entry of rector in Substation A potentially broke the line arrester, because there is a phase with transient voltage until 1180 kV which exceeds the nominal operation voltage for line arrester.

5 Conclusions and Recommendation

The model and simulation for the EHV transmission line from Substation A to Substation B is able to demonstrate the electrical resonance phenomena which significantly increase the voltage values for each phase in de-energized line. The resonance is triggered by the presence of reactor in Substation A and the capacitive and inductive reactance along the transmission line from Substation A to Substation B which match the resonance frequency of this system. Furthermore, electrical resonance phenomenon will potentially fail the auto recloses procedure which needs recalculation and resetting of reactor parameter to avoid this phenomenon. As recommendation for the further work, recalculation of the inductance value of the reactor is a key important to avoid the electrical resonance on the transmission line.

References

1. B. Filipović-Grčić, A. Župan, I. Uglešić, D. Mihalic, D. Filipović-Grčić, "Transients Caused by Switching of 420 kV Three-Phase Variable Shunt Reactor", Paper submitted to the International Conference on Power Systems Transients (IPST2015) in Cavtat, Croatia June 15-18, 2015.
2. D. A. N. Jacobson, L. Marti, R. W. Menzies, "Modeling Ferroresonance in a 230 kV Transformer-Terminated Double-Circuit Transmission Line", IPST '99 International Conference on Power Systems Transients, June 20-24, 1999, Budapest – Hungary.
3. Hazlee Azil Illias, Ab Halim Abu Bakar, Hazlie Mokhlis, Syahirah Abd Halim, "Calculation of Inductance and Capacitance in Power System Transmission Lines Using Finite Element Analysis Method", PRZEGLĄD ELEKTROTECHNICZNY (Electrical Review), ISSN 0033-2097, R. 88 NR 10a/2012, 2012.
4. Kamlesh K. Wadhwani, "Simulation of Parallel Resonance Condition in Electrical Network Using EDSA", International Conference on, Computer Modeling and Simulation, pp. 708-713, 2008
5. M. Bradt, B. Badrzadeh, E. Camm, D. Mueller, J. Schoene, T. Siebert, T. Smith, M. Starke, R. Walling, "Harmonics and Resonance Issues in Wind Power Plants", Technical Paper.
6. Roger C. Dugan, "Examples of Ferroresonance in Distribution Systems", Technical Paper.
7. Salil S. Sabade, Himanshu J. Bahirat, Mythili Chaganti, Bruce A. Mork, "Shunt Reactor Switching Transients at High Compensation Levels", Technical Paper.
8. Tugay, Y, "10th International Conference on Electrical Power Quality and Utilisation, 2009. EPQU 2009", 15-17 Sept. 2009, Lodz.
9. V. Valverde, A.J. Mazón, I. Zamora, G. Buigues, "Ferroresonance in Voltage Transformers: Analysis and Simulations", Technical Paper.

MATEC Web of Conferences **55**, 06001 (2016) DOI: 10.1051/matecconf/20165506001
ACPEE 2016

Review of the Optimal Design on a Hybrid Renewable Energy System

Yuan-Kang Wu[1,a] and Shih-Ming Chang[1]

[1]*Department of Electrical Engineering, National Chung-Cheng University, 62102 Chiayi, Taiwan*

Abstract. Hybrid renewable energy systems, combining various kinds of technologies, have shown relatively high capabilities to solve reliability problems and have reduced cost challenges. The use of hybrid electricity generation/storage technologies is reasonable to overcome related shortcomings. While the hybrid renewable energy system is attractive, its design, specifically the determination of the size of PV, wind, and diesel power generators and the size of energy storage system in each power station, is very challenging. Therefore, this paper will focus on the system planning and operation of hybrid generation systems, and several corresponding topics and papers by using intelligent computing methods will be reviewed. They include typical case studies, modeling and system simulation, control and management, reliability and economic studies, and optimal design on a reliable hybrid generation system.

1 Introduction

The limited reserves of fuel oils and their unstable prices have significantly increased the interest in renewable energy sources. The alternative energy sources are non-polluting, free in their availability, and continuous. The motivation behind the use of renewable energy sources is the reduction of CO_2 emissions. This is especially true in isolated, standalone, small islands where the access to renewable energy sources is the only solution to meet their energy needs. Today, wind and PV generators are utilized in such applications as water pumping, lighting, electrification of remote areas and telecommunications. For remote systems like radio telecommunication, satellite earth stations, or at sites that are far away from a conventional power system, the hybrid generation systems have been considered as preferred [1]. At present, standalone solar photovoltaic and wind systems have been promoted on a comparatively larger scale; however, these independent systems cannot provide continuous source of energy. Therefore, energy storage systems will be required for each of these systems in order to satisfy the power demands. Usually storage system is expensive and the size has to be reduced to a minimum possible for the renewable energy system to be cost effective. Hybrid power systems can be used to reduce energy storage requirements [2].

In the world, a significant number of villages may never be connected to the national grid due to their remoteness, or communities are far from the conventional electrical grid due to natural obstacles and environmental constraints. For the people living in these communities, access to renewable energy sources may be the only solution to meet their energy needs. Therefore, the design and operational

[a] Corresponding author : allenwu@ccu.edu.tw

experience of hybrid power systems for small micro grids are important. In order to meet sustained load demands during varying natural weather conditions, different renewable energy sources need to be integrated with each other. Therefore, dynamic modeling of various components in isolated systems and steady-state/transient simulation works are required. Furthermore, an overall control strategy for power management among different energy sources in a multisource energy system is needed. The controller should determine the online operation modes for each generation subsystems and switch from power regulation to maximum power conversion. Reliability evaluation and economic analysis are also very important in the planning of hybrid energy system for a particular site. In order to efficiently and economically utilize the renewable energy resources, one optimum sizing method considering system reliability and economical benefit for the hybrid generation system is necessary. The optimum sizing method can help to guarantee the lowest investment with full use of the PV array, wind turbine and energy storage system, so that the hybrid system can work at the optimum conditions in terms of investment and system reliability. Different sizing methods, such as probabilistic approach, iterative approach and artificial intelligence method, can be applied to reach a techno-economically optimum hybrid renewable energy system. Whichever sizing and optimization technique are used, they must ultimately search for an optimum combination of the system reliability and cost.

This paper will concentrate on reviewing significant research issues and current state about the hybrid renewable generation system. These topics include several typical case studies of renewable-based micro-grids, modelling and simulation for renewable energy systems, control and management of hybrid renewable energy systems, reliability and economic studies, and optimal design of a reliable hybrid generation system. A number of intelligent computing methods have been utilized in these researches and most of them will be summarized in this paper.

2 Case Studies on Small-Scale Hybrid Generation Systems

In the beginning of designing a small-scale hybrid generation system, the historical power consumption of load, load pattern, solar radiation values, and wind character have to be investigated. The availability of renewable energy resources is an important factor to develop the hybrid power system; that is, weather data are important factor for pre-feasibility study of renewable hybrid energy system for any particular site. For example, hourly solar radiation, clearness index, wind speed, and wind direction for several years should be collected so that the long-term average annual resource can be evaluated. Moreover, from the above given data, the probability functions of wind speed and solar radiation can be obtained. In [3], the design and sizing of hybrid power system for small base transceiver stations (BTS) in Indonesia are presented. Combination of wind and solar energy sources are used as the main source and utility line is used as a backup. The average wind speed in this area is about 4-5 m/s, and the average sunshine hour in this site is between 7 to 8 hours a day during the dry season and between 5 to 6 hours a day during the rainy season. The design of this hybrid generation system includes a 4.8kW PV array, 2.5kW wind turbine, and 1200Ah battery. It will provide electricity to a BTS with an average power demand of 4kW. By using this hybrid power system, about 90 percent of the required power can be supplied by renewable energy sources. Paper [4] proposes the most feasible configuration of a stand alone PV/Wind Hybrid Energy System with diesel generator as a backup for cellular mobile telephony base station site in isolated areas of Central India. In this case, the power requirements for the GSM load are about 2kW continuous, and the proposed hybrid system consists of two 7.5kW horizontal-axes type wind turbines, a 5kW Photovoltaic array, a 1kW diesel generator, 1.83kWh battery bank and 2kVA Inverter. Paper [5] proposes a photovoltaic/wind based hybrid power system connected to a common bus with battery storage and conventional backup source to supply a 2kW telecommunication load in Malaysia. The hybrid system consists of 4.5kW PV system, 2kW wind system, and 5kWh battery. Paper [6] also presents an innovative wind/PV/diesel hybrid system implemented in three remote islands in the Republic of Maldives. In this micro-grid system, twenty four 1.8 kW wind turbines are coupled to the micro-grid, and 2.5 kWp of amorphous silicon PV modules are connected to AC grid through single phase grid connected inverter. This paper

[6] undertakes renewable energy system planning using the software tool HOMER and to analyses the various options paying particular attention to the cost per unit of electricity consumed, fuel saved and initial capital requirements. In [7], a hybrid system of the pilot project in Lençóis Island located in the north area of Brazil is proposed. It composed of 21kW solar PV power, 22.5kW rated wind power, 120 batteries of 150Ah each, and a 53-kVA/48-kW diesel generator as a backup unit. This paper also addressed the field operation results for this system, including power quality and reliability analyses. In [8], several types of projects developed in different small islands and remote villages were reviewed. Reference [9] analyzed the potential implementation of hybrid photovoltaic (PV)/wind turbine/diesel system in southern city of Malaysia, in which seven different system configurations were studied. Reference [10] also investigated the potential operation of hybrid energy system in the northernmost city in Africa. In which there were several system configurations with and without battery storage elements were studied.

3 Modelling and Simulation of Renewable Energy Systems

In order to meet sustained load demands during varying natural conditions, different renewable energy sources need to be integrated with each other. In addition, dynamic modeling of various components of this isolated system should be presented. Several papers have described dynamic modeling and simulation results of a renewable energy based hybrid power system. For example, in [11], a detailed dynamic model and simulation of a solar cell/wind turbine/fuel cell/ultracapacitor hybrid power system is developed. In [12], a PV/Wind hybrid electric power system interface with electric utility for solving modeling and simulation problems by using Matlab/Simulink environment have been proposed. Solar panels are the medium to convert solar power into the electrical power. A number of solar cell models have been developed, but the one diode electrical equivalent circuit is commonly used for cell based or module based analysis. The current source generates the photo-current that is a function of the incident solar cell radiation and temperature [13-14]. Wind turbines are used to convert the wind power into electric power. Based on Betz theory, the maximum converting efficiency of a wind generation system is 59%. The developed wind power can be evaluated at a given wind speed through the power curve of the wind turbine. Power curves are usually provided by manufacturers of wind turbines. In terms of fuel cell, there are two typical types: low-temperature proton-exchange membrane FC (PEMFC) [15] and high-temperature solid oxide FC (SOFC) [16]. Both of them show great potential in hybrid energy system applications. The PEMFC model is an autonomous model operated under constant channel pressure with no control on the input fuel flow into the FC. The model was validated by experimental data measured from an Avista Labs SR-12 500 W PEMFC stack [15]. The FC will adjust the input fuel flow according to its load current to keep the channel pressure constant.

4 Control and Management of Hybrid Renewable Energy Systems

An overall control strategy for power management among different energy sources in a multisource energy system is needed. The objectives of the control and management of a hybrid energy system are to satisfy the changeable load power demand and to maintain the state of charge of the battery bank to prevent blackout and to extend the life of the batteries. For these purposes, the controller should determine the online operation modes for each generation subsystems and switch from power regulation to maximum power conversion. In [17], a comprehensive supervisor control for a hybrid system that comprises wind and photovoltaic generation subsystems, a battery bank, and an ac load is developed. Robust sliding-mode control laws have been considered in this paper to fulfill the different control objectives of the wind and solar subsystems. Paper [18] proposed an overall control strategy for a hybrid renewable energy system, in which the wind generation system is controlled by a pitch angle controller and a PV electricity generation unit is controlled by a maximum power point tracking (MPPT) controller; furthermore, dynamic models have been used for all the components of the system. At any given time, any excess wind and PV-generated power is supplied to the electrolyzer to generate hydrogen that is delivered to the hydrogen storage tanks. When there is a deficit in power generation, the fuel cell stack begins to produce energy for the load using hydrogen from the storage tanks. Paper [19] introduces the hardware realization of the energy management and control subsystem of a grid-connected wind/solar hybrid power system. The system is composed of several modules: programmable logic controller(FBs-40MAT from FATEK), AC multi-function electric power meters, Grid-connection control module, human-machine interface (HMI), dc electric power meters, and RS485/TCP converter to control and manage the operation of multi-source. Communication plays an important role in the system safe, reliable and precise operation for a hybrid renewable energy system. In [19], the communication protocol in the subsystem and between subsystems is Modbus RTU, while the communication with computer is implemented by RS485/TCP converters.

5 Reliability Studies of Hybrid Renewable Generation Systems

Reliability evaluation is very important in the planning of hybrid energy system for a particular site. The reliability is estimated in terms of the number of the critical load interruptions over a certain period of time; this choice is because the reliability of a system depends on the complexity of the system itself and not only on the number and the failure rates of its components. Typically, the reliability evaluation methods can be broadly classified into two types namely deterministic and probabilistic techniques[20]. Probabilistic techniques can be further classified into two types. They are analytical and Monte Carlo Simulation (MCS) methods. Several reliability indices are introduced in literature. Loss of Load Expected (LOLE), Loss of Energy Expected (LOEE) or Expected Energy not Supplied (EENS), Loss of Power Supply Probability (LPSP) and Equivalent Loss Factor (ELF) are some of the most common used indices in the reliability evaluation of generating systems. In general, the reliability assessment of hybrid generation system can be done by the following procedure:

- To collect required atmospheric data such as hourly solar radiation, hourly ambient temperature and hourly wind speed.
- To estimate hourly output power available from each type of source by using power generation models and considering outage schedules generated randomly using failure rates of units.
- To obtain total available generation with reliability index.

Reliability evaluation of PV/Wind hybrid energy system using well-being approach and Monte Carlo Simulation (MCS) method is discussed in [21], in which the "well-being" approach is implemented by combining deterministic and probabilistic approaches for applying to hybrid energy systems. Furthermore, the effects of different energy types, effects of changes in peak load and renewable capacity on system health and fuel savings are also studied in that paper. Paper[22] presented an evaluation model and applied it to analyze optimum generation expansion of small isolated systems using PV and wind energy sources. Paper [23] studied reliability and cost implications of PV and Wind Energy utilization in small isolated power systems. A Monte Carlo simulation approach has been utilized to incorporate the numerous random variables and their interactions. Paper [24] developed an analytical method for well-being assessment of small autonomous power systems having solar and wind energy systems.

6 Optimal Design of a Reliable Renewable Energy System

One of the most important issues in hybrid renewable systems is to optimally size the hybrid system components with possible minimum investment and operating costs. There are many studies about the optimization and sizing of hybrid renewable energy systems [25-29]. For example, References [25] reviewed different issues related to the modeling, analysis, optimal sizing, control and energy management strategies for hybrid system that are interconnected with wind power, solar photovoltaic and fuel cells. Reference [26] presented different sizing techniques and optimization techniques for off-grid PV–Wind hybrid systems. It also demonstrated that hybrid energy systems can significantly reduce the total life cycle cost of stand-alone power supplies and may constitute the most economical solution in many applications. Reference [27] designed and optimized (multi-objective optimization) the size of hybrid energy systems under smart grid theories that divide the load into high and low priorities. High priority load that should be supplied at any time but low priority load will be supplied only in case of the generation from wind/PV are greater than high priority load and the battery are full, otherwise this load can be shifted to next hour. Reference [28] applied the preference-inspired coevolutionary algorithm to the design of multi-objective hybrid renewable energy system. Three objectives, including the annualized cost of system, the loss of power supply probability and the fuel emissions, are minimized. Reference [29] modeled an optimal design of a PV/battery/micro-turbine hybrid energy system to supply power to remote communities by using genetic algorithm, in which the objective function minimizes the cost of energy production while covering the load demand with a specified value for the loss of load probability. Reference [30] designed techno-economically optimized systems by simulating behavior of various combinations of renewable energy systems with different sizing, including wind turbine (WT), photovoltaic (PV), fuel cell (FC), and battery banks.

7 Conclusions

The limited reserves of fuel oils and their unstable prices have significantly increased the interest in renewable energy sources. At a particular site, it is an important economical task for electrification of villages in rural areas or telecommunications. Therefore, the design of hybrid renewable power systems has received considerable attention in the last decade. In any hybrid generation system, sizing represents an important part of these system designs, including the optimal sizes of various renewable energy sources and storage systems. This paper has reviewed several typical case studies on hybrid renewable energy systems, including their modeling and simulation, control and management, reliability and economic studies, and the optimal size design by using intelligent computing methods.

References

1. C.H. Liu, K. T. Chau, and X.D. Zhang, "An Efficient Wind–Photovoltaic Hybrid Generation System Using Doubly Excited Permanent-Magnet Brushless Machine," *IEEE Trans. on Industrial Electronics*, vol.57, no.3, March 2010, pp.831-839
2. Francesco Bonanno, Alfio Consoli, Salvatore Lombardo and Angelo Raciti "A Logistical Model for Performance Evaluations of Hybrid Generation Systems," *IEEE TRANS. on Industry Applications*, vol.34, no.6, 1998, pp.1397-1403
3. P. A. Dahono, M. F. Salam, F. M. Falah, G. Yudha, Y. Marketatmo and S. Budiwibowo,"Design and Operational Experience of Powering Base Transceiver Station in Indonesia by Using a Hybrid Power System,"*Telecommunications Energy Conference,* Bandung, Bandung, Indonesia, 2009, pp.1-4.
4. P. Nema and S. Rangnekar,"Pre-feasibility study of PV-solar / wind hybrid energy system for GSM type mobile telephony base station in central India," *Computer and Automation Engineering*, Bhopal, India, 2010, pp.152-156.
5. D. B. Chia, B. L. Yong, R. K. Rajkumar, V. K. Ramachandaramurthy,"Design of a PV/Wind hybrid system for telecommunication load in Borneo region," *Environment and Electrical Engineering*, Malaysia, 2010, pp.73-76.
6. C. Nayar, M. Tang, W. Suponthana,"Wind/PV/Diesel Micro Grid System implemented in Remote Islands in the Republic of Maldives,"*Sustainable Energy Technologies*, Perth, WA, 2008, pp.1076-1080.
7. L. A. de Souza Ribeiro, O. R. Saavedra, S. L. de Lima, J. Gomes de Matos ,"Isolated Micro-Grids With Renewable Hybrid Generation: The Case of Lençóis Island,"*IEEE Trans. on Sustainable Energe*, vol. 2, no.1, January 2011,pp. 1-11.
8. Diana Neves, Carlos A. Silva and Stephen Connors, "Design and implementation of hybrid renewable energy systems on micro-communities: A review on case studies," *Renewable and Sustainable Energy Reviews*, vol. 31, 2014, pp.935-946.

9. Mei Shan Ngan and Chee Wei Tan, "Assessment of economic viability for PV/wind/diesel hybrid energy system in southern Peninsular Malaysia," *Renewable and Sustainable Energy Reviews*, vol. 16, no. 1, 2012, pp.634-647
10. Taher Maatallah, Nahed Ghodhbane and Sassi Ben Nasrallah, "Assessment viability for hybrid energy system (PV/wind/diesel) with storage in the northernmost city in Africa, Bizerte, Tunisia," *Renewable and Sustainable Energy Reviews*, vol. 59, 2016, pp.1639-1652.
11. H. C. Chen, J. C. Qiu and C. H. Liu, "Dynamic Modeling and Simulation of Renewable Energy Based Hybrid Power Systems,"*Electric Utility Deregulation and Restructuring and Power Technologies*, Taichung, 2008, pp.2803-2809.
12. H. H. El-Tamaly and A. A. E. Mohammed, "Modeling and Simulation of Photovoltaic/Wind Hybrid Electric Power System Interconnected with Electrical Utility,"*Power System Conference*, Middle-East, 2008, pp. 645-649.
13. M. J. Khan and M. T. Iqbal, "Dynamic Modeling and Simulation of a Small Wind-Fuel Cell Hybrid Energy System,"*Renewable Energy*, vol. 30, no. 3, March 2005, pp.421-439.
14. S. Arul Daniel and N. Ammasai Gounden, "A Novel Hybrid Isolated Generating System Based on PV Fed Inverter-Assisted Wind-Driven Induction Generators,"*IEEE Trans. on Energy Conversion*, vol. 19, no. 2, June 2004, pp.416-422.
15. C. Wang, M. H. Nehrir, and S. R. Shaw, "Dynamic models and model validation for PEM fuel cells using electrical circuits," *IEEE Trans. on Energy Conversion*, vol. 20, no. 2, Jun. 2005, pp.442-451.
16. C. Wang and M. H. Nehrir, "A physically-based dynamic model for solid oxide fuel cells," *IEEE Trans. on Energy Conversion*, vol. 22, no. 4, Dec. 2007, pp. 887-897.
17. F. Valenciaga and P. F. Puleston, "Supervisor Control for a Stand-Alone Hybrid Generation System Using Wind and Photovoltaic Energy,"*IEEE Trans.* on *Energy Conversion*, vol. 20, no. 2, June 2005, pp.398-405.
18. C. Wang and M. H. Nehrir,"Power management of a stand-alone wind/photovoltaic/fuel cell energy system," *IEEE Trans. on Energy Conversion*, vol. 23, no. 3, September 2008, pp. 957-967.
19. G. LI , Y. Chen and T. LI , "The Realization of Control Subsystem in the Energy Management of Wind/Solar Hybrid Power System," *Power Electronics Systems and Applications*, China, May 2009, pp.1-4.
20. Billinton. R and R. Allan, *Reliability evaluation of power systems*, 2nd Edition, Plenum Press, New York, 1996.
21. L. N. Kishore and E. Fernandez, "Reliability Well-Being Assessment of PV-Wind Hybrid System Using Monte Carlo Simulation," Roorkee, India, March 2011, pp.63-68.
22. R. Billinton and R. Karki, "Capacity expansion of small isolated power systems using PV and wind energy," IEEE *Trans. on Power Systems*, vol. 16, no. 4, Nov 2001, pp.892-897.
23. R. Karki and R. Billinton, "Reliability/cost implications of PV and wind energy utilization in small isolated power systems," *IEEE Trans. on Energy Conversion*, vol. 16, no. 4, Dec 2001, pp.368-373.
24. D. K. Khatod, V. Pant and J. Sharma, "Analytical Approach for Well-Being Assessment of Small Autonomous Power Systems With Solar and Wind Energy Sources," *IEEE Trans. on Energy Conversion*, vol. 25, no. 2, June 2010, pp.535-545.
25. O. Erdinc and M. Uzunoglu, "Optimum design of hybrid renewable energy systems: Overview of different approaches," *Renewable and Sustainable Energy Reviews*, vol. 16, no. 3, April 2012, pp.1412-1425.
26. Ahmed Said Al Busaidi, Hussein A Kazem, Abdullah H Al-Badi and Mohammad Farooq Khan, "A review of optimum sizing of hybrid PV–Wind renewable energy systems in oman," *Renewable and Sustainable Energy Reviews*, vol. 53, 2016, pp.185-193.
27. Ali M. Eltamaly, Mohamed A. Mohamed and Abdulrahman I. Alolah, "A novel smart grid theory for optimal sizing of hybrid renewable energy systems," *Solar Energy*, vol. 124, 2016, pp.26-38.

28. Zhichao Shi, Rui Wang and Tao Zhang, "Multi-objective optimal design of hybrid renewable energy systems using preference-inspired coevolutionary approach," *Solar Energy*, vol. 118, 2015, pp.96-106.
29. M.S. Ismail, M. Moghavvemi and T.M.I. Mahlia, "Genetic algorithm based optimization on modeling and design of hybrid renewable energy systems," *Energy Conversion and Management*, vol. 85, 2014, pp.120-130.
30. Hosseinalizadeh Ramin, Shakouri, Hamed G. Amalnick and Mohsen Sadegh, "Economic sizing of a hybrid (PV-WT-FC) renewable energy system (HRES) for stand-alone usages by an optimization-simulation model: Case study of Iran," *Renewable and Sustainable Energy Reviews*, vol. 54, 2016, pp.139-150.

MATEC Web of Conferences **55**, 06002 (2016) DOI: 10.1051/matecconf/20165506002
ACPEE 2016

Impact of Wind Speed Correlation on the Operation of Energy Storage Systems

Mathieu Davril[1,a],Xiaoyuan Xu[1] and Zheng Yan[1]

[1]*Key Laboratory of Control of Power Transmission and Conversion, Ministry of Education, Department of Electrical Engineering, Shanghai Jiao Tong University, Shanghai 200240, China*

Abstract. Renewable resources technologies such as wind power currently demonstrate a worldwide popularity thanks to their environmental friendly status and their economic potential. The variability of the wind power output implies the use of practical solutions such as energy storage systems in order to retain the power system stability and reliability. The wind geographical correlation between different wind farms also impacts the system reliability. This paper studies the effects of the wind speed correlation level on the performance of the associated energy storage system (ESS). Wind correlated model using Weibull probability distribution and Nataf transformation is presented. Energy storage system model and energy management algorithm are developed. Both are applied to a modified IEEE-RTS power generation and load model. The case simulation results indicate that the wind speed correlation level between two wind farms impacts the power distribution inside energy storages and that it needs to be considered in order not to overestimate ESS benefits on the system.

1 Introduction

Renewable energies are seen as a way to fight against environmental issues. They are mainly subject to intermittency and randomness due to resource uncertainty. That introduces technical challenges, such as power quality, control and reliability [1]. With the increase of the renewable energies penetration rate, it becomes important to develop methods to evaluate the impact of these resources generation on power system reliability and to manage this power output more efficiently.

The use of the energy storage system (ESS) has shown promising results in this area, and many publications discuss about this topic. Analytical procedures were proposed to evaluate operational reliability and energy utilization efficiency of power systems with high wind power or solar power penetration [2]. The impact of the initial stored energy and the rated capacity of energy storage on the system reliability has been shown [3]. Operation and control methodology of battery ESS were investigated [4]. Energy management algorithm and rule-based control method for the integration of wind and solar power were proposed [5, 6].

Wind farms at relatively close geographical locations can have their wind speeds strongly correlated. A multitude of methods has been developed to simulate correlated wind speed [7-9]. The important impact of a high correlation among wind farm on locational marginal prices has been studied [8].Taking the correlation level into account has also been found relevant on the study of

[a] Corresponding author: davril.mathieu@gmail.com

economic dispatch [9].The significant impact of the wind geographical correlation level on reliability studies has been demonstrated [10].However no study has been realised about the impact of the correlation level among wind farm on the energy storage system associated.

The added value of this paper is the consideration of an ESS working together with wind farms of different correlation degrees. The model of wind farms with different correlation levels and the ESS model are descripted. There are applied to an updated version of the IEEE-RTS system. Correlated wind speeds are obtained thanks to an algorithm based on the Weibull probability distribution and the Nataf transformation. An energy management algorithm is also elaborated. System adequacy assessments indices as well as energy storage characteristics are studied. Simulations are carried on the MATLAB software. The impact of the correlation on the system adequacy assessment indices and on the characteristics of the ESS is finally analysed.

2 System Modelling

2.1 Generating and Load Model

Two-state models are used to simulate the generating units operating cycles. A unit can move from an available state to an unavailable state following a failure rate (λ) and returns to an available state with a repair rate (μ). The time to failure (TTF) and the time to repair (TTR) are then used to produce an up-and-down operating cycle of the unit.Peak load models are usually used in reliability analysis. The load model used in this paper is an hourly varying load curve over a simulation period of 1-year. Although more accurate models for short operating periodslike 5 min exist, it is more useful to simulate the effect of unit ramp rates, a part that is not considered in this paper. It is assumed that load level within an hour is constantand the hourly time varying load model has been adopted.

2.2 Correlated wind Speed Model

Given as input data the Weibull distribution of random variables $v = (v_1, \dots, v_N)$ and their associated correlation matrixR_v:

$$R_v = \begin{pmatrix} 1 & \rho_{1,2} & \cdots & \rho_{1,N} \\ \rho_{2,1} & 1 & \cdots & \rho_{2,N} \\ \vdots & \vdots & \ddots & \vdots \\ \rho_{N,1} & \rho_{N,2} & \cdots & 1 \end{pmatrix} \quad (1)$$

where N is the number of wind farms and $\rho_{i,j}$ the correlation coefficient between farm i and farm j.It is possible to generate a set of correlated standard normal values $w = (w_1, \dots, w_N)$ from a set of independent standard normal values $y = (y_1, \dots, y_N)$. First, a Gaussian quadrature technique is used to generate the correlation matrix R_w of the set of correlated normal values from the correlated matrixR_v of the random variable following a Weibull distribution. Then the set of correlated normal values w is generated through an orthogonal transformation, i.e.:

$$w = Ly \quad (2)$$

where L is the lower triangular matrix of the Cholesky decomposition of the correlation matrixR_w.

From the set of correlated normal values$w$, it is then possible to generate samples of correlated wind speed $v = (v_1, \dots, v_N)$ thanks to the inverse of the Nataf Transformation[11, 12]as follow:

$$v_q = F_q^{-1}(\phi(w_q)); q = 1, \dots, N \quad (3)$$

where $F_q^{-1}$is the inverse of the cumulative distribution function (cdf) of v_q, and $\phi(.)$ is the cdf of the correlated standard normal variables. The samples of correlated wind speed values finally obtained follow the correlation matrix R_v and a Weibull probability distribution.

2.3 Power Output Model of the Wind Turbine Generators

Wind speed values can be converted in power output values by using the power curve of the installed wind turbine generator (WTG). The power curve of a WTG can be easily modelled as follows

$$P_{wt} = \begin{cases} P_r \times (v_t - v_{in}) \div (v_l - v_{in}) & v_{in} < v_t < v_r \\ P_r & v_r < v_t < v_{out} \\ 0 & \text{otherwise} \end{cases} \tag{4}$$

where P_{wt} is the WTG power output at hour t, v_t is real wind speed at hour t, P_r is the WTG rated power, v_r is the rated wind speed, and v_{in} and v_{out} are the WTG cut-in and cut-out wind speed.

2.4 Energy Storage Model

In this paper, atypical energy storage is studied: the battery. Battery energy storage can be characterized by four parameters: the charging/discharging rate, the maximum storage capacity, the discharging limit and the state of charge (SOC). The SOC is the available energy remaining in the battery unit. In fact, one of the main advantages of the battery is its fast charging/discharging rate. Our simulations do not take in account the ramp rate of conventional generators (CGs). The charging/discharging rate of battery units will not be considered in our study. The charging/discharging cycle will be modelled as a step function.

3 Evaluation Method

The operating strategy instigated in this paper is described as follows. Each wind farms possess the same number n of WTGs with the same parameters. The maximum total wind farm power output will coincide with a fixed percentage F_{pl} of the peak load. This percentage will be considered as the integration percentage of wind power into the energy production of our system and will be equally dispatched between wind farms. In other words, for a total number m of wind farms connected to the grid, each wind farm WTGs rated power $P_r$is given by equation (5).

$$P_r = F_{pl}/(m \times n) \tag{5}$$

The coordination strategy of the ESS can be described as follows: when the power amount of the generating system is more than the load at time t, the ESS is in charging mode. The power output of wind farms is used to charge the ESS. If the SOC is less than the amount of power available, the energy storage (ES) will charge at its maximum capacity and the remaining energy will be added as surplus power in the system. If the power amount of the generating system without wind farm is less than the load at time t, but it becomes more when we take in account the wind farms production, there is a surplus of power. Then the ESS is still in charging mode. The excess of power from wind farms is used to charge up the ES to the limit of batteries maximum capacity. Finally if the power amount from the generating system including wind farm is less than the load at time t, the ESS is in discharging mode. The remaining energy in ES is dischargeduntil the minimum energy allowed in the storage. The model of ES following this operating strategy is givenby

$$P_{mt} = P_{gent} + P_{wft} - P_{lt} \tag{6}$$

$$P_{est} = \begin{cases} -\min(P_{mt}, P_{wtf}, E_{SOCt-1}) & \text{if } P_{mt} \geq 0 \\ -\min(P_{mt}, E_{SOCt-1} - E_{msc}) & \quad\; P_{mt} < 0 \end{cases} \tag{7}$$

 DOI: 10.1051/matecconf/20165506002

where $P_{mt}$the power margin at hour t is, P_{gent} the generating system power output at hour t, P_{wft} the total amount of power generated by all wind farms at hour t, and P_{lt} the load at hour t. P_{est} represents the amount of power provided or saved at hour t by the ESS. It can be either positive when the ES is discharging, or else negative. $E_{SOCt}$is the available energy in the energy storage at time t and E_{msc} represents the maximum storage capacity.

It is then possible to evaluate if the system is at failure state or provide a surplus of power at hour t thanks to the followingequation:

$$P_{s/ft} = P_{mt} + P_{est} \tag{8}$$

where $P_{s/ft}$ stands for the power lack or surplus at hour t.

In order to show the impact of the correlation level on the energy storage improvement for the wind power management,we will analyzethe system reliability by using adequacy assessment indices: theloss of load expectation (LOLE) and the loss of energy expectation (LOEE). Another index used is the expected energy stored in the energy storage (EESES) [13]. This last index enables us to determine how much energy the ESs will be able to save from the surplus of power coming from the global generating system.

4 Case Study

The simulations are carried on a generating system created on the basis of IEEE-RTS. The original IEEE-RTS is a 24 bus system with 32 conventional units.The peak demand had been increased to 3150MW in order to simulate the impact of WTGs and ESSs.In order to study the impacts of wind speed correlation level, two wind farms are considered in this study.Wind speeds follow the Weibull models. For one Weibull model, the scale parameter is set to 7.5 and the shape parameter is 3. For the other model, the scale parameter is set to 7 and the shape parameter is to 2. The wind speed correlation level variation is studied from 0 to 0.99. ES capacities are taken from 50MW to 175MW. The wind power integration percentage varies from 5% to 25% of the load demand.

4.1 System Adequacy with ES and Wind Speed Correlation

The variation of LOLE and LOEE indices with the increase of wind speed correlation level is shown in Figure1. The ES capacity is 150MW and the wind power integration percentage is taken at 20%. It can be observed that the LOLE and the LOEE will increase with the increase of the correlation level.

Figure 2 shows the variation of the LOLE and the LOEE function of ES capacity and wind speed correlation level. It can be observed that the decreasing rate for both the LOLE and the LOEE is slightly higher with the increase of the wind speed correlation level. Results show that the wind speed correlation level impacts lightly the beneficial contribution of the ES capacity to the system adequacy. However it has to be noticed that for a fixed ES capacity, the LOLE and the LOEE are always higher with a greater wind speed correlation level. If the wind speed correlation level is not taken into account, there will be a clear underestimation of the LOLE, the LOEE and the benefit of adding ES.

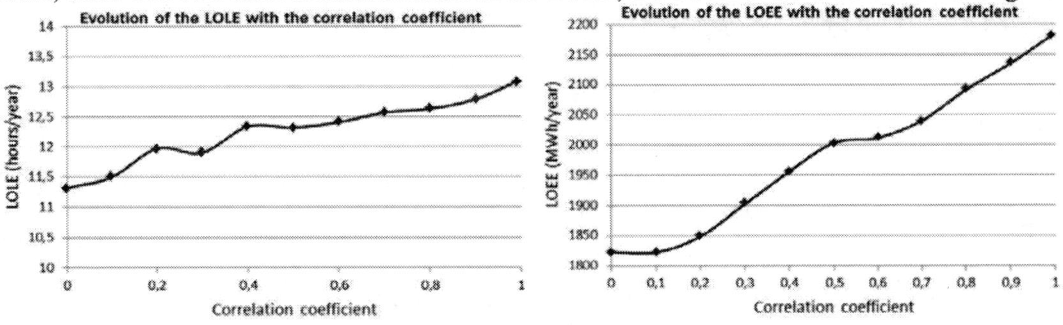

Figure 1. Effects of the wind speed correlation level on LOLE and LOEE.

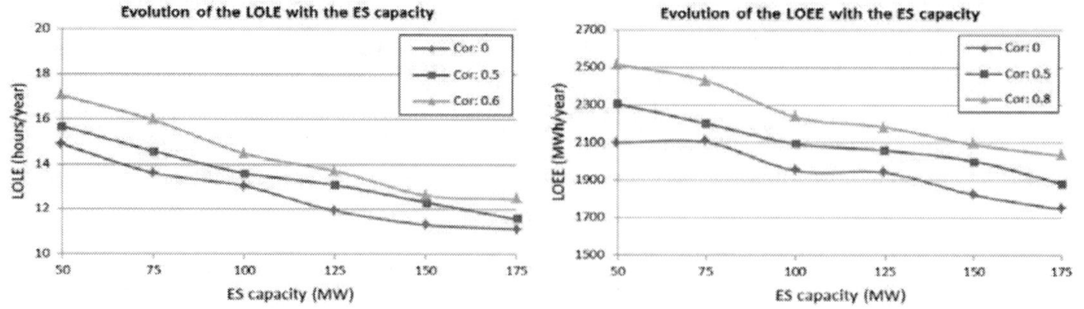

Figure 2. Effects of ES capacity on LOLE and LOEE.

4.2 Evolution of the Energy Stored in ES with Wind Speed Correlation

Figure 3 gives the effects of wind power integration percentage and ES capacity on EESES. For the figure in the left, the ES capacity is fixed at 100MW. It can be seen that although in the case of a wind speed correlation level of 0 the EESES noticeably decrease with the integration percentage, it decreases much slowly with a correlation level of 0.5 or 0.8. At a low correlation level,the variance of the power output of the two wind farms added together can be considered as low.Hence,the combination of low wind power variance and more power available lead to a decrease of EESES. In the case of a high correlation level, the probability of wind power output peak increases.With the increase of wind power integration, those peaks involve a bigger power surplus and consequently a growth of the EESES. This increase implies more charging/discharging cycles for the batteries and a more frequent utilization of the ES capacity.

For the figure on the right, the evolution of the EESES with the increase of ES capacity is depicted. The wind power integration percentage is taken at 20%. The EESES increases with the augmentation of ES capacity for each of the three correlation levels. It can be explained by the increase of storage available for wind power peaks. But it can be observed that the increasing rate of the EESES becomes higher with the growth of the wind speed correlation level. From the smallest ES capacity to the largest, the EESES increases of 484MWh, 547MWh and 617MWh for respectively the case of a correlation level of 0, 0.5 and 0.8. This result is justified by the wind power output peaks that are likely to be largest in the case of a strong correlation compared to a weaker one.

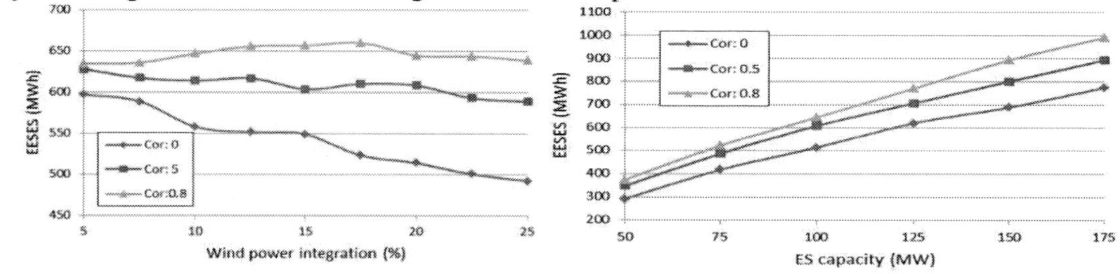

Figure 3. Effects of wind power integration percentage and ES capacity on EESES.

4.3 Evolution of the Allocation of the Energy Inside the ESwith Wind Speed Correlation

The allocation of the energy stored inside the ES is studied, and results are shown in Figure 4 with respectively wind power integration percentage of 10% and 20%. The ES capacity is chosen to be 100MW. The utilization frequency of the ES capacity globally decreases with the augmentation of wind power integration percentage thanks to the increase of available power capacity ensuing. It can be observed that for both studies, the utilization frequency of the ES being charge between 20% and 30% of its capacity increase with the correlation level. It means that ES are more frequently put in a state of deep-discharge in the case of a high correlation level than on a lower. We can also notice that the gap between the three correlation level increases with the integration percentage for the same charging state of 20/30%. For an integration percentage of 10%, the difference between the case of a correlation level of 0 and 0.8 is 0.5 hours/year. For an integration percentage of 20%, it becomes 1.8 hours/year. The frequency of ES deep-discharge state increases with the wind power integration growth. This result is in accordance with the one extracts from Figure3.

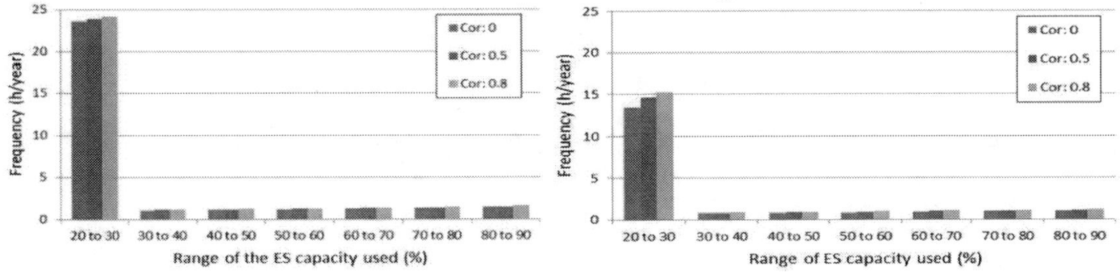

Figure 4. Allocation of the energy stored inside the ES for 10% and 20% of wind power integration.

5 Conclusion

This paper proposes a study of the effects of wind speed correlation on operation of the energy storage system (ESS) associated with wind farms. Generating system and load, wind speed and EES models have been presented. The Weibull probability distribution and the Nataf transformation have been used to simulate correlated wind speeds. The effects of wind speed correlation on energy storage characteristics are studied through the evolution of the energy stored inside the energy storage and the allocation of this energy. The results show that not considering the wind speed correlation could lead to an underestimation of adequacy assessment indices and an overestimation of the profit of adding ESS. It also shows that a strong correlation level will put more pressure on energy storage (ES): it will cause an increase of the number of charge/discharge cycles and of deep-discharge states. Besides not considering the wind speed correlation could lead to an overestimation of the lifetime of the ESs as well as underestimatethe associated costs.

References

1. S. Teleke, M. E. Baran, A. Huang, S. Bhattacharya, L. Anderson, IEEE Trans. Energy Convers., **24**, 725–732 (2009)
2. P. Wang, Operational adequacy studies of power systems with wind farms and energy storage, *Power and Energy Society General Meeting (PES)*, 1 (2013)
3. Z. Y. Gao, P. Wang, J. Wang, Impacts of energy storage on reliability of power systems with WTGs, *Probabilistic Methods Applied to Power Systems (PMAPS)*, 65-70 (2010)
4. C. A. Hill, M. C. Such, D. Chen, J. Gonzalez, W. M. Grady, IEEE Trans. Smart Grid, **3**, 850-857 (2012)
5. S. Teleke, M. E. Baran, S. Bhattacharya, A. Q. Huang, IEEE Trans. Sustain. Energy, **1**, 117-124 (2010)
6. R. Zamora, A. K. Srivastava, Energy management and control qlgorithms for integration of energy storage within microgrid, *Industrial Electronics (ISE)*, 1805-1810 (2014)
7. K. Xie, R. Billinton, Electr. Pow. Syst. Res., **79**, 687-693 (2009)
8. J. M. Morales, A. J. Conejo, J. Perez-Ruiz, IEEE Trans. Power Syst., **26**, 820-828 (2011)
9. D. Villanueva, A. Feijoo, J. L. Pazos, IEEE Trans. Sustain. Energy, **3**, 142-149 (2012)
10. F. Vallee, J. Lobry, O. Deblecker, IEEE Trans. Power Syst., **22**, 2232-2239 (2007)
11. P. L. Liu, A. Der Kiureghian, Probab. Eng. Mech., **1**, 105–112 (1986)
12. J. M. Morales, R. Mínguez, A. J. Conejo, Appl. Energy, **87**, 843–855 (2010)
13. H. Wang, X. Bai, Adequacy assessment of generating systems incorporating wind, PV and energy storage, *Innovative Smart Grid Technologies- Asia (ISGT Asia)*, 1-6 (2012)

MATEC Web of Conferences **55**, 06003 (2016) DOI: 10.1051/matecconf/20165506003
ACPEE 2016

Short-term Electricity Load Forecasting Model and Bayesian Estimation for Thailand Data

Kamal Chapagain[1,a] and Somsak Kittipiyakul[1]

[1] *Sirindhron International Institute of Technology, Thammasat University, PathumThani, Thailand*

Abstract. This paper proposes multi-equation linear regression model with autoregressive AR(2) method for modelling and forecasting a day ahead electricity load. AR(2) is used to show the dependency of next data on its previous two days data because the nature of electricity load consumption for the next day follow the pattern of previous days. Since, we allocate one equation for particular half hour, we need 48 separate equations to predict for one complete day. Parameters of model are estimated based on two different approaches- (i) Classical approach, and (ii) Bayesian approach. Classical or Ordinary Least Square approach estimates the parameters in terms of single value and hence its forecast is also single value, where as Bayesian approach includes the predictive distribution of electricity load due to multiple values of parameters. So, we can forecast electricity load not only from mean, but median and with other percentile values. In this paper, we use 70-percentile value for forecasting because the performance for all models accounts better in this percentile than that of mean and median forecast. Finally, we compare the performances where Bayesian estimation provides better and consistent performance than that of OLS estimation.

1 Introduction

Accurate short term load forecasting is important and crucial because electricity distribution grid requires proper balancing between electricity supply from production companies and its consumption at every instant. If the electricity load is not balanced, it affects the electricity distribution system with unnecessary blackouts or load shedding and huge loss of revenue for utility company. So, proper forecasting helps to maintain power system more secure, avoid blackout risk and provide adequate electricity supply. Therefore modelling and forecasting of electricity load for short horizon is alarming issue today.

Normally, forecasting system consist the several models, and we normally provide historical observed data as main input variables. Many authors build univariate time series model without exogenous variable. For example, paper [1] follows a double seasonal exponential smoothing for half hour data and predict with mean average percentage error (MAPE) 1.25 to 2 percent. In paper [2] overall forecast result is about MAPE 1.5 to 3 percent for the dataset of 10 European countries. But, for proper prediction only the historical load data may not sufficient because there are several other important factors that cause instant variation on demand of load. Paper [3] investigatesthoroughly on the electricity load and temperature issue showing the relationship between electricity demand and

[a] Corresponding author: kamal_chapagain@wrc.edu.np

temperature in Great Britain. Harvey and Koopman [4] has used structure time series model and include temperature using cubic spline to get forecast MAPE up to 3 percent, where as same model extended by [5] with Bayesian semi-parametric regression model on new South Wales data gives forecasting MAPE around 2 percent. Since, the load of special days, and public holidays deviates significantly from normal load and need to be treated separately.

The multiple linear regression method is still an interesting forecasting option because of its simplicity. Paper [6-9] use regression based method for short term load forecasting of normal days and use dummy variable for special days. Time series ARIMA process also perform forecasting for non-stationary mean by a differencing process [10] and is the popular model for forecasting. Paper [9] present a model for hourly electricity load forecasting state space model based on stochastically time-varying process that are designed to account for changes in customer behaviour and in utility production efficiency. For long term forecasting, Paper [11-12] forecast electricity demand with different forecasting methods-ARIMA, ANN, and MLR also compare their performance for Thailand data. Similarly, [13] propose new Gene Expression Programming (GEP) approach for the prediction of electricity demand. Multivariate weather- like temperature, wind speed, cloud cover, humidity, with historical load based models are employed for modelling electricity consumption load [12] [3]. Classical approach for short-term forecasting normally employs regression methods with dummy variables for example [6-8].

The paper [9] develops a regression model for each hour of the day, and incorporates dummies for special days-like bridge days as well. Bridge day is the special day if there is a normal day like Monday before holiday or Friday after holiday. In such special days,consumption of electricity load is lower than other normal days, so they treated separately. Paper [6] build a separate regression model for each of the hour and include special day, weather effects on load using dummy variables. Paper [7] use multiple equation regression model vector auto-regression model for 48 coefficients for each dummy variable to reflect intraday seasonality using half hourly load and estimated using Markov chain MonteCarlo technique. Paper [13] builds a dynamic multivariate periodic regression model for hourly data. Since, unrestricted model contains many unknown parameters and effective methodology is developed with the state-space framework where as [8] developtwo-stage model for each hour of the day for load, such that special days load in the first stage using dummy variable. Any unexplained component in the load is then modelled in the second stage using either AR or ANN. But, we adapt the combination of [6] and [7] techniques as a base paper for this research.

The plan of the paper is as follows. Section 2 presents the model and modelling strategy. Two different approaches for parameter estimation are described in Section 3. Section 4 presents the forecasting result, and their performance is analysed in Section 5. A final remark as conclusion is found in Section 6 and abbreviations used in this paper is listed in Section 7.

2 Model Specifications

We develop multiple linear regression (MLR) model with autoregressive AR(2) inspired by the paper [6] who develop multiple regression model with separate equations for each half hour of the day. Since, we develop the model to predict the electricity load for whole day with every half hour interval, we develop 48 separate prediction equations. The model is formulated as a AR(2) regression with its own coefficients and error terms,

$$Y_{dh} = X_{dh} + \epsilon_{dh} \tag{1}$$

where $\epsilon_{dh} = N(0, \sigma^2)$. In equation 1,$d$ represents the day and h for every half hour. The some of the variables used are perfectly predictable like day of the week, and month of the year. The Weather variables such as temperature and square of temperature are also used with some lagged load variables in this AR(2) model.The complete weekday/holiday model for particular hour can be written as,

$$Y_d = \beta_0 + \beta_1 Y_{d-1} + \beta_2 Y_{d-2} + \beta_3 Temp_d + \beta_4 Temp_d^2 + \beta_5 DMon_d + \beta_6 DTue_d + \cdots + \beta_8 DThu_d + \beta_9 DJan_d + \beta_{10} DFeb_d + \cdots + \beta_{19} DNov_d + \epsilon_d \tag{2}$$

where, Y_{d-1}, and Y_{d-2} explains the AR(2) model that accounts dependency of current value to previous two days. $Temp_d$, and$Temp_d^2$ use the current value of temperature and temperature square. Both weekday and holiday model equation 2 consist same 20 coefficients for weekday and holiday that we have to estimate. And similarly, weekend model for particular hour is expressed as

$$Y_d = \beta_0 + \beta_1 Y_{d-1} + \beta_2 Y_{d-2} + \beta_3 Temp_d + \beta_4 Temp_d^2 + \beta_5 DSat_d + \beta_6 DJan_d + \beta_7 DFeb_d + \cdots + \beta_{16} DNov_d + \epsilon_d \quad (3)$$

This equation 3 represents the model for weekend prediction and we use separate historical weekend data for the prediction.

3 Estimation of Parameters

3.1 Classical Approach (OLS Technique)

We use ordinary least square estimation technique as a classical method for the estimation of parameters. This method minimizes the sum of squared residuals (SSE) to obtain the parameters. The load model for this method can be written as,

$$\boldsymbol{Y} = \boldsymbol{X\beta} + \boldsymbol{\epsilon} \quad (4)$$

where, $\boldsymbol{Y}$ is$n \times 1$observations, $\boldsymbol{X}$ is a$n \times k$ matrix of predictors, $\boldsymbol{\beta}$ is a$k \times 1$ parameters, and $\boldsymbol{\epsilon}$ is an $n \times 1$ random distributions.Since, parameters β_i are unknown and estimated from the observation data of $\boldsymbol{Y}$ and variables $X_i$as $(\boldsymbol{\beta}) = (\boldsymbol{X}'\boldsymbol{X})^{-1}\boldsymbol{X}'\boldsymbol{Y}$

3.2 Bayesian Approach

Unlike to OLS, we draw burst of $\hat{\beta}$ and corresponding $\hat{\sigma}^2$ from the posterior distributionwhere, posterior distribution is the product of prior belief and likelihood function. Prior belief is the initial assumption of $\hat{\beta}$ with some value of variation and likelihood function is the observational data with some mean and variance. As, we consider the prior belief is normally distributed so that our posterior distribution is also normally distributed. So, $P(\hat{\beta}, \hat{\sigma}^2 | Y) = N(Mean^*, Var^*)$ so that, $Mean^* = \left(\Sigma_0^{-1} + \frac{1}{\sigma^2} X'X\right)^{-1} \left(\Sigma_0^{-1}\beta_0 + \frac{1}{\sigma^2} X'Y\right)$ and $Var^* = \left(\Sigma_0^{-1} + \frac{1}{\sigma^2} X'X\right)^{-1}$ where, β_0, Σ_0^{-1} *are* the initial assumption of $\hat{\beta}$ value and its variance $\hat{\sigma}^2$ as prior belief. From this $Mean^*$ and Var^* we draw the number of values using Gibbs sampling. These samples are stored after crossing the burn in stage of Markov chain and discard other samples. In our algorithm only 1000 samples are saved out of 5000 samples. So, the series of samples are stored as, $\hat{\beta} = Mean^* + \left[\hat{\beta} * (Var^*)^{\frac{1}{2}}\right]$ and for $\hat{\sigma}^2$ we sample scalar value from inverse Gamma distribution with degree of freedom $\frac{T}{2}$ and scale parameter $\frac{\theta}{2}$as $\hat{\sigma}^2 = \Gamma^{-1}\left(\frac{T}{2}, \frac{\theta}{2}\right)$

4 Forecasting Result

From the historical data set, 80 percentages of available data was used for training and remaining 20 percentage data was separated for validation purpose, but here we predict only for five days and present the result as in figure 4.

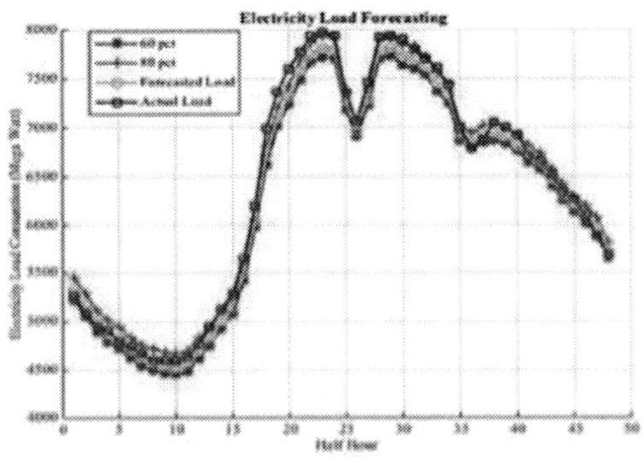

Figure 1. Weekdayloadforecast for day1

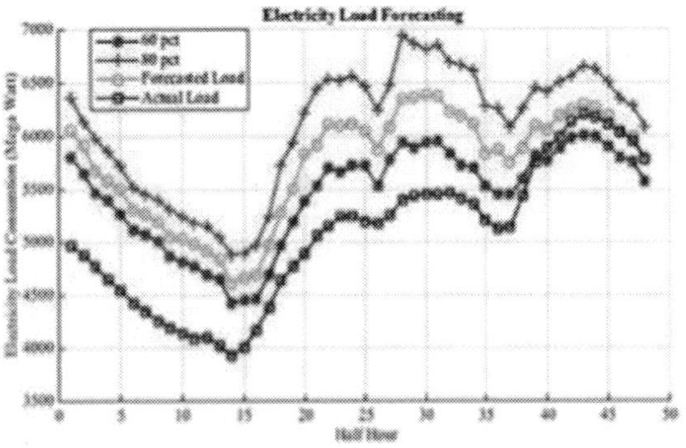

Figure 2. Holiday loadforecast for day1

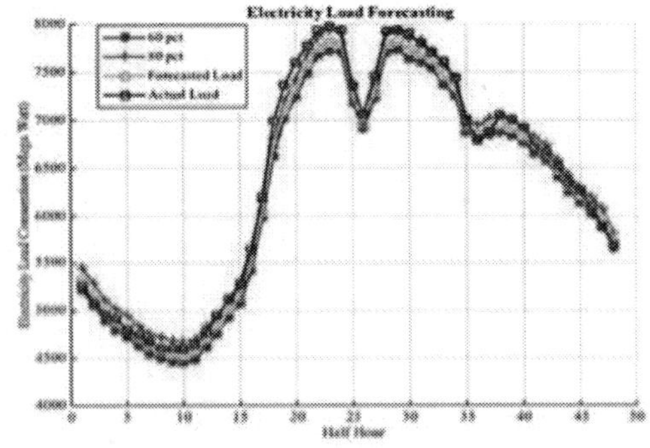

Figure 3. Weekend loadforecast for day1

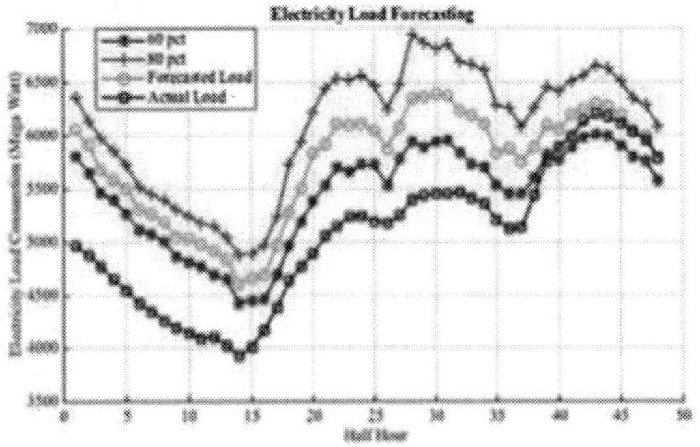

Figure 4. Consistancy of retaindraws for parameters

In Figures 1-3, prediction of electricity load from Bayesian estimation is plotted for each half hour and compared with actual load. We utilize the advantage of probabilistic approach and predicted the next day electricity load taking 70-percentile value from the distribution. Because, the prediction from 70 percentile providing here a consistent value with actual load for each model than that of mean and median prediction value. Although, this performance is still not enough to satisfy for weekends and holidays forecast.We forecast the electricity load for same horizon from OLS and compare with Bayesian approach and tabulated in performance analysis section. The consistency of estimation in parameter can be examined from the sequence of retained draws during Gibbs sampling. The visual inspection in Figure 4 is easier to check from its random fluctuation of all 20 coefficients of weekdays without any trends.Since, they are fluctuating randomly around the stationary mean, we can conclude that our estimation of parameters from Gibbs sampling is converge.

5 Performance Analysis

The performance is based on the capability of model that can predict the future value with better accuracy. Mean Absolute Percentage Error (MAPE), and Root Mean Squared Error (RMSE) are widely used techniques to measure performance in forecasting.

$$MAPE = \frac{1}{N}\sum_{i=1}^{N}\left|\frac{(Y_i - \hat{Y}_i)}{Y_i}\right| \times 100 \quad (5)$$

$$RMSE = \sqrt{\frac{1}{N}\sum_{i=1}^{N}(Y_i - \hat{Y}_i)^2} \quad (6)$$

Table 1. Overall performances of our models (weekdays, weekend, and holiday)

Method	Estimate	Forecast for	Weekdays		Weekends		Holidays	
			MAPE (%)	RMSE (MW)	MAPE (%)	RMSE (MW)	MAPE (%)	RMSE (MW)
MLR-AR (2)	OLS	Day1	4.86	369.88	4.83	260.6	6.16	313.15
		Day2	2.81	189.29	33.45	2054.3	28.62	1843.8
		Day3	3.63	259.87	6.23	342.37	8.71	539.6
		Day4	1.75	119.57	27.23	1718.7	6.59	363.75
		Day5	5.62	401.5	5.46	315.66	5.69	332.37
MLR-AR (2)	Gibbs sampling	Day1	1.442	95.91	3.52	176.23	13.93	670.74
		Day2	1.207	79.93	5.84	374.2	9.98	607.56
		Day3	0.848	58.48	2.05	114.5	11.75	767.84
		Day4	0.865	53.72	7.13	451.3	23.06	1229.3
		Day5	4.831	258.45	3.71	195.9	4.52	255.31
MAPE Range	OLS	Hourly	0.99-5.37		11.59-14.0		6.66-12.89	
	Gibbs	Hourly	0.54-3.68		2.30-6.49		4.92-17.46	

Since, MAPE measures the deviation of predicted data from our real load in terms of percentage. We also account RMSE, which measures the deviation in terms of Megawatts. Performance of weekdays prediction from both OLS and Bayesian approach (Gibbs sampling) is better than that of the weekends and holidays predictions. In five days prediction horizon, Gibbs sampling estimation technique gives better performance for all five days with the MAPE range from 0.54 to 3.68 for every half hour.

For weekends, though the forecast performances from both techniques do not look good. Bayesian approach provides more consistent MAPE and RMSE value than that of OLS. In real data analysis, the consumption of electricity load on Sunday is much less than that of Saturday, but here we treat within a same model. So, this may be one possible cause of inconsistent performance. So, the performance of weekdays and holiday forecast is really inconsistent from both techniques.

6 Conclusion

This paper presentsMLR with AR(2) model for one day ahead forecasting which can predict for every half hour electricity load consumptions. The model is treated separately according to weekdays, weekends, and holidays. Since, we estimate the coefficients in two different techniques and performance is compared for individual days. The overall MAPE calculation for weekdays using least square method is 3.73 percent where as Gibbs sampling result 1.77 for same time horizon. Some inconsistency for the performance of weekend is due to the different average load consumption on two consecutive days.

7Abbreviations

Table 2. List of abbreviations used in the paper

Full form	Abbreviation		Full form	Abbreviation
Autoregressive	AR		Ordinary Least Squares	OLS
Mean Absolute Percentage Error	MAPE		Multiple Linear Regression	MLR
Artificial Neural Network	ANN		Moving Average	MA
Autoregressive Integrated Moving Average	ARIMA		Neural Network	NN
Gross Domestic Product	GDP		Electricity Generating Authority of Thailand	EGAT
Sum of Square Residual	SSE		Root Mean Squared Error	RMSE

References

1. J W Taylor,*Short-term electricity demand forecasting using double seasonal exponential smoothing.* J Oper Res Soc, **54(8)**:799–805, (2003).
2. J.W. Taylor and P.E. McSharry,*Short-term load forecasting methods: An evaluation based on european data.* Power Systems, IEEE Transactions on, **22(4)**:2213–2219, (2007).
3. John Peirson and Andrew Henley,*Electricity load and temperature: Issues in dynamic specification.* Energy Economics, **16(4)**:235–243, (1994).
4. Andrew Harvey and Siem Jan Koopman,*Forecasting hourly electricity demand using time-varying splines.* Journal of the American Statistical Association, **88(424)**:1228–1236, (1993)
5. Michael Smith, *Modeling and short-term forecasting of new south wales electricity system load.* Journal of Business & Economic Statistics, 18(4):465–478, (2000).
6. RamuRamanathan, Robert Engle, Clive W. J. Granger, FarshidVahid-Araghi, and Casey Brace. *Short-run forecasts of electricity loads and peaks. International Journal for Forecasting,***(13)**:161–174, (1997)
7. R. Cottet and Michael Stanley Smith,*Bayesian modeling and fore-casting of intraday electricity load.*Journal of the American Statistical Association, **98**:839–849, (2003).
8. Lacir J. Soares and Marcelo C. Medeiros, *Modeling and forecasting short-term electricity load: A comparison of methods with an application to brazilian data.* International Journal of Forecasting, **24(4)**:630–644, (2008).
9. V. Dordonnat, S.J. Koopman, M. Ooms, A. Dessertaine, and J. Collet. *An Hourly Periodic State Space Model for Modelling French National Electricity Load.*Tinbergen Institute Discussion Papers 08-008/4, Tinbergen Institute, (2008).
10. J.W. Taylor and R. Buizza,*Neural network load forecasting with weather ensemble predictions.* Power Systems, IEEE Transactions on, **17(3)**:626– 632, (2002).
11. H.S. Hippert, C.E. Pedreira, and R.C. Souza,*Neural networks for short-term load forecasting: a review and evaluation.* IEEE Trans. Power Systems, **16(1)**:44–55, (2001).
12. Karin Kandananond, *Forecasting Electricity Demand in Thailand with an Artificial Neural Network Approach.* Energies, **4(8)**:1246–1257, (2011).
13. Seyyed Mohammad Mousavi, Elham Sadat Mostafavi, and FaribaHos-seinpour. *Gene expression programming as a basis for new generation of electricity demand prediction models.*Comput. Ind. Eng., **74**:120–128, (2014).

MATEC Web of Conferences **55**, 06004 (2016) DOI: 10.1051/matecconf/20165506004
ACPEE 2016

Absorber Tube with Internal Pin-Fins for Solar Parabolic Trough Collector

Kalidasan B [1,a], Shankar R [2] and Srinivas T [2]

[1]*Department of Mechanical Engineering, Bannari Amman Institute of Technology, Satyamangalam-638401, India.*
[2]*Energy Division, Vellore Institute of Technology, Vellore-632014, India*

Abstract. Solar parabolic trough collectors exploit solar energy for power generation in solar thermal power stations. These systems require long arrays of reflective troughs with absorber tube running along the axis of parabolic dish. A successful attempt to reduce the length of arrays was accomplished by experimentally analysing the modifications done in absorber tube. Two out of three tubes were fabricated and they were employed to obtain the performance parameters through experimentation conducted at VIT University, Vellore, India. Distilled water was used as the working fluid. Maximum efficiency of 39.12% was obtained at 451.6 W/m^2 of direct normal irradiance (DNI) for absorber tube with internal pin-fins and without glass tube ($AFWG_t$) compared to 8.15% obtained at same value of DNI and other conditions for simple absorber tube without glass cover (AWG_t). Cylindrical parabolic trough available at the university was utilized, providing the basis for designing and fabrication of the tubes. Plots for varying mass flow rate at interval of 10 minutes were made against instantaneous thermal efficiency and heat utilized, for direct normal irradiance vs. temperature difference across the tubes and instantaneous thermal efficiency. Through the experimentation conducted, better performance was procured compared to earlier works. Thus, the proposal infers that absorber tube with internal fins has good scope for its application, both domestically as well as industrially. It also calls for further research and development of proposed techniques so as to achieve better performance curves.

1 Introduction

As by Fernandez-Garcia et al. (2010) [1] the concentrating systems can be classified into three major categories, point focus concentration, line focus concentration and linear concentration. Concentrating paraboloid concentrators are types of point focus concentrations, whereas parabolic trough collectors lies in category of line focus concentrators and Fresnel collectors lies in the category of linear collectors. Solar parabolic trough collector (SPTC) provides an effective way to harness solar energy providing "one-axis concentration" such that fluid flowing through the absorber tube, gains the concentrated energy and accomplish useful work. Parabolic trough collectors, with water as fluid, are basically used for two major applications, industrial steam generation and water heating application. Former application requires heating of water up to a temperature of range 300-400 °C, while the later requires temperature range of around 100-250 °C.

[a] Corresponding author : kalidasan@bitsathy.ac.in

Since many years, researchers are working to optimize the operating performance of SPTC so that rapid heat transfer can take place and overall length of the trough collector can be reduced. Supporting the above statement, the review paper on state of parabolic trough collectors by Price et al. (2012) [2] conveys that intensive research and development is required in this field of SPTC to reduce cost and augment their operation. Work done by Tao et al. (2011) [3] introduces design method and working principle for new type of trough collector with efficient performance of SPTC by introducing widely opened concentrating collectors. However, increase in the width of concentrating apparatus may result to unbalanced design and higher probability of back reflection of the incident solar radiations. To avoid such hindrances, requirement of precise tracking and high cost engrossment will be needed, which add some more minuses to the proposed design. Effort on solar selective coatings by Farooq and Raja (2008) [4] and cobalt electrodeposition on absorber tubes by Barrera et al. (1998) [5] enhance the efficiency for performance of solar devices. But radiation thermal losses may increase with increase in coating temperature, thus results in development of high thermal stresses in the collector tube.

Valenzuela et al. (2006) [6] showed the implementation and development of recirculating operation mode for SPTC, increasing the output temperature and performance of the apparatus. But, the system possesses minor drawbacks concerning the use of steam at high pressures, risk for steam leakage and higher stresses on tube. Experimental work with bimetallic tube by Flores and Almanza (2004) [7] is appreciable, posing a new field for further research. However, employment of bimetallic tube requires high investment and also the proposed tube has very large diameter (31.7 mm), which affect heat transfer rate. Performance analysis of selectively coated solar collector tube with U-tube by Ma et al. (2010) [8] employing reticulation of working fluid presented results employing very less mass flow rate through the tube, while the tube had unexploited and considerable space to fit in another U-tube. Utilising the available space, performance of the tube could have been optimised. An innovative numerical model evaluation of heat transfer characteristics for porous disc receiver by Ravi Kumar and Reddy, 2009 [9] presents an efficient design of absorber tube, but flow across the tube is very high and also the system will require only pure fluid to avoid accumulation of particles on the tube walls than can be caused by highly interrupted flow through the receiver tube with sequential porous discs.

Although there may be an extensive list of researches, developments and innovations for improving effectiveness of absorber tube, the authors of the paper present another innovative research to enhance efficiency nearly by two to three times compared to that of conventional absorber tubes. The fabrication procedure and experimental analysis for the modifications done in the absorber tube are presented. The proposed solar absorber tube can be employed for application in both line and linear focus concentration apparatuses with numerous advantages as well. With VIT University, Vellore (12.92 °N, 79.13 °E) as the experimentation site, the experiments were performed on August 23rd, 2014 with Absorber tube with glass cover (AGt), and on August 25th, 2014 with Absorber tube without glass cover (AWG_t) and Absorber tube with internal fins and without glass cover ($AFWG_t$). Because of unavailability of the glass tube, the experiment on August 25th was performed without evacuated glass cover on both the tubes. Nevertheless, the obtained results are very much accurate and can be compared with available studies as well.

2 Materials and methods

Among the three tubes used for the experiments, two tubes were designed and fabricated, excluding absorber tube with glass cover (AG_t) that was available from the university. The specifications of cylindrical axis trough collector and the absorber tubes are given in Table 1 respectively. All the tubes were made of copper. Although the total length of the two fabricated tubes was kept as 2 m, effective length of tubes receiving reflected radiations from parabolic trough was 1.5 m and arrangements were made such that the temperature differences for the fluid across the effective length of the tubes were measured accurately. Parabolic trough available at the university, with rim angle of 120° and focal length 26.26 cm, was used which has reflective surface of aluminium alloy and reflectivity of about

80%. Details for fabrication processes of the tubes and pictures taken during the experiments have been included.

Table 1. Specification of Apparatus.

Sl. No	Component	Dimension
1	Length, L	1.5 m
2	Width, W	0.91 m
3	Concentration ratio, C	19
4	Specific heat of fluid, C_p	4182 J/kgK
5	Tilt factor for beam radiation, R	1.00
6	Inner diameter, d	0.013 m
7	Outer diameter, D	0.015 m
8	Inner diameter of glass tube	0.030 m
9	Thickness of glass tube	0.005 m
10	Vacuum pressure in glass tube	0.8 bar

AG_t, was the conventional tube of specified dimension with glass cover. The enclosed volume in the glass tube was evacuated to develop a low pressure vacuum, thus reducing the heat transfer losses. Due to unavailability of more number of glass tubes, the experiment on other two tubes was performed without evacuated glass cover. AWG_t was similar to AG_t with the only difference of the glass tube covering the absorber tube, and thus the tube was subjected to minor heat transfer losses. For $AFWG_t$, drill holes were made in helical pattern all over the tube at pitch distance of 25 mm and extrusion using a solid copper wire, 1.5 mm diameter, was provided at each hole with gas welding on the surface. The length of the extruded wires, functioning as internal pin-fins for the tube, was kept constant at 5 mm. The 3D models for simple absorber tube figure 1 and absorber tube with internal fins figure 2 are presented showing internal design. Feature 'a' presents the isometric view of a section of absorber tube while feature 'b' presents the front view.

Figure 1. Simple Absorber Tube (AG_t / AWG_t)

Figure 2. Absorber Tube with Internal Fins ($AFWG_t$)

Initially, absorber tube was fixed on SPTC's axis and the trough was adjusted such that axis of focus and the absorber tube accords. Then the tube was connected to the pump through a hose and distilled water, used as working fluid, was collected in tank. The pump was switched on, leaving the system for 10 minutes with specified volume flow rate, controlled by valves. Parabolic trough collector was adjusted frequently to align the absorber tubes with focus of reflected radiations to maintain tilt factor with the value of 1.00 ± 2%.Pyrheliometer, with sensitivity of 4.95 mV/W/m^2, was used to measure the beam radiation. The required performance parameters were noted, such as DNI

Figure 3. Experimental Set-Up for AG_t

Figure 4. Experimental Set-Up for $AFWG_t$

value from multimeter, temperature values for atmospheric condition, inlet and outlet, and volume flow rates. Then, flow rate was altered at an interval of 10 minutes and data for same parameters was collected at different flow rates. The experiment for each tube was conducted for a span of 50 minutes and total of 5 readings were obtained. The experimental setup for AG_t is shown in figure 3 and for $AFWG_t$ in figure 4.

A tank of 15 liters capacity was employed to store distilled water that was pumped across the tubes through a valve and a rotameter. Thus, volume flow rates could be varied and experimentation was done with varying mass flow rates. Digital thermocouple was used to measure the temperature values and stop watch for measuring time required to fill a liters capacity collecting jar.

3 Results and discussions

Based on the experimental results and observations, plots were obtained for mass flow rate vs. energy utilized in figure 5, mass flow rate vs. instantaneous efficiency in figure 6, DNI vs. temperature difference in figure 7 and DNI vs. instantaneous efficiency in figure 8.

The figure 5 shows that for almost same mass flow rates, the simple tubes, both with and without glass tube, could utilize very less amount of energy ranging from about 50-75 W, while $AFWG_t$ utilized nearly four-fold of their utilization ranging from 200-250 W. The extrusions or pin-fins provided turbulence in flow with extended surface area for heat transfer and thus higher energy was used. The simple tubes neither provided turbulence nor additional heat transfer resulting to their lowest transfer rates. Hence, in figure 6, the instantaneous efficiencies follow the same increasing trend with AWG_t having the lowest performance efficiency, almost equal to that obtained by AG_t and $AFWG_t$ resulted to highest performance efficiency of range 35-40 %, about three times increment from AWG_t.

It can be inferred that both efficiency and energy utilized increases with decrease in mass flow rate of the working fluid. The procured results can be very well compared to the results presented by[8].The authors maintained very low mass flow rates of 0.001 kg/s, 0.002 kg/s and 0.003 kg/s and obtained efficiency of about 35-40% with higher radiation densities of about 950 W/m^2. In the present work, it should be noticed that AG_t and AWG_t utilizes the same amount of energy and delivering almost same thermal instantaneous efficiencies. But, as per theory, energy utilized by AG_t must be higher than that of AWG_t with availability of glass cover. This shift can be explained using the fact that performance test for AWG_t was done on August 25th, when higher radiation intensity was available.

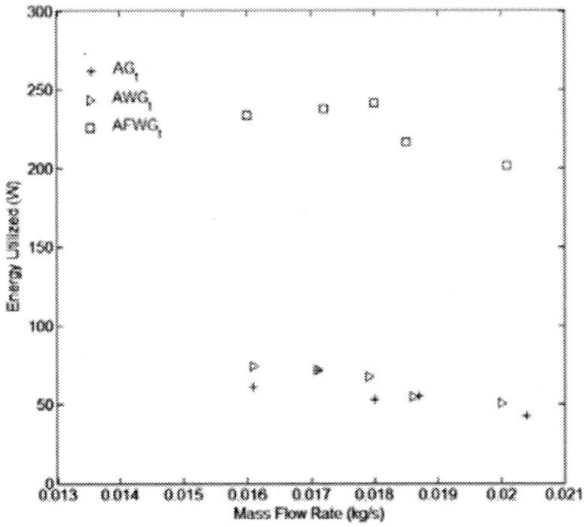

Figure 5. Mass Flow Rate vs. Energy Utilized by Absorber Tubes

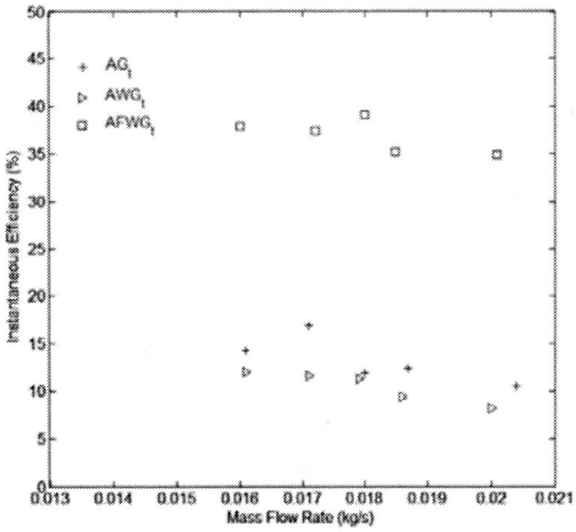

Figure 6. Mass Flow Rate vs. Instantaneous Efficiencies of the Tubes

Variation of temperature difference across the absorber tubes with varying DNI in figure 7 presents the direct proportionality relation amongst the two parameters i.e. with increase in DNI, temperature difference for the working fluid also increments. Again because of difference in incident radiation intensities, the difference of temperatures for both AG_t and AWG_t is almost constant. But, this difference of temperature augments to four times for $AFWG_t$. The higher temperature verifies the better efficiency of $AFWG_t$ compared to other absorber tubes, as presented in figure 8. It should be noted that this temperature difference was obtained only for 1.5 m long SPTC and at low DNI values.

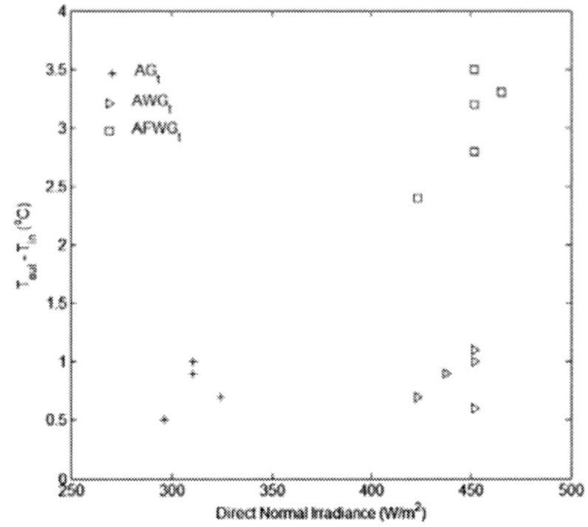

Figure 7. DNI vs. Temp. Difference of Working Fluid Across Absorber Tubes

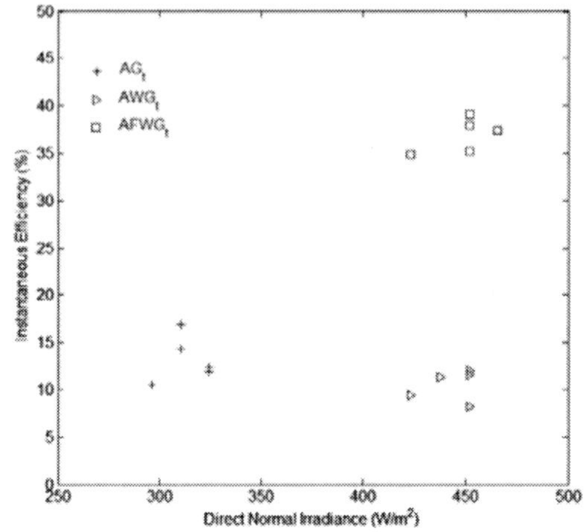

Figure 8. DNI vs. Instantaneous Efficiencies of Absorber Tubes

The experimental data provided by Dudley et al. (1994) [10] for SEGS LS-2 collector, about 4 m long, used DNI values varying from 900 W/m^2 to 930 W/m^2 with very high inlet temperature of working fluid ranging from 100 °C to 350 °C and achieved temperature increment of about 20 °C in all cases. Whereas, the experimental values provided by the authors for $AFWG_t$ have DNI values ranging from about 400 W/m^2 to 450 W/m^2 and almost constant inlet temperature of 31.5 °C, achieving temperature rise of 3.5 °C within the absorber tube length of only 1.5 m. For direct steam generation using parabolic troughs with collector length of 600 m and inlet temperature of 190 °C, Odeh et al. (1998) [11] presented analysed and developed model achieving maximum efficiency of about 70%, whereas efficiency close to 40% was achieved for $AFWG_t$, without evacuated glass cover.

The proposed design appears to have substantial application in the developing fields as well. It can be used for thermal applications, involving heat transfer, organic rankine cycles and even in absorption power and cooling cycles, making performance of system efficient.

4 Conclusion

Vellore is 216 m above sea level and receives two periods of monsoons mostly during June – August and October – December. As the experiments were conducted during the second phase of monsoon, intensified solar flux was not available compared to that obtained in summer season, affecting the performance curves of the proposed absorber tubes. Also the arrangement of aluminium alloy sheet over the parabolic trough had minor irregularities, resulting to lessened reflective and optical efficiency compared to mirror reflectors. The unavailability of glass tube for $AFWG_t$ also contributed to the procured less efficient results. Reduction of overall length of arrays required for harnessing of solar energy, especially for application of direct steam generation in solar thermal power plants, was the prime aim of this project. With the obtained results, it can be concluded that the modifications of internal fins in the absorber tubes of solar parabolic concentrating collector can deliver highly efficient performance compared to that of traditional tubes. Average instantaneous thermal efficiencies of 36.89 % were obtained for internally finned tube. Such modifications can be revolutionary in the field of solar energy, hence cutting off the overall cost of the projects to a great extent. It can be inferred that absorber tube with internal pin-fins proves to be most efficient, although there was a con for the experimentation that use of evacuated glass cover was avoided due to lack of sufficient funding. If evacuated glass cover are used, better and more efficient results could have been obtained compared to the various referred works.

References

1. A. Fernandez-Garcia, E. Zarza, L. Valenzuela, M. Perez, Renew. and Sustain. Energy Rev. **14**, 1695–1721 (2010)
2. H. Price, E. Lupfert, D. Kearney, E. Zarza, G. Cohen, R. Gee, R. Mahoney, J. of Sol. Energy Eng. **124**, 109-125 (2012)
3. T. Tao, Z. Hongfei, H. Kaiyan, A. Mayere,. Sol. Energy **85**, 198–207 (2012)
4. M. Farooq, I.A. Raja, Renew. Energy **33**, 1275–1285 (2008)
5. E. Barrera, I. Gonzalez, T. Viveros, Sol. Energy Mater. and Sol. Cells **51**, 69-82 (1998)
6. L. Valenzuela, E. Zarza, M. Berenguel, E.F. Camacho, Sol. Energy **80**, 1–17. (2006)
7. V. Flores, R. Almanza, Energy **29**, 645–651. (2004)
8. L. Ma, Z. Lu, J. Zhang, R. Liang, Build. and Environ. **45**, 1959-1967. (2010)
9. K. Ravi Kumar, K.S. Reddy, Appl. Energy **86**, 1804-1812. (2009)
10. V.E. Dudley, G.J. Koib, M. Sloan, D. Kearney, D., Test results SEGS LS-2 solar collector, Sandia National Laboratories, 1994
11. S.D. Odeh, G.L. Morrison, M. Behnia,. Sol Energy 62, **No. 6**, 395–406. (1998)

MATEC Web of Conferences **55**, 06005 (2016) DOI: 10.1051/matecconf/20165506005
ACPEE 2016

A Real-Time Load Distribution Method of Hydro-Thermal Power Grid Based on Working Conditions Self-Matching

Fang Chen[1], Jianzhong Zhou[1,a] and Xuemin Wang[1]

[1]*School of Hydropower and Information Engineering, Huazhong University of Science and Technology, Wuhan, 430074, P R China*

Abstract. Short-term hydro-thermal power optimal dispatch is a classical problem. Traditional scheduling methods cannot meet the requirements of different dispatch periods separately, and can hardly ensure the safety operation of cascade hydropower stations and reduce wasting water. So the problem how to ensure the safety operation of power grid and the rational utilization of water resources should be settled urgently. In order to solve the problem, a real-time load distribution method of hydro-thermal power grid based on working conditions self-matching is proposed in this paper. In the proposed method, working conditions self-matching is proposed to meet the requirements of different dispatch periods separately for ensuring the efficiency and economy of the power grid. Finally, the proposed method is applied to Sichuan power grid to verify its feasibility and effectiveness. Compared to traditional scheduling methods, the proposed method not only has excellent performance in the utilization of water resources but also can meet all constrains of each scheduling period, which provides a new method to solve the real-time load distribution of hydro-thermal power grid.

1 Introduction

The real-time load distribution of hydro-thermal power system [1]refers to distribute the load among hydropower and thermal power stations according to certain economic operation[2] mode. However, when the load changes, dispatchers usually make real-time load adjustment based on their scheduling experience. The adjustment plan which they made is always not conducive to the economic dispatch of cascade hydropower stations. There are some research on short-term hydro generation scheduling of Xiluodu-Xiangjiaba cascade hydropower stations and Three Gorges–Gezhouba[3, 4] cascaded hydropower plants. The results are excellent and powerful. But they neglected the adjustment ability of thermal power and did not take the security and stability operation of the power grid fully into consideration. So a new real-time load distribution method of hydro-thermal power grid, which can not only meet all constrains but also ensure the efficiency, economy and security of the power grid, is needed urgently.

2 Real-time Load Distribution of Cascade Hydropower Stations

[a] Corresponding author : 1103448274@qq.com

The real-time load distribution of hydro-thermal power grid can be divided into two parts. The real-time load distribution among hydropower stations and thermal power stations is one part. The real-time load distribution of cascade hydropower stations is another part. The real-time load distribution of cascade hydropower stations is aimed to maximize the total energy storage of cascade hydropower stations over the current scheduling period, while meeting kinds of equality and inequality constrains.

2.1 Objective Function

$$obj = \max \sum_{i=1}^{M} \left(Z_i^{up} - Z_i^{down} \right)\left(Q_i^{in} - Q_i^{out} \right) \tag{1}$$

The above formula is aimed to maximize the total energy storage of cascade hydropower stations over the current scheduling period. Where obj is the total energy storage of the hydropower stations over the current scheduling period; M is the number of hydro plants; Z_i^{up} is the average upstream water level of the i-th hydro plant over the current scheduling period; Z_i^{down} is the average downstream water level of the i-th hydro plant over the current scheduling period; Q_i^{in} is the inflow of the i-th hydro plant over the current scheduling period; Q_i^{down} is the outflow of the i-th hydro plant over the current scheduling period.

2.2 Constrains

In the process of real-time load distribution, various complex equality and inequality constrains, such as restrictions of water levels, outputs and hydraulic connections, should be taken into account for restricting the total energy storage optimization. The constraints are described as follows:

(1) Water level constrains

$$Z_{i,j}^{\min} \le Z_{i,j} \le Z_{i,j}^{\max} \tag{2}$$

In the above inequality, $Z_{i,j}$ is upstream water level of the i-th hydro plant at the end of the j-th period; $Z_{i,j}^{\min}$ and $Z_{i,j}^{\max}$ represent the lower and upper water level limits of the i-th hydro plant of the j-th period, separately.

(2) Output constrains

$$N_{i,j}^{\min} \le N_{i,j} \le N_{i,j}^{\max} \tag{3}$$

In the above equation, $N_{i,j}$ is output of the i-th hydropower station in the j-th period; $N_{i,j}^{\min}$ and $N_{i,j}^{\max}$ represent the minimum and maximum output limits of the i-th hydropower station in the j-th period, separately.

(3) Water discharge constrains

$$Q_{i,j}^{\min} \le Q_{i,j} \le Q_{i,j}^{\max} \tag{4}$$

In the above equation, $Q_{i,j}$ is water discharge of the i-th hydropower station in the j-th period; $Q_{i,j}^{\min}$ and $Q_{i,j}^{\max}$ represent the minimum and maximum water discharge limits of the i-th hydropower station in the j-th period, separately.

(4) Hydraulic connection

$$I_{i,j} = \sum_{l \in \Omega_i} Q_{l,j} + B_{i,j} \tag{5}$$

In the above equation, $I_{i,j}$ and $B_{i,j}$ represent the inflow and local inflow of the i-th hydropower station in the j-th period, separately; $Q_{l,j}$ is water discharge of the l-th hydropower station in the j-th period; Ωi is upper hydro plants set of the i-th plant.

(5) Water balance equation

$$V_{i,j} = V_{i,j}^{\ 0} + \left[I_{i,j} - Q_{i,j} \right] \Delta t \tag{6}$$

In the above equation, $V_{i,j}^0$ is the storage of the i-th hydropower station at the beginning of the j-th period; $V_{i,j}$ is the storage of the i-th hydropower station at the end of the j-th period; $I_{i,j}$ and $Q_{i,j}$ are the inflow and water discharge of the i-th hydropower station in the j-th period, separately; Δt is interval of scheduling term.

3 Strategies of HTPC and WCSM

The hydro-thermal power combined adjustment (HTPC) is proposed to enhance the safety and stability operation of the power grid. When the total load decreases, thermal power output reduces firstly. However when the load increases, hydropower output raises in flood or falling off season and thermal power output raises in dry or impounding season firstly. Different scheduling periods have different inflow and water level requirements. If scheduling according to the same rules, it will not only lead to the waste of water resources, but also a certain threat to the safety operation of plants. Then different strategies are proposed based on each scheduling period, which is called working conditions self-matching (WCSM). Differential evolution (DE)[5, 6] is suitable to solve the load optimal allocation model of cascade hydropower stations in chapter 2. The detailed illustrations are as follows:

Assuming that ΔP is the D-value between the real-time load and the planned load.

3.1 Strategies for Flood Seasons When the Load Increases

In the flood season[7], the inflow is plentiful. In order to ensure the safety operation of the hydropower plants, the upstream water level can rise in a small range only when the plants have spilled water. This measure allows the reservoir to save more energy. Otherwise, the upstream water level keeps unchanged.

At first, if there is no spilled water and the hydropower output has been the maximum output under the current conditions, the hydropower output keeps unchanged, and the increment of thermal power output is ΔP. And if there is no spilled water and the hydropower output has not reached the maximum output under the current conditions, the upstream water levels keep unchanged and just adjust the load distribution to reach the maximum output under the current conditions. Furthermore, if there is lots of spilled water and the hydropower output has been the maximum output under the current conditions, the upstream water level can rise to the allowable maximum to storage water until there is no spilled water. Then the maximum hydropower output $P_s^{\max}$ under the current conditions can be obtained by checking the table which describes the relationship between the maximum output and the inflow. And then the maximum increment of hydropower output $\overline{\Delta P_s}$ can be calculated according to formula (10). If $\overline{\Delta P_s}$ is more than or equal to ΔP, the increment of hydropower output is ΔP and the thermal output remain unchanged. Otherwise, the increment of hydropower output is $\overline{\Delta P_s}$ and the increment of thermal power output ΔP_h can be calculated according to formula (11). Meanwhile, the thermal power output turns to $P_h^{\max}$ if it cannot meet formula (12). And the increment of thermal power output turns to UR_h if it cannot meet formula (13). Then the adjusted outputs of cascade hydropower stations and thermal power stations are obtained. Finally, according to Section 2 and 3, the real-time load distribution of hydro-thermal power grid can be completed.

$$\overline{\Delta P_s} = P_s^{\max} - P_s \tag{7}$$

$$\Delta P_s + \Delta P_h = \Delta P \tag{8}$$

$$\alpha P_h^{\max} \leq P_h \leq P_h^{\max} \tag{9}$$

$$DR_h \leq P_h^t - P_h^{t-1} \leq UR_h \tag{10}$$

P_s is the current hydropower output; $P_s^{\max}$ is the maximum hydropower output under the current conditions; $\overline{\Delta P_s}$ is the maximum increment or decrement of hydropower output; $\overline{\Delta P_h}$ is the maximum increment or decrement of thermal power output; P_h is the current thermal power output, $P_h^{\max}$ is the maximum thermal power output, UR_h is the maximum increment of thermal power output, and DR_h is the maximum decrement of thermal power output. α is a constant between 0 and 1. It is a compression ratio of thermal power output.

3.2 Strategies for Falling off Seasons When the Load Increases

In the falling off season, when the total load increases, the upstream water level of hydropower stations can fluctuate within a certain range. If the inflow is plentiful, the upstream water level can rise to increase the storage of energy. Otherwise, the upstream water level can drop to increase the output. If the hydropower output cannot increase under the current conditions, the hydropower output keep unchanged, and the increment of thermal power output is ΔP . Otherwise, the upstream water level at the end of the period Z_1 can be calculated according to the planned load and forecasted inflow.

3.3 Strategies for Impounding Seasons or Dry Seasons When the Load Increases

In impounding season or dry season, the inflow is insufficient. So it is necessary to save water. When the total load raises, increasing thermal power output can save water to ensure the hydropower stations run at high water levels. Thus the overall principle is that thermal power output increases firstly.

3.4 Strategies for the Load Decreases

When the total load decreases, decreasing thermal power output can reduce waste water to improve the efficiency of hydropower and enhance the safety and stability of cascade hydropower stations. Furthermore reducing coal consumption is conducive to environmental protection. So the overall principle is that thermal power output decreases firstly. According to formula (12) and (13), the maximum decrement of thermal power $\overline{\Delta P_h}$ can be calculated. If ΔP is less than $\overline{\Delta P_h}$, the decrement of thermal power output is ΔP , and the hydropower output keep unchanged. Otherwise, according to formula (10) and (11), the decrement of thermal power output ΔP_h and the decrement of hydropower power output ΔP_s can be calculated. Then the adjusted output of cascade hydropower stations and thermal power stations are obtained. Finally, according to Section 2 and 3, the real-time load distribution of hydro-thermal power grid can be completed.

4 Case Study

For the Xiluodu-Xiangjiaba cascade hydropower stations, the load is assigned by the dispatching centre of power grid. And the load is made up of two parts, one part is assigned by State Grid, another part is assigned by Southern Power Grid. The fluctuation of the load is generated mainly from the Sichuan Power Grid. Meanwhile, the load assigned by Southern Power Grid is provided by the right bank units of Xiluodu. And the load assigned by State Grid is provided by Xiangjiaba and the left bank units of Xiluodu. The case took 15 minutes as a scheduling period. Assuming that the thermal power output of Sichuan Power Grid has been the minimum. The basic parameters of Xiluodu and Xiangjiaba is shown in Table 1. The real-time load and planned load is shown in Fig.1 and Fig.2.

Table 1. The basic parameters of Xiluodu and Xiangjiaba.

Reservoir	Normal water level (m)	Dead water level (m)	Initial water level (m)	Water level amplitude (m)	Regulation storage(m^3)	Installed capacity (MW)
Xiluodu	600	540	580	0.5	64.6×10^8	13500
Xiangjiaba	380	370	375	1	9.03×108	6400

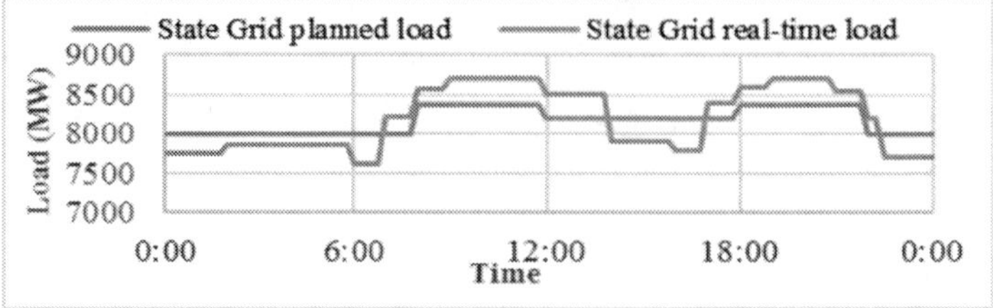

Figure 1. The Load of State Grid.

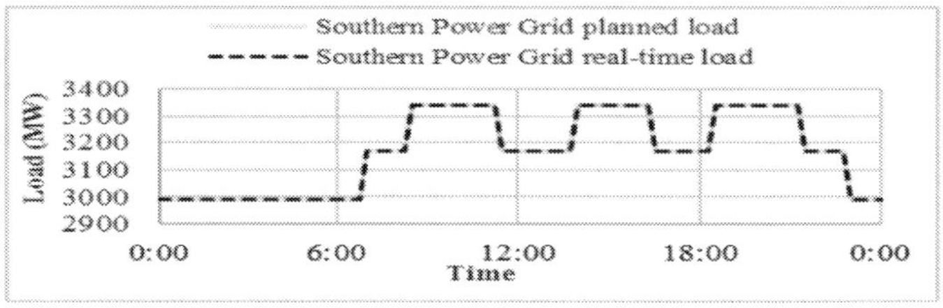

Figure 2. The Load of Southern Power Grid.

And the results are shown in Fig.3 and Fig.4.

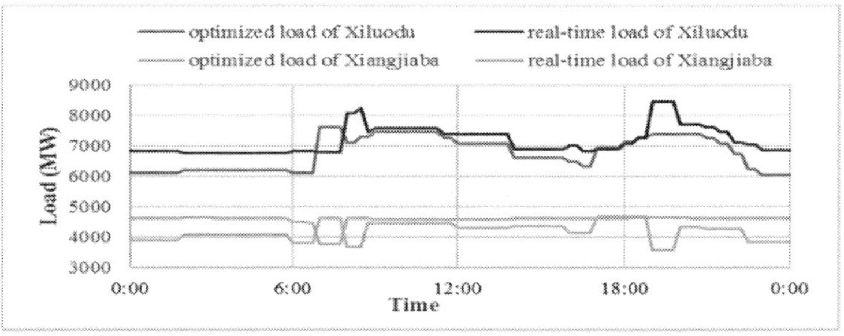

Figure 3. The Load of Xiluodu and Xiangjiaba.

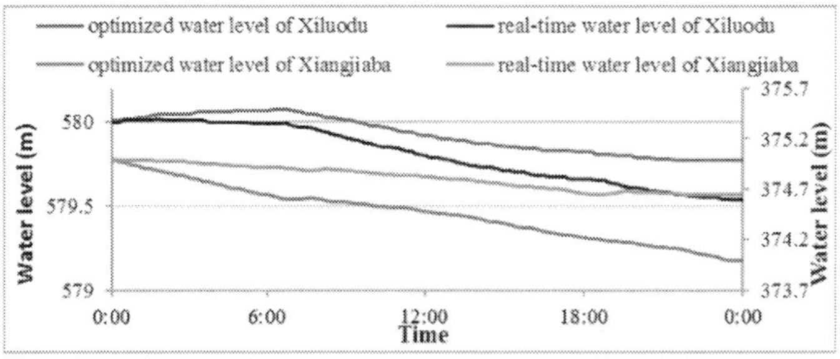

Figure 4. The Water Level of Xiluodu and Xiangjiaba.

According to Fig.3 and Fig.4, in the day, the total water consumption of Xiluodu was 2.61×108 m^3, and the total water consumption of Xiangjiaba was 2.24×108 m^3. During the 96 scheduling periods, the real-time energy storage was 27.367×10^8kWh, and the optimized energy storage was 27.797×10^8kWh. So the energy storage increased 0.43×10^8kWh in the day. Furthermore, the two hydropower stations both have no waste water all day. Moreover, the variation of Xiangjiaba water level was less than 1m, then the optimized results cannot have influence on the economic operation of the next day.

5 Conclusions

The results of Case study show that the proposed method can obtain more total energy storage in falling off season. And the comparison between the optimized outputs and actual outputs shows that the proposed strategies can be more in line with the requirements of the grid and have effective utilization of water resource while ensuring the security and stability of the system. Finally, the proposed HTCA and WCSM can generate reasonable load distribution as well as meet all kinds of hydraulic and electrical constraints. And the two proposed strategies have effective utilization of water resource while ensuring the security and stability of the power grid.

References

1. S.E. Humphris, M.K. Tivey, M.A. Tivey. The Trans-Atlantic Geotraverse hydrothermal field: A hydrothermal system on an active detachment fault. Deep Sea Research Part II Topical Studies in Oceanography. (2015)
2. A.H. Elbatran, O.B. Yaakob, Y.M. Ahmed, H.M. Shabara. Operation, performance and economic analysis of low head micro-hydropower turbines for rural and remote areas: a review. Renewable & Sustainable Energy Reviews. 43 (2015) 40-50
3. L. Mo, P. Lu, C. Wang, J. Zhou. Short-term hydro generation scheduling of Three Gorges–Gezhouba cascaded hydropower plants using hybrid MACS-ADE approach. Energy Conversion & Management. 76 (2013) 260–73
4. W. Chao, J. Zhou, L. Peng, Y. Liu. Long-term scheduling of large cascade hydropower stations in Jinsha River, China. Energy Conversion & Management. 90 (2015) 476–87
5. M. Hou, C. Luo, L. Wei, L.I. Peng. Study on short-term hydrothermal scheduling based on differential evolution. Power System Protection & Control. (2015)
6. J. Brest, A. Zamuda, B. Bošković. Adaptation in the Differential Evolution. Adaptation Learning & Optimization. (2015)
7. J.X. Yuan, D.Q. Zhang, Z. Zhang. Study on Optimal Operation for Recession Flood of Hydropower Reservoir. South-to-North Water Transfers and Water Science & Technology. (2007)

MATEC Web of Conferences **55**, 06006 (2016) DOI: 10.1051/matecconf/20165506006
ACPEE 2016

Research of Charging(Discharging) Orderly and Optimizing Load Curve for Electric Vehicles Based on Dynamic Electric Price and V2G

Shuai Yang[1,a], Zhaowen Luan[2] and Zhen Qin[3]

[1,2,3]*Electrical Engineering College, Shandong University, 250061 Jinan, Shandong Province, China*

Abstract. Firstly, using the Monte Carlo method and simulation analysis, this paper builds models for the behaviour of electric vehicles, the conventional charging model and the fast charging model. Secondly, this paper studies the impact that the number of electric vehicles which get access to power grid has on the daily load curve. Then, the paper put forwards a dynamic pricing mechanism of electricity, and studies how this dynamic pricing mechanism guides the electric vehicles to charge orderly. Last but not the least, the paper presents a V2G mechanism. Under this mechanism, electric vehicles can charge orderly and take part in the peak shaving. Research finds that massive electric vehicles' access to the power grid will increase the peak-valley difference of daily load curve. Dynamic pricing mechanism and V2G mechanism can effectively lead the electric vehicles to take part in peak-shaving, and optimize the daily load curve.

1 Introduction

Compared with traditional vehicle, electric vehicle has advantages of low pollution and abundant fuel supply. According to analysis, by 2020 the number of electric vehicle ownership in china will exceed 5 million[1]. Nowadays, many scholars are applying themselves to researching on the subject of large scale of electric vehicles connecting to the power grid[2-5].

This paper is organized as follows. In section Ⅱ, we use Monte Carlo method to model the conventional charging behaviour of electric vehicles and use analogy simulation to model the fast charging behaviour of electric vehicles. In section Ⅲ, we propose a kind of dynamic electricity price method which can guide electric vehicles to charge orderly. In section Ⅳ, we propose a kind of V2G method which can guide electric vehicles to charge(discharge) orderly. In the last section, we look back the full text and tally up textual point.

2 Model the Behaviour of Electric Vehicles

According to the forecast[1], by 2020 the number of electric vehicle ownership in china will exceed 5 million, of which the personal cars are more than 4.3 million, almost 86% of the total number. Electric vehicles mainly have two kinds of charging models, which are conventional charging model and fast

[a] Corresponding author : jinandiqu@163.com

charging model. Conventional charging facilities are generally located in residential, corporation, parking lot and so on. In general, charging power of this model is about 10kW, its charging time is longer and its charging speed is slow. Conventional charging is used to meet the needs of daily use, usually being conducted in the evening or at night. Fast charging facilities are generally located in a special charging station, mainly to meet the needs of temporary charging. Its charging speed is relatively fast and its charging power is about 40kW. When being charged half an hour in the fast charging station, electric vehicles can drive almost 100 kilometers.

We use the Monte Carlo method and analogy simulation to model the behaviors of electric vehicles.

2.1 Conventional Charging Modeling

We use *DENZA* charging facility in this paper which produces 10kW of power. According to related data, we can get the last trip ending and return time distribution probability density[6] and electric vehicle driving distance distribution probability density[7][8]. Continuous charging time is calculated as follow:

$$t_i = \frac{s_i}{100\times(p_c\times\eta)}\times k \tag{1}$$

t_i is the charging time of the i_{th} datum, s_i is the driving distance of the i_{th} datum. The two share the same distribution probability density. p_c is the rated power of charging facility. $\eta=0.9$ is charging efficiency, k is power consumption per hundred kilometers. The $k=18.6kwh$ refers to *Qin* of *BYD*. The result are shown in Figure 1 and Figure 2

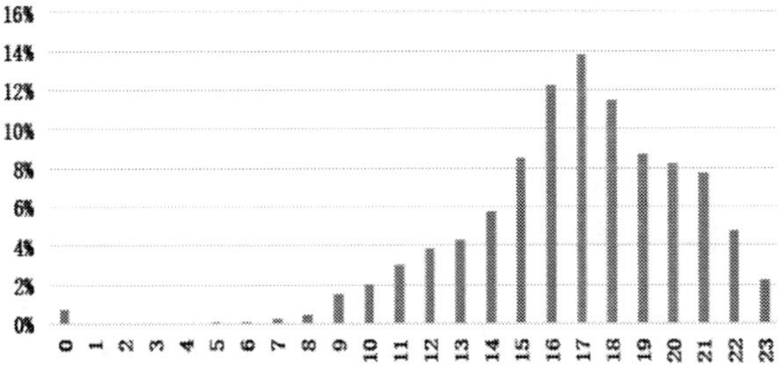

Figure 1. Last trip ending and return time distribution probability density

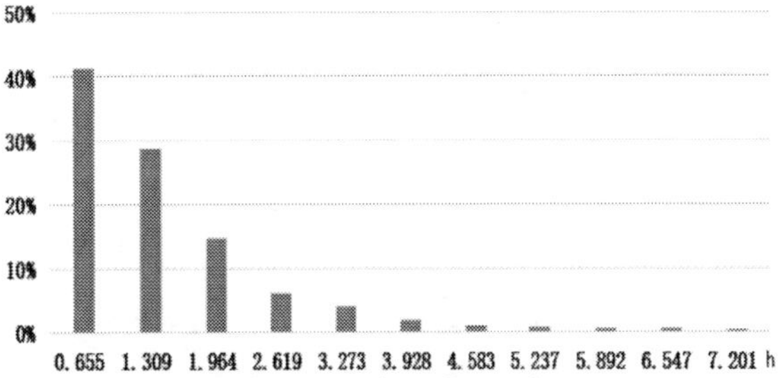

Figure 2. Continuous charging time distribution probability density

2.2 Monte Carlo simulation

Monte Carlo simulation is based on Probability and Mathematical Statistics. Under this method, the problem is solved with a certain probability model, and the approximate solution of the problem is obtained by using large scale simulation[9].

The mathematical and empirical evidence show that the random variables which are influenced by plenty independent homogeneous-small effects are subject to normal distribution. That is to say, if a variable is influenced by many small independent random factors, then it can be considered to be subject to normal distribution. Charging starting time and duration of charging are related to personal preference and mutual independence. Therefore, the charging time of electric vehicle and the duration of charging can be used for the curve fitting of normal distribution[5]. The results of curve fitting are as follows:

$$f_1(\mathrm{t}) = a_1\times\exp\left(-\left(\frac{t-b_1}{c_1}\right)^2\right), \mathrm{t}\in[0,24] \tag{2}$$

$$f_2(\mathrm{t}) = a_2\times\exp\left(-\left(\frac{t-b_2}{c_2}\right)^2\right), \mathrm{t}\in[0,\infty) \tag{3}$$

a_1=0.1195, b_1=17.32, c_1=4.721; a_2=0.5283, b_2=-0.3791, c_2=2.113; 95% confidence interval. $f_1(t)$ is the probability distribution of charging start time, $f_2(t)$ is the probability distribution of charging duration. Then daily charging expectation of single electric vehicle can be gotten by Monte Carlo simulation. The calculation is shown in Figure 3 and the result is shown in Figure 4.

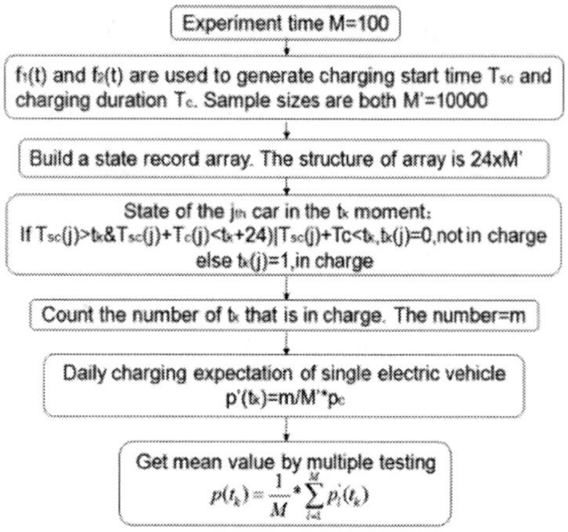

Figure 3. Calculation of Monte Carlo simulation

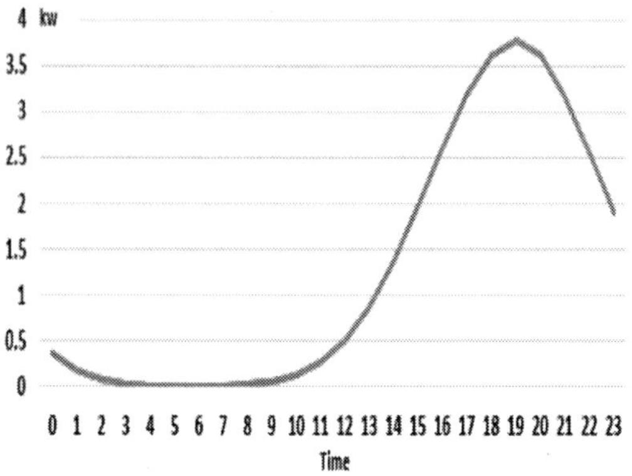

Figure 4. Daily load curve of single electric vehicle

2.3 Fast charging modeling

Because it is hard to estimate the fast charging behaviour, we get the load curve of fast charging through modeling the business of fast charging station. The vehicle flow of charging station is not the same in different time[10]. The power of charging facility is 40kW, which is referred to *BYD*'s charging facility. The modeling of fast charging behaviour is shown in Table 1.

Table 1. Model of fast charging behavior

	Station Numbers	**Charging Facility Number in Each Station**	**Charging Manners**	**Charging Time per Car**	**Business Hours**
Charging Station Status	220 sites	5 sites	1 facility 1 charging interface	30min	5:00-24:00
Arrival Interval	**5:00-6:00**	**6:00-10:00**	**10:00-14:00**	**14:00-17:00**	**17:00-20:00**
	15min	10min	8min	6min	4min
	20:00-21:00	**21:00-22:00**	**22:00-24:00**		
	6min	10min	15min		

According to calculation, the daily load curve of fast charging is obtained as Figure.5.

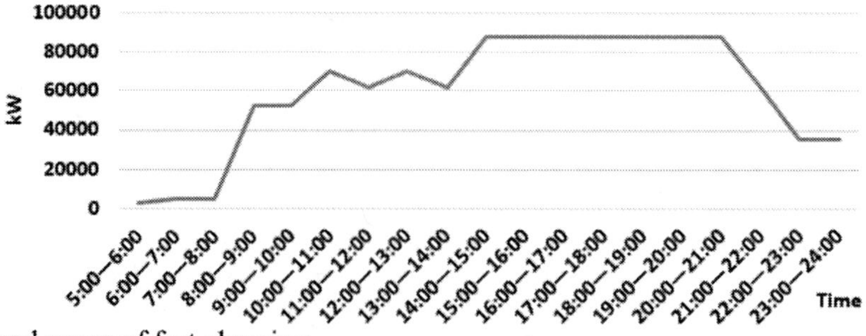

Figure 5. Daily load curve of fast charging

2.4 Fast charging modeling

 DOI: 10.1051/matecconf/20165506006

We choose the daily load of a northern city in China as the basic load. When the number of electric vehicles is 100,000, 200,000 and 500,000 respectively, the daily load curves are shown in Figure 6.

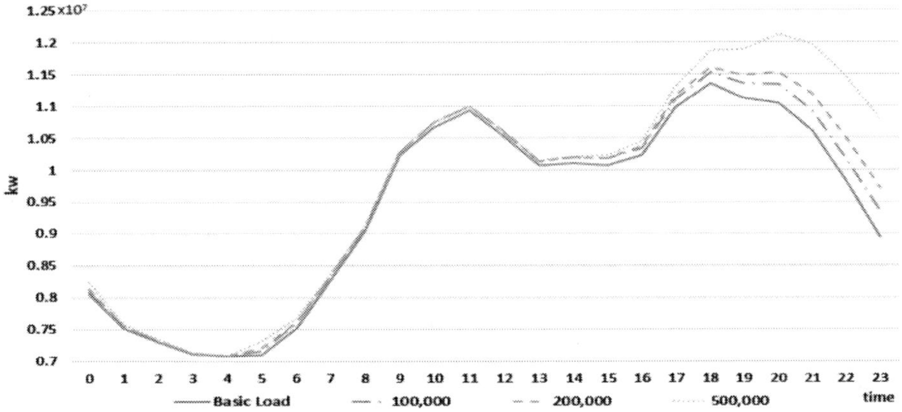

Figure 6. Daily load curve after electric vehicles connecting to the grid

Definite peak valley ratio:

$$\delta = \frac{(P_{max} - P_{min})}{P_{max}} \times 100\% \tag{4}$$

In the formulation above, δ is peak valley ratio, P_{max} is peak load, P_{min} is valley load.

Definite peak superposition and valley superposition:

$$\gamma_g = \frac{\left(p'_{min} - p_{min}\right)}{p_{min}} \times 100\% \tag{5}$$

$$\gamma_f = \frac{\left(p'_{max} - p_{max}\right)}{p_{max}} \times 100\% \tag{6}$$

In the formulations above, γ_g and γ_f are peak superposition and valley superposition respectively. p'_{min} is valley load which contains both basic load and electric vehicle load; p'_{max} is peak load which contains both basic load and electric vehicle load. p_{min} is basic valley load and p_{max} is basic peak load. The result of calculation about load curve that electric vehicles connect to the grid is shown in Table 2.

Table 2. The calculation about the electric vehicles connecting to the grid

	p'_{min} (MW)	γ_g	p'_{max} (MW)	γ_f	δ
Basic Load	7,075	0	11,350	0	37.67%
100,000	7,076.82	0.03%	11,525.23	1.544%	38.60%
200,000	7,078.72	0.05%	11,534.14	1.62%	38.63%
500,000	7,084.3	0.13%	13,056.35	15.03%	45.74%

According to Figure.6 and Table.2, the load characteristics of electric vehicles are similar to the city's electricity load characteristics. The reason is that most users tend to charge electric vehicles after stopping work and arriving at home in the evening in the meanwhile the charging period is also the peak period. This characteristic makes the difference between peak load and valley load bigger. In addition, it can be inferred from Figure.1 and Figure.2 that the electric vehicle is usually used as a tool replacing the foot, so the driving distance is not so long. This characteristic makes it unavailable to fill the valley by charging in the midnight. The characteristics discussed above may waste the power, increase the power loss and increase the burden of power grid. From what has been discussed above, we may safely arrive at the conclusion that due attention has to be paid to guide the charging behaviors of electric vehicles.

3 Dynamic Power Price

Different power prices in different time can guide charging behaviors of electric vehicles. When the price is high, users tend to charge at other times. Conversely, when the price is low, users tend to charge in this time. The price elasticity function can explain this phenomenon:

$$\varepsilon = \frac{\Delta D_t / D_t}{\Delta C_t / C_t} \tag{7}$$

In formulation(7), ε is price elasticity index; ΔD_t and ΔC_t are demand changes and price changes respectively in t_{th} moment; D_t and ΔC_t are demand and price in t_{th} moment.

In this paper, one day is separated into 24 electric price ranges. Each range starts at the beginning of this hour and the price continues one hour. The electric price of each range is related to the total load of this range and the mean load of the whole day. The models are as follows:

$$D_t^{'} = D_t + \Delta D_t \tag{8}$$

$$\Delta D_t = D_t \times \frac{\Delta C_t}{C_t} \times \varepsilon \tag{9}$$

$$\frac{\Delta C_t}{C_t} = \frac{D_t - \bar{D}}{D_t} \tag{10}$$

$$\bar{D} = \sum_{t=1}^{T} D_t \tag{11}$$

D'_t is the load adjusted by dynamic electric price, $\bar{D}$ is the mean demand of the day, T is the time range, $\varepsilon=-0.3$[11], the number of electric vehicles connecting to power grid is 500,000. Comparison of load curves before and after being adjusted by dynamic electric price are shown in Figure 7 and Table 3.

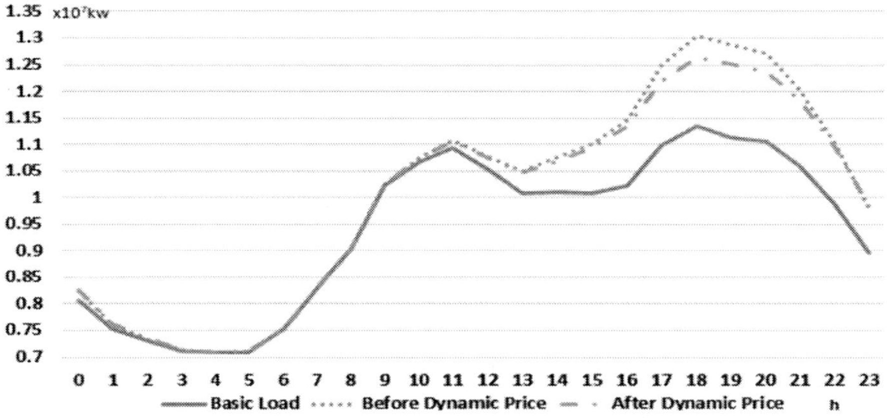

Figure 7. The effect of dynamic electric price 1

Table 3. The effect of dynamic electric price 2

	γ_g	γ_f	δ
Basic Load	0	0	37.67%
Before Dynamic Price	0.131%	15.030%	45.74%
After Dynamic Price	0.117%	11.381%	43.97%

It can be seen from Figure.7 and Table.3 that load curve condition is improved by dynamic electric price. Under this dynamic electric price, although γ_g has a little decrease, γ_f has a significant reduction. γ_f decreases from 15.030% to 11.381%, having 3.649% fall. In addition δ decreases from 45.74% to 43.97%, having 1.77% fall. So it can be concluded that it is useful for this dynamic electric price to guide the charging behaviors of electric vehicles and smooth the load curve.

4 V2G Regulation

Vehicle to Grid is called V2G for short[12]. V2G describes a system where plug-in electric vehicles communicate with the electric power grid to sell demand response[13]. Hence, V2G application is transformational and revolutionary in that it precipitates the confluence of transportation and electric power system, thereby introducing electricity as a meaningful automotive fuel[14]. Under V2G regulation, electric vehicles require charging and discharging batteries.

Its model is as followed:

- The electric vehicle can obtain power from power grid when it need to charge;
- The electric vehicle can release power energy to power grid when it need to discharge.

It can be inferred from Figure.1 and Figure.2 that most users drive their electric vehicles not so far. So their charging time is not so long. Charging time is less than 4 hours in a day for most users. It means that when the battery is full, there will be a long time without any activities for electric vehicles. The free time can be used to fill the valley of load curve in order to reduce the burden of power grid and conserve the energy. Under V2G regulation, electric vehicles are discharged in peak period and charged in valley period.

Based on Figure.1, Figure.7 and the experience, the users who get home from 16:00pm to 23:00pm can be regarded as main force to participate in V2G program. Referring to the step electric price mechanism of Beijing[15], we set the peak load period from 18:00pm to 23:00pm and the valley load period from 23:00pm to 7:00am. Under the prerequisite of fully charging in electric vehicles, the cars participating in V2G program discharge in peak load period and charge in valley load period. The main mission of V2G program set by us in this paper is to clip the peak load and fill the valley load. The car that begins to discharge in i o'clock. The model is as follows:

$$t_{fi} \rightarrow max \tag{12}$$

$$T_i + 2t_{fi} + t_{ci} \leq (24-23+7) \tag{13}$$

$$0 \leq t_{fi} \leq (23-16) \tag{14}$$

t_{fi} is the maximum discharging time of electric vehicles that begin to discharge in i o'clock, t_{ci} is the charging time of electric vehicles which begin to discharge in i o'clock. After calculation, the maximum discharging time of electric vehicles that begin to discharge in i o'clock is shown in Table 5.

Table 4. Maximum discharging time

	0.655h	**1.309h**	**1.964h**	**2.619h**	**3.273h**	**3.928h**	**4.583h**	**5.237h**
16:00	7h	6.691h	6.036h	5.381h	4.727h	4.072h	3.417h	2.763h
17:00	6h	6h	6h	5.381h	4.727h	4.072h	3.417h	2.763h
18:00	5h	5h	5h	5h	4.727h	4.072h	3.417h	2.763h
19:00	4h	4h	4h	4h	4h	4h	3.417h	2.763h
20:00	3h	3h	3h	3h	3h	3h	3h	2.763h
21:00	2h	2h	2h	2h	2h	2h	2h	2h
22:00	1h	1h	1h	1h	1h	1h	1h	1h

In Table 4, the first row represents duration and the first column represents starting time.

Definition: a_{ij} is the discharging element exiting in Table.5. i is the beginning time of discharging, j is the duration of charging. The total number of electric vehicle accessing the grid is A=500,000, p_i is the probability of charging in i o'clock, q_j is the probability of that electric vehicle charge for j hour(s), D_i is the power that is injected to the grid by electric vehicles in i o'clock. Search the a_{ij} above the Table.5 one by one and calculate the total contribution every a_{ij} makes to D_i. Sum of D_i is the power injection. The calculation and results are shown in Figure.8 and Table 5 respectively.

 DOI: 10.1051/matecconf/20165506006

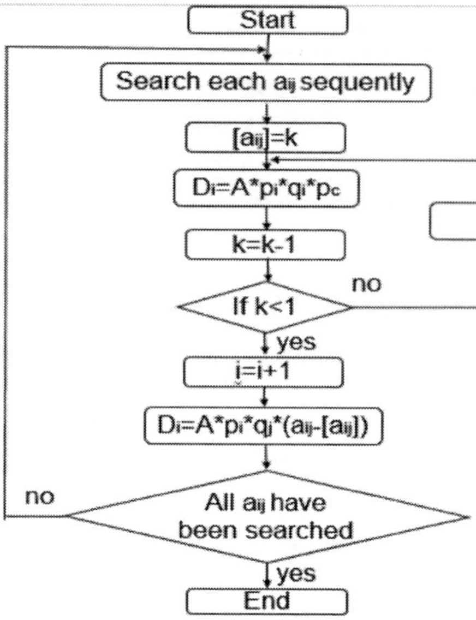

Figure 8. The calculation of discharging

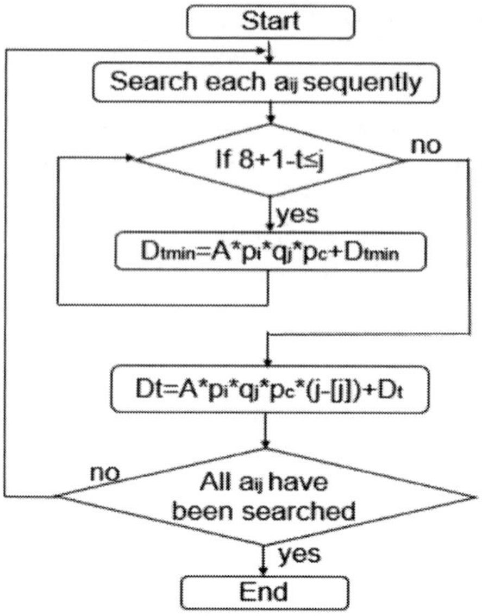

Figure 9. The calculation of charging

Table 5. Power of discharging

	16:00-17:00	**17:00-18:00**	**18:00-19:00**	**19:00-20:00**	**20:00-21:00**	**21:00-22:00**	**22:00-23:00**
16:00	611,750 kw	611,750 kw	610,552 kw	603,465 kw	584,159 kw	542,239 kw	377,856 kw
17:00		690,000 kw	690,000 kw	688,649 kw	680,656 kw	658,880 kw	601,186 kw
18:00			575,000 kw	575,000 kw	573,874 kw	567,213 kw	540,390 kw
19:00				437,500 kw	437,500 kw	436,643 kw	431,575 kw
20:00					412,500 kw	412,500 kw	387,500 kw
21:00						387,500 kw	387,500 kw
22:00							237,500 kw

In Table 5, the first row represents the time period of discharging and the first column represents the start time of discharging.

After getting t_{fi} and t_{ci}, we can calculate t'_{ci}, the value of charging time in valley load period.

$$t'_{ci} = t_{fi} + t_{ci} \tag{15}$$

We separate the valley load period into 8 time periods. Each a_{ij} select one period that has the minimum load. Generally speaking, from 23:00-7:00, the later time is, the lower the loads are. So the charging rule we propose is that charging in later time period is prior than charging in the early time period. If a_{ij} can charges only 1.3h, for example, it charges 1h in the time period of 6:00-7:00 and charges 0.3h in the time period of 5:00-6:00. In other words, it charges from 6:00 to 7:00 at rated power and charges from 5:00 to 6:00 at 30% rated power. The calculation of charging is shown in Figure.9 and Table 6.

Table 6. Power of charging

	23:00-0:00	**0:00-1:00**	**1:00-2:00**	**2:00-3:00**	**3:00-4:00**	**4:00-5:00**	**5:00-6:00**	**6:00-7:00**
16:00	524,536	611,750	611,750	611,750	611,750	611,750	611,750	611,750
17:00	263,811	591,630	690,000	690,000	690,000	690,000	690,000	690,000

18:00	54,871	219,842	493,025	575,000	575,000	575,000	575,000	575,000
19:00	19,731	49,711	160,669	375,128	437,500	437,500	437,500	437,500
20:00	5,586	18,604	46,870	151,488	353,692	412,500	412,500	412,500
21:00	758	5,247	17,476	44,029	142,307	332,256	387,500	387,500
22:00	0	464	3,216	10,711	26,985	87,220	203,640	237,500

Based on Figure 8, Figure 9, Table 5 and Table 6, we can get the daily load curve adjusted by dynamic electric price and V2G. The comparison between load curve and load curve adjusted by dynamic electric price and V2G is shown in Figure.10 and Table 7. The number of electric vehicles connecting the grid is 500,000.

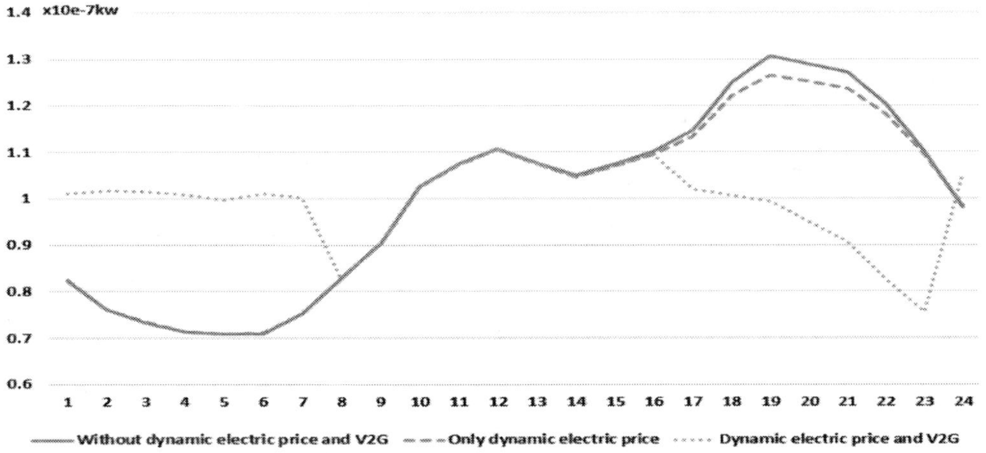

Figure 10. The effects of V2G and dynamic electric price 1

Table 7. The effects of V2G and dynamic electric price 2

	Valley Time	p'$_{min}$	Peak Time	p'$_{max}$	δ
No dynamic electric price	4:00-5:00	7081 MW	18:00-19:00	13056 MW	45.76%
Only dynamic electric price	4:00-5:00	7083 MW	18:00-19:00	12642 MW	43.97%
Dynamic electric price and V2G	22:00-23:00	7553 MW	11:00-12:00	11046 MW	31.62%

It can be known from Figure.10, and Table.7 that the method of V2G is valuable for improvement of power load characteristics. But it should be noticed that the valley time has been changed after V2G method adopted. If the number of electric vehicles remain elevated, V2G maybe be not enough. The problems will be more complicated. If we just make use of V2G simply when the number of electric vehicles is large enough, the new valley load period may appear.

5 Conclusion

When the number of electric vehicles connecting to power grid is small, the influence to the grid is limited. When the number of electric vehicles connecting power grid is big, the influence to the grid is tremendous. Dynamic electric price is available to optimize the daily load curve. It can guide electric vehicle users to charge their cars orderly. V2G is more useful to optimize the daily load curve. It allows users of electric vehicles to participate in peak/valley load regulating of power system. But it should be noticed that the valley time has been changed after V2G method being adopted. If the number of electric vehicles remain elevated, V2G will be not enough. If we just make use of V2G simply when the number of electric vehicles is large enough, the new valley load period may appear.

Many problems still need to be further studied, such as charging/discharging period segmentation, the subsidy rule to V2G users, coordination between renewable energy and electric vehicles and so on.

References

1. "Electric Vehicle Charging Infrastructure Development Guide (2015-2020)", China National Development and Reform Commission,2015:3-5.
2. Lopes, J.A.P.; Soares, F.J.; Almeida, P.M.R., "Integration of Electric Vehicles in the Electric Power System," in *Proceedings of the IEEE* , vol.99, no.1, pp.168-183, Jan. 2011
3. Madawala, U.K.; Thrimawithana, D.J., "A Bidirectional Inductive Power Interface for Electric Vehicles in V2G Systems," in *Industrial Electronics, IEEE Transactions on* , vol.58, no.10, pp.4789-4796, Oct. 2011
4. Rahman, S.; Shrestha, G.B., "An investigation into the impact of electric vehicle load on the electric utility distribution system," in *Power Delivery, IEEE Transactions on* , vol.8, no.2, pp.591-597, Apr 1993
5. Shaoyun Ge; Liang Feng; Hong Liu, "The planning of electric vehicle charging station based on Grid partition method," in *Electrical and Control Engineering (ICECE), 2011 International Conference on* , vol., no., pp.2726-2730, 16-18 Sept. 2011
6. A. M. William and A. M. Martin, "Travel estimation techniques for urban planning," Transport Research Board National Research Council, U.S., NCHRP 365 Rep. B8-29.6,1998.[online]. Available: http://ntl.bts.gov/lib/21000/21500/21563/PB99126724.pdf.
7. Taylor M J, Alexander A. Evaluation of the impact of plug-in electric vehicle loading on distribution system operations. IEEE Power & Energy Society General Meeting, Calgary, Canada, 2009: 1-6.
8. Vyas A, Santini D. Use of national surveys for estimation 'full' PHEV potential for oil user reduction, 2008-07-21. Available: http://www.transportation.anl.gov/pdfs/HV/525.pdf.
9. Xiao Gang, "Monte Carlo method in system reliability analysis", published by Science Press,2003.
10. Bai Gaoping, "Research on the Adaptability of Large-scale Electric Vehicle Charging(discharging) Stations and the Grid", Master's thesis , Beijing Jiaotong University, 2011.
11. Lamont, J.W.; Obessis, E.V., "Emission dispatch models and algorithms for the 1990s," in *Power Systems, IEEE Transactions on* , vol.10, no.2, pp.941-947, May 1995.
12. Chukwu, U.C.; Mahajan, S.M., "V2G electric power capacity estimation and ancillary service market evaluation," in *Power and Energy Society General Meeting, 2011 IEEE* , vol., no., pp.1-8, 24-29 July 2011.
13. Available: http://en.wikipedia.org/wiki/Vehicle-to-grid
14. Chukwu, U.C.; Mahajan, S.M.; Spina, I.; Rizzo, R., "A nomogram for estimating energy loss in a distribution network due to penetration of V2G," in Clean Electrical Power (ICCEP), 2013 International Conference on , vol., no., pp.476-481, 11-13 June 2013.
15. National Development and Reform Commission, "*Notice of the national development and Reform Commission on the implementation of the linkage of coal and electricity prices in North China Power Grid*", China public document,2005.

MATEC Web of Conferences **55**, 06007 (2016) DOI: 10.1051/matecconf/20165506007
ACPEE 2016

Digital Metering System of New Generation Smart Substation in China

Jingfen Bai [1,a], Fantao Lin[1], Xiangjiang Yang[1], Jia Zuo[1], Yingying Cheng[2] and Chunping Yu[1]

[1]*China Electrical Power Research Institute, 100192 Beijing ,China*
[2]*Chongqing Electric Power Research Institute, 400000 Chongqing, China*

Abstract. The new generation of smart substation is a summary and improvement of the earlier intelligent substations within the past 5 years. As a part of the substation operation function, the electric energy metering system must adapt to the requirements of the new generation smart substation which is more intelligent, full digitalization, networked and integrated. With the features of digitized, easy information sharing, and hardware shared, the digital measurement system is the inevitable choice for the new generation smart substation. This paper is the first literature to introduce the concept and structure of the digital metering system. The composition and technical characteristics of the new generation Smart Substation measurement system are introduced in detail.

1 Introduction

As one of the important parts in developing smart grid, the smart substation uses advanced, reliable, integrated, low- carbon and environmentally friendly equipments and design. With qualities of digitization of whole station information, networking of communication platform, standardization of information-sharing, integration of system function, compact design of structure, intelligent of high-voltage equipment and visualization of running status, smart substation can support power grid to realize real-time online analysis and control decision, and improve the operation reliability and economy.

The digitized, networked, standardized and integrated features of smart substation not only affect the design and configuration of secondary systems such as protection, measurement and control systems, but also affect the metering systems of the substation. Since 2009, the construction of smart substation has started in china [1] , the early metering system of smart substation used traditional transformer, the digital level is not so high, and the integration with other secondary systems is not so deep, which limits the information-sharing between different services of the smart substation, and affects the level of intelligence. Compared with the earlier smart substations, the new generation smart substation puts forward higher requirements to intelligent, digitized and integrated level. The digital metering system should meet the overall requirements of the new generation smart substation with more comprehensive intelligent, digitization and integration.

2 Digital Metering System

[a] Corresponding author : jfbai@epri.sgcc.com.cn

With the feature of digitization, the digital metering system aims at metering and collecting electric energy in someone measuring point, it consists of digital electric energy metering device for power metering and acquisition terminal for power collection.

2.1 Digital Electric Energy Metering Device

The digital electric energy metering device consists of transformer, merging unit and digital input electricity meter, it is different from the traditional ones, and it adopts optical cable instead of electric cable to connect merging unit and digital input electricity meter, so the voltage drop caused by secondary cable can be reduced and result in composite error reduction [2-4]. This paper holds different views with previous records and believes that electronic transformer is not a distinguishing feature of the digital metering device, electronic transformer and traditional transformer are just in different way of sensing and different degree of digitization. In this paper, the digital electric energy metering device is classified into full digital and half digital in accordance with different signal sensors.

The signal sensing unit of full digital metering device is the electronic transformer. Between the transformer and the merging unit, and between the merging unit and the digital input electricity meter, voltage and current are transferred in the form of network packets which contain the discrete sampled values via different protocols, the structure is as shown in Figure 1:

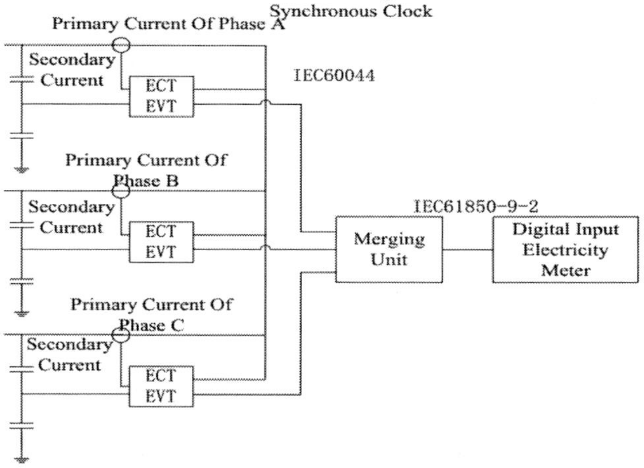

Figure 1. Structure of the full digital metering device.

The electronic transformer senses primary voltage and current, and transfers the sampled value to merge unit in the form of packet via the protocol of IEC60044, after the packet being collected and processed by merging unit, it will be assembled into new network packet via the protocol of IEC61850-9-2, and then be transferred to the digital input electricity meter of bay level by point-to-point or high -speed Ethernet mode, after a series of data processing and calculation, the electric energy accumulation of measurement point is completed.

The signal sensing unit of half digital metering device is the traditional electromagnetic transformer, traditional analog voltage and current signals are transferred between the transformers and merging unit, the structure is as shown in Figure 2:

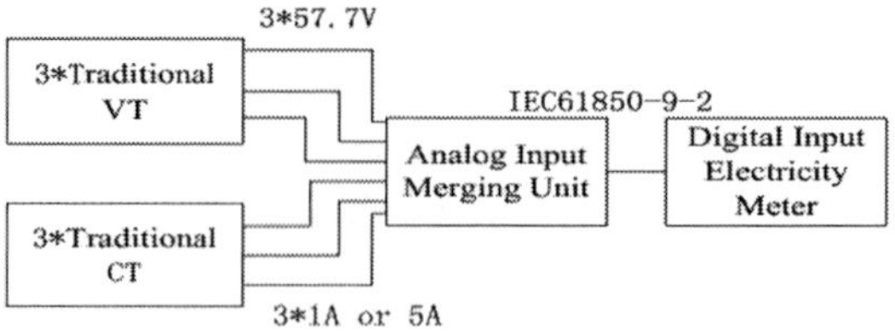

Figure 2. Structure of the half digital metering device.

The traditional transformer converts high current and voltage into small signals of rated voltage 57.5V, and rated current 1A or 5A, the analog input merging unit samples data uniformly. Then device transfers the signals and accumulates power in the same way as the full digital one.

2.2 Electric Energy Acquisition Mode

Smart substation is based on IEC61850 uniformly, in the early construction of smart substation, the digital input electricity meter only realized the digital input, but didn't completely follow the 61850 protocol to realize modelling and networking. The electric energy acquisition still used half duplex communication mode based on DL/T 645 communication protocol, which not only affects the internal sharing of data and the standardization of substation, but also affects the timeliness and integrity of data acquisition due to the limitation of the protocol itself, that the information related to primary power energy data and sampled value can't be transferred. The electric energy acquisition terminal of digital metering system acquires positive and negative active power, four quadrant reactive power, demand, real-time voltage/current/power, abnormal events and other information of the digital input electricity meter according to IEC61850-8-1MMS protocol, and then transfers the information to the main station in the form of packet according to IEC60870-5-102 protocol, which is as shown in Figure 3:

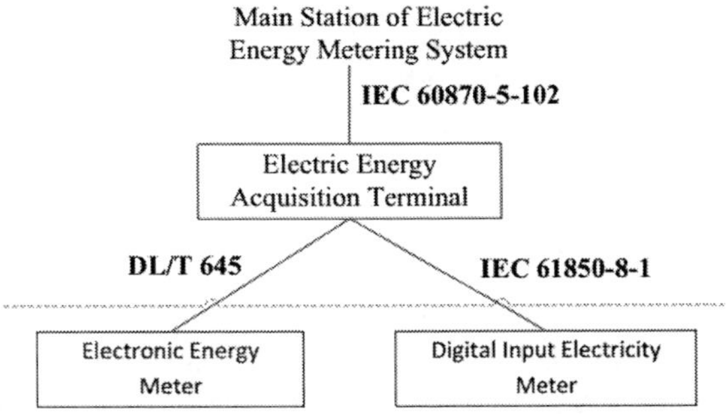

Figure 3. Acquisition structure of digital metering system.

The first version of IEC 61850 has only one logical node MMTR designed for electric energy, even the second version distinguishes three phase data from single phase data and just adds one logic node MMTN, which could not fully meet the requirement of electric energy data, so when use in China, it should be developed so as to adapt to the domestic business situation [5-9].

2.3 Advantages of Digital Metering System

(1)Digital signal is transferred by the digital metering system, so there is no error caused by secondary voltage drop due to the secondary cable connection, the error of the electric energy meter is reduced for the accumulation of electrical energy is a pure numerical calculation process, therefore, the composite error of the digital metering device only depends on the transformer and the merging unit, and thus will be greatly reduced.

(2)The digitization and networking features, the unified IEC 61850 protocol, the same data source and acquisition path with other secondary systems such as protecting, measuring and controlling system, all these make the digital metering system easy to realize data sharing and software intelligence, as well as hardware resource sharing.

(3)For full digital metering system, electronic transformer with the advantages of small size, unsaturated, wide band and so on, can expand the spectrum width of the system, and will be more suitable for harmonic measurement and impact load measurement.

3 Configuration Schemes of New Generation Smart Substation Electric Energy Metering System

3.1 Electric energy metering system of earlier smart substation

The typical configuration of earlier smart substation electric energy metering system is shown as Figure 4:

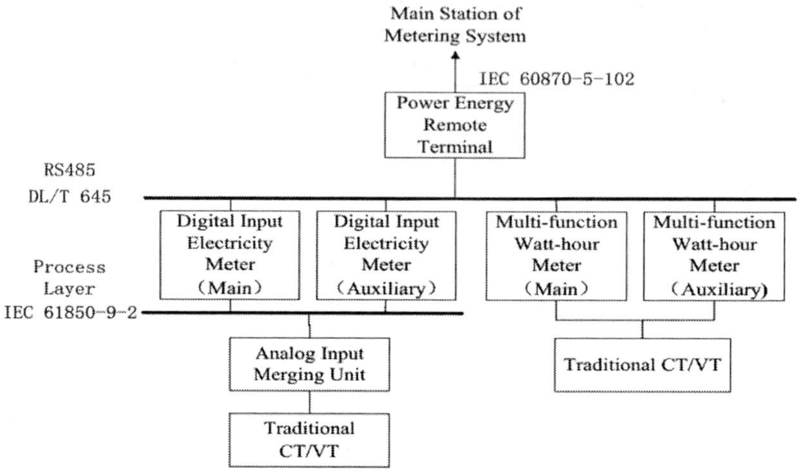

Figure 4. Typical configuration diagram of metering system.

Because of the electronic transformer exposed some stability and reliability problems in the early stages of development, the earlier smart substation did not use electronic transformer, but use the half digital metering device which was composed by analog input merging unit and digital electric energy meter. This composition can only be applied to the assessment measurement point where electric energy measurement is only used for the assessment of internal economic indicators or the assessment of power balance.

Measurement points for settlement belong to the category of national legal management. At present, in the case of merging unit and digital electric energy meter national metrological verification regulations have not been published, traditional electric energy measurement methods have still been used. The data acquisition method of assessment and settlement measurement points is to report power data according to DL/T 645 protocol by electric energy acquisition terminal through RS485 data bus.

3.2 Electric Energy Metering System of New Generation Smart Substation

Compared to the earlier smart substation, the new generation smart substation faces the demand of smart grid, places more emphasis on advanced features, integrated applications and interactive, strengthens the function applications of intelligent equipment, condition monitoring, advanced function of substation automation system and so on.

Compared with the earlier smart substation, the new generation smart substation has more prominent features of digitalization, networking. It uses electronic transformers as voltage and current sensing devices uniformly. Information accesses the substation automation system by merging unit through network, so as to realize the substation real-time information acquisition and security control functions. Compared to traditional transformer, electronic voltage/current transformer does not have many secondary windings for different function systems, the voltage/current digital sampler is included in the transformer, the merging unit only need to complete the merging and synchronization of the sampled value which from different voltage and current transformers. The output of the merging unit can also be the source of public information for the relay protection, measurement control and metrology devices, and so on.

In order to solve the problems such as overlapping network, multiple devices, taking multi screen cabinet, high construction cost, complex operation and maintenance in the earlier smart substation, one of the design targets of the new generation smart substation metering system is the integration with other secondary systems. And different services among metering, protection, and measurement control can share the network resources and the data source in order to simplify the network structure. Further, hardware resource can also be shared, the assessment measurement function which been undertook by independent meter originally can be incorporated into the measuring and controlling device, in such a way, the number of internal equipments and screen cabinets in station can be reduced, and the construction cost can be cut down. The configuration schemes of new generation smart substation measurement system are as follows.

Assessment measurement points share the output of merging units with other equipments in the station. Such kind of measurement points do not need independent digital electric energy meter, the metering function is integrated with other multifunction devices which have the same data source. For 110kV and above voltage levels, electronic transformers are used, measurement function is integrated in the digital input multi-function measuring and controlling device or device having integration of protection and measuring functions, for 10kV/35kV voltage levels, analog transformers are used, measurement device is integrated in the multifunctional device. For metering integrated multifunction device, the positive/negative, active/reactive energy metering, demand metering, freeze, meter cleared, demand cleared, event logging and other basic measurement functions should be realized, the power pulse output port should be with, which facilitates the detection of energy measurement error, algorithm of energy measurement was improved to make sure the measurement and control device could meet the requirements of 0.5S level, and meets the electric energy metering accuracy need of assessment measurement.

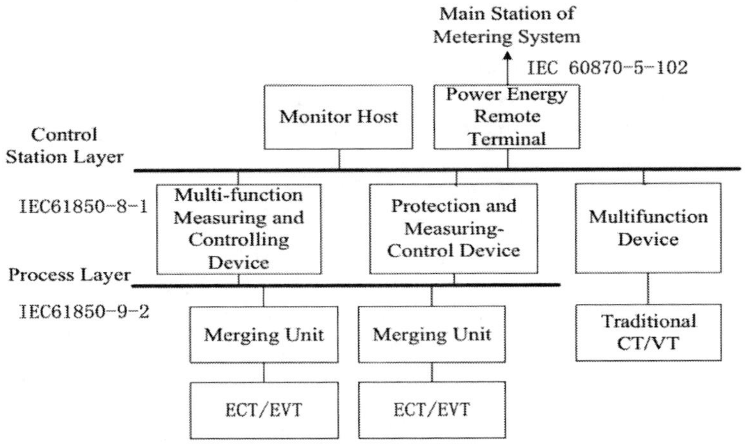

Figure 5. Assessment measurement point configuration scheme.

For settlement measurement points, the relevant equipment verification regulations haven't yet been published at present, the traditional transformer and electricity meter are configured independently according to DL/T 448. If the tracing and transmitting system of the digital electric energy meter, electronic transformer, merging unit and other equipment is established in the future, the digital metering system will be used for trade settlement, according to the reliability of different components and the metering needs of settlement measurement points, the following settlement measurement point configuration scheme is predicted, as shown in Figure 6:

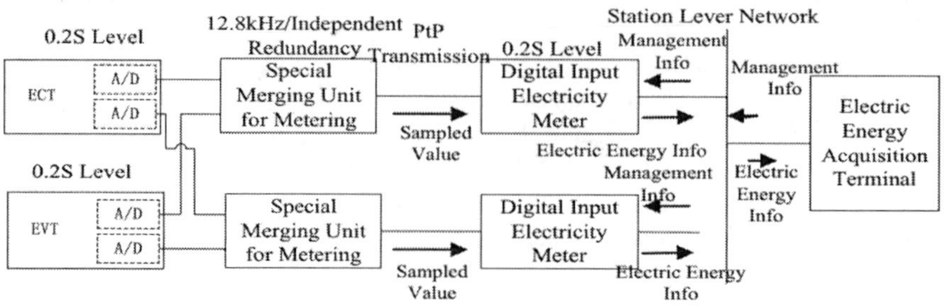

Figure 6. Advise of settlement measurement points configuration scheme.

The A / D sampling modules of ECT/EVT, merging unit, and digital input electricity meter are all electron devices, the reliability of which is lower than the transformer sensor modules. Therefore, the main/standby redundant configuration method needs to be prepared. The bay level needs to be equipped with special merging unit for metering with hardware protection function. The sampling data of merging unit is transferred point-to-point to the digital input electricity meter, which guarantees the security and independence of settlement measurement system. The electronic current transformer, the accuracy level of which is 0.2S, can meet the needs of small current measurement. Meanwhile, the merging unit, the output sampling frequency of which is 12.8 kHz, can support less than 40th harmonic metering. In addition to meeting the basic energy metering, the digital input electricity meter should also have the harmonic metering function, moreover, coordinating with other software functions such as measurement data timing freeze so as to ensure that the settlement measurement system is absolutely safe, reliable and fair.

There is a kind of measurement point that usually for assessment, but in special circumstances may be transformed into a settlement point. For example, if in the opposite side is the settlement point, in this side is such kind of measurement point, it will be transformed into settlement one when the device in the opposite side is invalid. For this kind of measurement point, the transformer, merging unit, etc, are designed according to the assessment measurement point, but the digital input electricity meter should be configured independently, as shown in Figure 7:

With the exception of the multi-function Walt-hour meter configured on settlement point still using DL/T 645, the digital input electricity meter and all kinds of measurement integration devices all use extended IEC 61850-8-1 protocol to acquire electric energy.

The devices such as multi-function measuring and controlling device, multifunction device, digital input electricity meter, merging unit, and electronic transformer, should meet the requirements of corresponding accuracy level in accordance with DL/T 448.

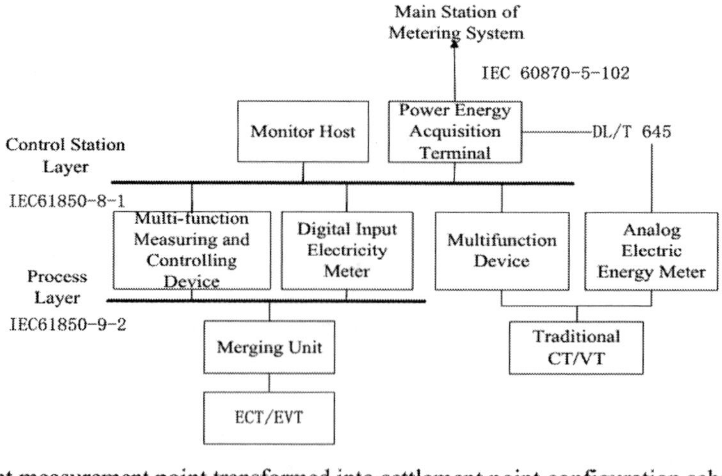

Figure 7. Assessment measurement point transformed into settlement point configuration scheme.

3.3 Metering Communication Model of New Generation Substation based on IEC 61850

Up till now, there is no metering communication data model based on IEC 61850 both in China and abroad. In order to achieve the standardization of measurement data, and to guarantee the interoperability and exchange of each equipment in the system, the following extension is carried out based on the analysis of the functions such as electric energy metering, demand metering, freeze, event reporting, and time consumption metering, as shown in Figure 8:

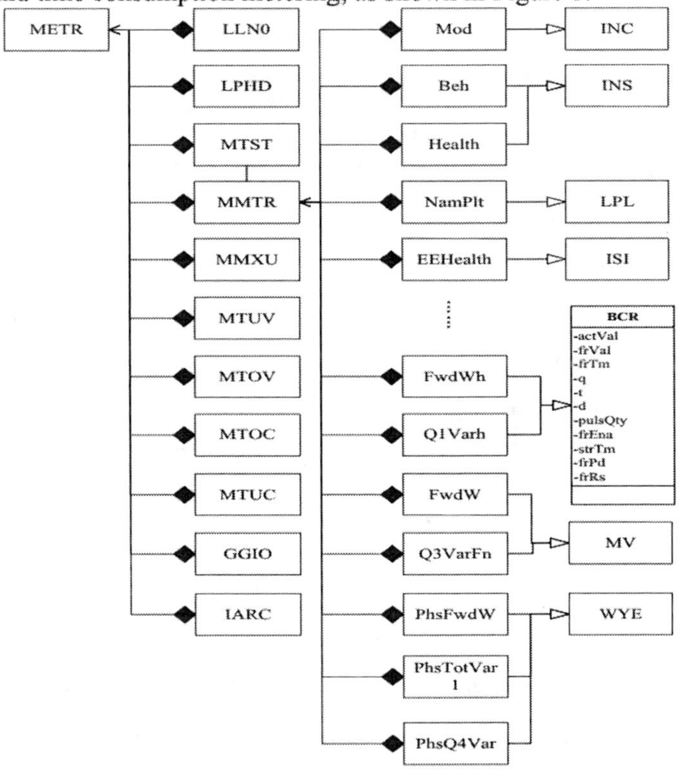

Figure 8. Energy Metering Communication Model based on IEC 61850

(1)The IEC61850-7-4 logic node MMTR can be extended for positive and negative active power metering, four quadrant reactive power metering and demand metering (MMTR for three-phase electric energy meter, and MMTN for single-phase electric energy meter). Different cases should be created for electric energy metering and demand metering, for instance, case MMTR1 for electric energy metering and case MMTR2 for demand metering, different cases are distinguished by Arabia digital suffix.

(2)The original logic node MMXU is used to measure voltage, current and so on (MMXU for three-phase electric energy meter, and MMXN for single-phase electric energy meter).

(3)The loss of voltage, phase failure and loss of current alarm limits for metering are different from the protection and control alarm limits, the MTUV、MTOV、MTUC、MTOC are created so as to monitor alarm. Different cases are created for different event alarms, even if they belong to the same type, for instance, case MTUV1 for under voltage, and case MTUV2 for loss of voltage, different cases are distinguished by Arabia digital suffix.

(4)The original logic node GGIO is used to report alarm events which needn't to be set or customized, such as abnormal sampling, electric energy meter failure, power fail, and voltage negative phase sequence, etc.

(5)The original logic node IARC is used to record programming events such as time zone /time consumption table programming, demand interval programming, meter cleared, demand cleared, event cleared, cover opening, and end button box opening, etc.

(6)The logic node MTST is created for saving and recording time zone table.

(7)The meter asset information, such as rated voltage, rated current, active/reactive power combination status word, active power pulse constant, and reactive power pulse constant, is taken as the data object of extended logic node LLN0.

3.4 Characteristics of the New Generation Smart Substation's Metering System

(1)Based on the design concept of unified data source and simplifying communication network in station, the full digital measurement system and the technical solution of integrating measurement service with other systems have been adopted, by incorporating the special communication network into the public communication network which based on IEC61850, data source and hardware resource sharing among measurement service and other systems will be realized to simplify the secondary system, that lays a good foundation for intelligent enhancing.

(2)According to the demand of measurement service, the measuring points can be sorted into three types: the assessment measurement points, the settlement measurement points and the assessment measurement points which can be transformed into settlement points in special circumstances. Different types of measurement points need different implementation schemes and different technical conditions.

(3)The IEC61850 node model and service model for the measurement and management of new generation smart substation have been established, furthermore, real time record and easy call after event of large number of frozen data have been achieved by use of IEC61850 file services, meanwhile, real-time active reporting of abnormal events has also been achieved by use of IEC61850 reporting services.

4 Measurement Technology of the New Generation Smart Substation Metering Device

Integrated equipments with measuring functions, such as digital input electricity meter, merging unit, multi-function measuring and controlling device are all new type of measurement devices, the using environment and application features are very different from the original analog electric energy meter[10-11]. On the basis of focusing on the test of basic error and measurement function, in the professional testing of measurement equipments in the new generation smart substation demonstration project, China Electric Power Research Institute adds anomaly detection in the network pressure and data packet loss situation, the added test items are as follow:

(1)The sampling value input interface receiving power test is to detect the receiving power range of the sampling value input port of the testing device. The optical attenuation meter is connected in series between a digital electric energy meter calibrator and the digital input electricity meter under test, slowly increase the attenuation of the optical attenuation meter from the beginning of 0 until the alarm lamp of the digital input electricity meter is lit, then pull off the tail fiber of digital input electricity meter, and access it to the optical power meter, read the power value at this time as the result of measurement, the test block diagram is shown as Figure 9:

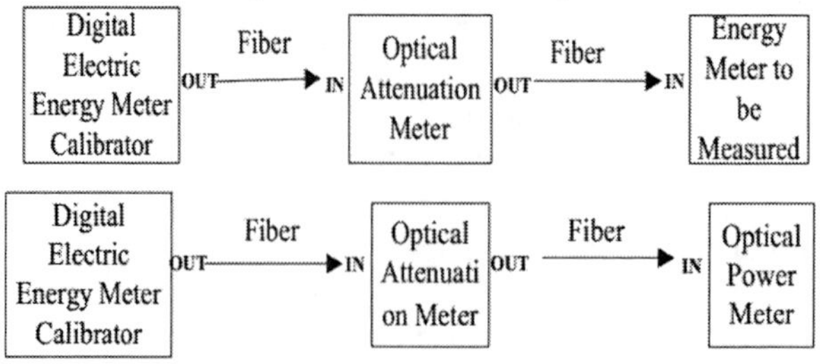

Figure 9. Sampling value input interface receiver power test block diagram.

(2)The sampling data loss effect test is to validate whether the error of energy meter meets the requirement of the grade index in case of abnormal sampling data. The digital input electricity meter testing device, in accordance with the lost point rate that not exceed the grade index, outputs exception signals of sampled value in a loop continuously, and records the measurement error, the error should be in accordance with the grade index.

(3)The network storm effect test is to validate whether the measurement performance of energy meter meets the requirement of the grade index when the sampled value input interface is in large network pressure. The digital input electricity meter testing device sends sampled value signals via switch port 1, the network tester join the broadcast storm to switch port 3 via port 2, the measurement error of the digital input electricity meter being tested should be in accordance with the grade index, the test block diagram is shown as Figure 10:

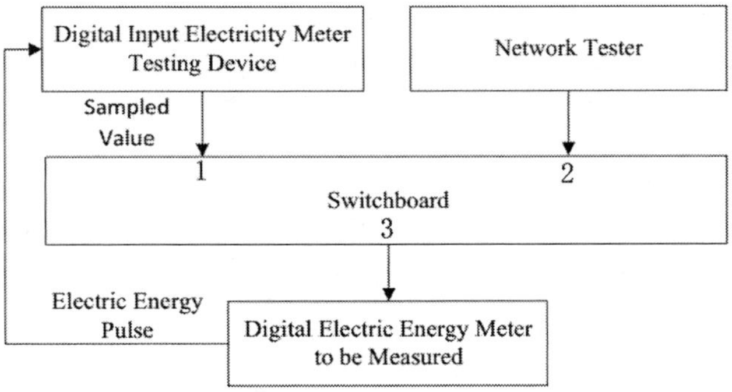

Figure 10. Network storm effect test block diagram

(4)The protocol conformance check includes sampled value protocol conformance test, configuration file check, and IEC 61850 communication function dynamic check, etc. The block diagram of dynamic check is shown as Figure 11. The testing software controls the digital electric energy meter calibrator with the pre edited test case by simulating normal state and different abnormal events according to the abnormal device duration, the output voltage and current value, the meter calibrator receives the MMS packet returned by the tested power meter, and analyses the MMS packet response node, so as to judge whether the tested power meter could respond, whether the action is consistent with the preset result of the test case, and whether the return time meets the standard requirements .

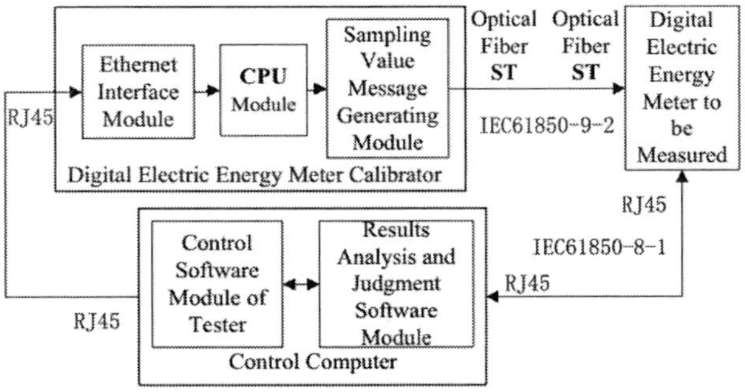

Figure 11. Block diagram of the IEC 61850 protocol conformance test system.

5 Conclusion

The new generation smart substation is more digitized, networked, integrated and intelligent. To adapting these improvements of the new smart substation, an even bigger evolution is carried on the internal metering system of the substation. On the basis of introducing the digital metering system, this article provides details of the system structure and related equipment detection technology of the digital metering system of the new smart substation. The new digital metering system is a business integration and promotion of the ordinary digital metering systems. The new metering devices, such as the testing and control device with metering function, and the new metering technologies, such as IEC 61850 meter reading application, are used for the first time. Whether these new devices and new technologies can meet the reliability, stability and security requirements of the system needs to be summarized and found in the operation process of the substation. At present, the digital metering system cannot be used as the basis for trade settlement yet. However, with the establishment of national digital metering system, in the near future, the digital metering system will be applied to all voltage levels and all types of measurement points thanks to its unique technical characteristics.

Acknowledgement

Supported by science and technology project of China State Grid Corp.(research of the on-site calibration technology for digital metering device and monitoring system of the new generation smart substation)

References

1. Q.Y. Zhang, Guizhou. Electr. Pow. Technol **14**, 1(2011)
2. S.J. Wu, J.F. Bai, H.B. Yu, et al, Power. Syst. Technol **38**, (supplement 1) (2014)
3. L.M. Cai, G.H. Liu, Electr. Pow **44**, 4 (2011)
4. K. Jiang, T. Shang, B. Chen, et al, J. Zhongyuan. Univers. Technol **23**, 2(2012)
5. J.F. Bai, F.T. Lin, S.J. Wu, Y.H. Xu, et al, ICMPMT 2013
6. L. He, *IEC 61850 Application* (CEPP, 2012)
7. IEC 61850 Communication networks and systems in substations (2003)
8. Y.L. Li, G. Li, Power. Syst. Technol **34**, 4 (2010)
9. J.M. Zhang, S.T. Jiang, N.Z. Jin, et al, Automat. Electr. Pow. Syst **34**, 11(2010)
10. Y. Xiao, B. Jiang, W. Zhao, B. Ai, S.L. Huang, Y. Tang, X.S. Chen, Electr. Meas. Instr **51**,1(2014)
11. Q. Li, S.H. Zhang, Y.B. Lu, D. Han, H.L. Hu, H. Li, D.Y. Li, Electr. Meas. Instr **47**, 10(2010)

Sustainable Development of Geothermal Industry in China: An Overview

Bang Xu [1,a], Menggang Li [1], Xiyu Pi [2,3]

[1] Postdoctoral research station of China Industry Safety Research Center, Beijing Jiao tong university, Beijing 100044, China
[2] State Key Laboratory of High-efficiency Mining and Safety of Metal Mines, Ministry of Education, University of Science and Technology Beijing, Beijing 100083, China;
[3] School of Civil and Environment Engineering, University of Science and Technology Beijing, Beijing 100083, China

Abstract. With a wide distribution, large reserves, low cost, sustainable energy use and environmental protection and other unparalleled advantages, geothermal energy resources is important for China's energy structure adjustment, energy conservation and environment improvement . Currently, geothermal utilization in China is still in its infancy, and Sustainable Development of the geothermal industry is also having a lot of problems. In this paper, the current research on sustainable development of geothermal industry focuses on two aspects: 1. the current situation of geothermal industry development and existing problems, 2. the current situation of sustainable development of the geothermal industry. On the basis of the review, some suggestions for further study on the sustainable development of geothermal industry are put forward.

1 Introduction

In the 21st century, with the rapid development of Chinese social economic and continuous growth of population, energy consumption increased significantly, traditional fossil energy resources such as coal, oil, gas are dwindling, and as a consequence the environmental degradation increasingly apparent. Adjustment of energy consumption structure, need for energy conservation and environmental protection, development and utilization of renewable energy is increasingly attract the attention of the whole society [1]. Geothermal resources, as one of the new energy resources, with a wide distribution, large reserves, low cost, Sustainable Utilization, environmental friendly and other unparalleled advantages, if can be fully exploited, it will save a lot of traditional fossil energy such as coal, oil and gas, which will be of great significance to improve Chinese energy structure and solve the increasingly serious environmental problems [2]. However, in the process of the development and utilization of geothermal resources in China exists many problems, such as the low degree of development and utilization, serious waste of resources, outdated management situation, and the thermal pollution, harmful gas pollution, chemical pollution, high fluoride, soil hardening and salinization during exploitation, the thermal reservoir water table drop caused by the unreasonable exploitation is also more prominent [3], doing research on sustainable development of geothermal

[a] Corresponding author : bxu2@uwyo.edu

DOI: 10.1051/matecconf/20165506008

resources in China, realizing the ecological, economic and social coordinated development, prolong the life of the development and utilization of geothermal resources are of great significance to promote the healthy development of geothermal energy industry in China [4].

At present, the academic research on the sustainable development of geothermal industry in China is mainly concentrated in two aspects: one is the present situation and problems in geothermal industry, the other is the evaluation of the sustainable development of the geothermal industry and the development strategy research. This article will systematically introduce the sustainable development of geothermal industry in China from the above two aspects.

2 Review of the Current Development of Geothermal Industry in China

Geothermal industry refers to enterprises and institutions aggregate that participate in a series of process activities such as geothermal resource exploration, development and application [5]. In general, the geothermal industry chains in China are as shown in figure 1. Geothermal exploration evaluation is the basis of geothermal industry, and geothermal resources exploitation and utilization way is the key of the geothermal industry development.

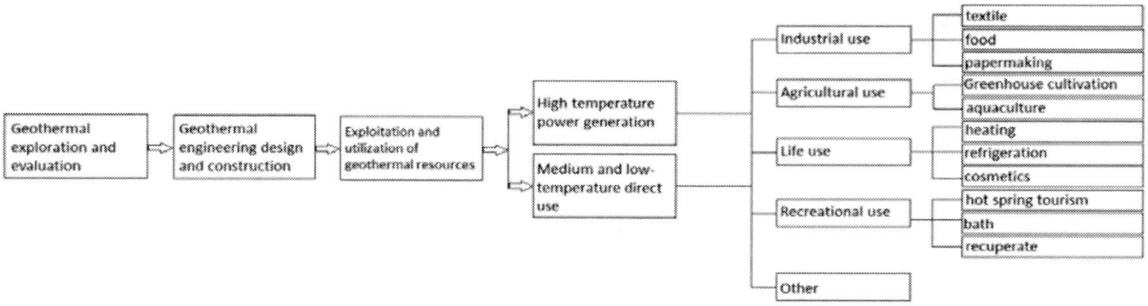

Figure 1. Geothermal industry chains in China[6]

Geothermal resource is the available geothermal energy, geothermal fluid and its useful components for development and utilization in shallow crust under the condition of current technology economy and geological environment, it is a kind of clean and renewable energy [6, 7]. According to its properties can be divided into three types: high temperature (> 150 ℃) convection type geothermal resources, mainly distributed in southern Tibet, western Sichuan, western Yunnan and Taiwan; Medium temperature (90 ~ 150 ℃) and low temperature (< 90 ℃) convection type geothermal resources, mainly distributed in Guangdong, Hainan, Fujian and other provinces; Medium and low temperature conduction type geothermal resources, mainly distributed in northern China, Sichuan, Songliao, Ordos, etc. [8-10]. The temperature distribution of land area 5.5 km deep geothermal resources in China are shown in figure 2 as follows.

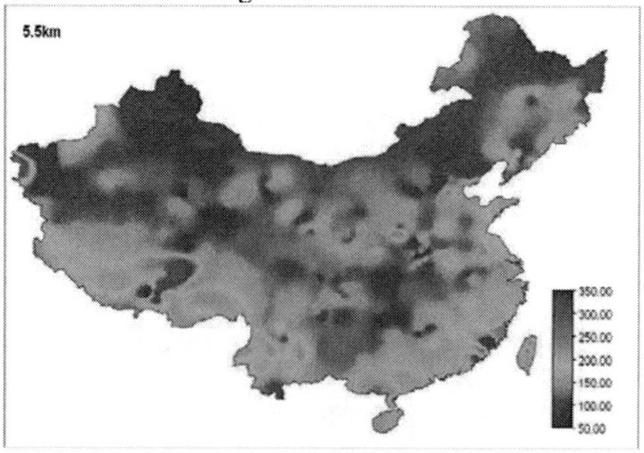

Figure 2. Land area 5.5 km deep geothermal resources distribution in China[10]

The geothermal resources in China is quite rich, according to the preliminary evaluation, the conventional geothermal resources equals to 853 billion tons of standard coal resources in China, and the recoverable resources equals to 256 billion tons of standard coal, is mainly the medium and low temperature geothermal resources [11]. There are only two high temperature geothermal field in China, they are Yangbajing and Yangyi geothermal field, the rest are medium and low temperature geothermal field [12]. In the provinces (autonomous regions and municipalities), the most abundant geothermal resources is in Tibet autonomous region, the proven recoverable heat of geothermal resources is 1732.2 MW, followed by Yunnan, Guangdong, Hebei, Tianjin and other provinces (municipalities). The proven recoverable heat of the above five provinces (autonomous regions and municipalities) geothermal resources is 3157.1 MW, accounts for about three-quarters of the total geothermal resources in China.

2.1 Geothermal resources exploration and evaluation status and problems

Rich geothermal resources provide the foundation for speeding up the development of geothermal exploration technology in China, the current method used in geothermal exploration mainly are geothermal geology, remote sensing, geophysical, geochemical and isotope geology, drilling engineering and laboratory analysis, etc. [13, 14], but each method its theoretical and technical defects and deficiencies, so different geothermal resources exploration stages need different effective combination of methods to reduce the risk, the rationalize investment and improve economic benefit [15].

Shallow geothermal resource exploration and evaluation in China is officially starts pilot in 2006, in 2009 the ministry of land and resources introduced the "shallow geothermal energy exploration technical specification". Early work in Beijing, Tianjin, Shanghai and Hebei has completed preliminary investigation and evaluation work [16]. Due to complicated geological conditions in China, there are problems such as deep carving, complicated structure, poor physical property, great difficulty with exploration technical about deep geothermal resource. The main areas of deep geothermal exploration are mainly in northeast, north and northwest area, most of them are basin (transmitting) geothermal field, although some achievements has been obtained, but the basic theory system of geothermal geological remains to be further study. And from geothermal resource prediction, recognition to the output, compared with the developed countries, the exploration and development technology in China also has great disparity [17,18].

In general, the geothermal resource exploration and evaluation in China mainly exists following problems: (a) the low degree of geothermal resource exploration and evaluation around the country, lacking of unified exploration evaluation system, dynamic monitoring system and exploration and development planning, failing to form a systematic exploration and evaluation agency [19]. At present, the total geothermal resources has always been an approximate number, the unified data has not been recognized, especially in the western region in China and north China plain area, the normal geothermal exploration work has not been carried out [20]. (b) the shallow geothermal exploration seriously lags behind its development. (c) the exploration evaluation of deep geothermal resources under 2000m has not been actualized and the resource extent is not clear. (d) the research on hot dry rock utilization technology has not been carried out [21, 22].

2.2 Geothermal resources development and utilization situation and existing problems

The utilization of geothermal resources is divided into two ways: one is geothermal power generation and its development is relatively slow. High temperature geothermal power is represented by Yangbajing, and medium-low temperature geothermal power generation only remaining in the Guangdong Fengshun and Hunan Huitang two power plants [23]. The other kind is the direct use of geothermal energy, including the direct use of geothermal water (such as geothermal heating, bathing, aquaculture, etc.) and ground source heat pump heating, cooling, [24, 25], among them bathing and spa account for 47.55%, heating accounts for 30.77%, other account for 21.68%. In China, directly using of geothermal equipment capacity was 8898 mw in 2010, ranking second in the world [26].

At present, the utilization of geothermal resources in China has formed geothermal power generation represented by the Tibet Yangbajing, geothermal heating represented by Tianjin, Xi 'an and Beijing, surface water source heat pump heating and refrigeration represented by Chongqing, sea water resource heat pump heating and refrigeration represented by Dalian, residents and tourism represented by southeast coast, as well as planting and breeding represented by the north China plain pattern of development and utilization.

Yangbajing Geothermal Power Plant, by utilizing recovered geothermal water, built up an 800MW screw expanding generator unit. The internal efficiency of such power generation systems has reached up to 70-80%; when its load has a big change in range, its performance indicators fluctuate quite small; in addition, its equipment inlet design parameters are matched to that of the geothermal energy resource; it is the best geothermal power generation mode at present. Fig. 3 is the diagram of the Thermal system for Yangbajing Dual Medium Circulating Screw Expanding Power Generator Unit. Hot heat réservoirs Fields of Yambajan include two parts, which named shallow and deep, belong to the same two different parts of the water heating system. The first generating unit is successful on September in 1977 and installed capacity was expanded to 25MW by 1991, annual output can reach 1 x 10^8 KWh [27].

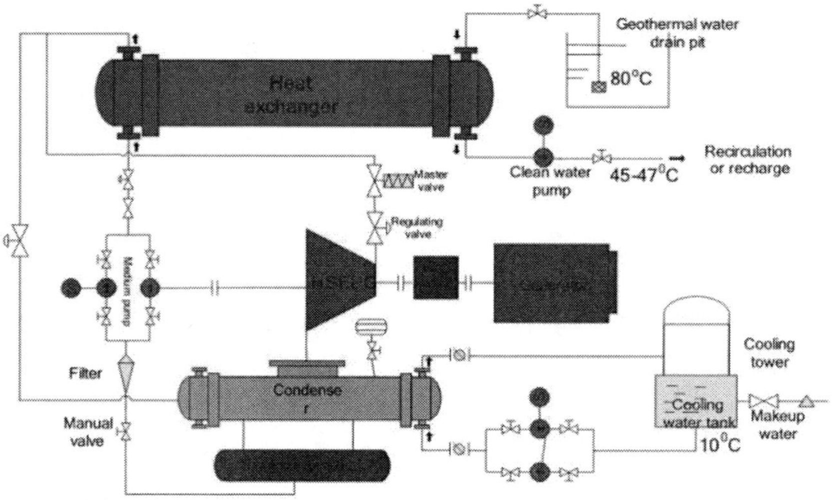

Figure 3. Diagram of the Thermal system for Yangbajing Dual Medium Circulating Screw Expanding Power Generator Unit [27].

Geothermal power is capital-intensive industry, so the initial investment is large enough, which is the total fee of all resource. Electric power generation which is based on the fossil fuels such as natural gas and coal cost too much fuel investments, but geothermal power once it is built down; there is no need any fee to support it. Table 1 is the the average cost of the budget for different power plants in 2016 [28].

Table1. The average cost of the budget for different power plants in 2016 [28]

Energy type	Energy efficency (%)	average cost budget for power plants in 2016 /[dollar (MW h) $^{-1}$]				
		The main body core average cost	Maintenance costs	Variable costs (including fuel)	The transmission input	System's overall cost of homogenization
traditional coal -fired	85	65.3	3.9	24.3	1.2	94.8
Advanced coal-fired	85	74.6	7.9	25.7	1.2	109.4
Advanced coal-fired With	85	92.7	9.2	33.1	1.2	136.2

carbon dioxide capture							
Natural gas	Traditional combined cycle	87	17.5	1.9	45.6	1.2	66.1
	Advanced combined cycle	87	17.9	1.9	42.1	1.2	63.1
	Advanced combined cycle With carbon dioxide capture	87	34.6	3.9	49.6	1.2	89.3
	Traditional gas turbine	30	45.8	3.7	71.5	3.5	124.5
	Advanced gas turbine	30	31.6	5.5	62.9	3.5	103.5
Advanced nuclear power		90	90.1	11.1	11.7	1.0	113.9
Wind energy		34	83.9	9.6	0	3.5	97.0
Coastal wind power		34	209.3	28.1	0	5.9	243.2
solar photovoltaic power generation		25	194.6	12.1	0	4.0	210.7
solar-thermal		18	259.4	46.6	0	5.8	311.8
geothermal energy		92	79.3	11.9	9.5	1.0	101.7
biomass energy		83	55.3	13.7	42.3	1.3	112.5
hydro (electric) power		52	74.5	3.8	6.3	1.9	86.4

Note: data from the U.S. energy information administration (EIA) annual energy outlook (2011), the cost budget through the existing capacity effectively in the grid can be calculated, not considering the preferential tax or incentives.

The main problems existing in the process of development and utilization of geothermal resource in China are as follows: (a) geothermal resource utilization rate is low, comprehensive benefit is not significant, waste of resources phenomenon is serious; (b) excessive mining in some areas and the phenomenon of environmental pollution and destruction [29].

2.3 The current situation and existing problems of geothermal resources management

In order to promote the development of geothermal technology and industry in China, a series of laws and regulations related to geothermal energy policy and technical specifications have been issued from national to local governments in recent years, mainly include: “mid long term development of renewable energy”, “geothermal resources geology exploration specifications”, “technical specification for the ground source heat pump system engineering”, “geothermal power plant connected to the power system technical regulations”, “technical specification for evaluation of the development of shallow geothermal energy” and "technical specification of geothermal heating project in town”, etc. [30].

The geothermal resource management still exist a series of problems: (a) hysteresis of legislation, management function has not been put in place. Now in addition to part of provinces, national laws and regulations of geothermal resources development and utilization has not been issued and failed to form a certain legal system, management style, methods, and the administrative subject is also not clear [31]. (b) the geothermal resource management related technical measures did not reach the designated position, which influenced the management functions [32]. (c) National attention is not enough, lack of prophase investment, and incentives and incentive measures has not been introduced to promote its development [33].

3 Current Research Review on Sustainable Development of The Geothermal Industry in China

Take a wide view of academic research for the sustainable development of the geothermal energy industry in China, domestic scholars mainly put forward some managerial and technical suggestions on sustainable development research from geothermal resource development situation our country or regions the development situation, development and utilization of geothermal resources current situation and existing problems.

From the perspective of management, Li xia [34] put forward that for the healthy development of geothermal industry in China, we should carry out reasonable planning of the development and utilization of geothermal resources, guide and standardize the industry development, actively develop the shallow geothermal energy resources exploration evaluation, promote the sustainable development of the industry, create a good policy environment, support the development of geothermal industry, increase the geothermal exploitation and utilization of the technology innovation, and improve the technical support system. Cheng Boyu [35] proposed that to realize the sustainable development of geothermal resources, we should exploit normatively, strengthen the construction of laws and regulations and the planning work and so on. XingQian [36] also thought we should reasonably planning the development and utilization of geothermal resources, strengthening the construction of relevant laws and regulations to realize the sustainable development and utilization of geothermal resources.

From a technical perspective, Xu junxiang [37] put forward to strengthen the exploration of geothermal resources evaluation, carry out "comprehensive utilization, step-by-step development", rely on high-tech, establish the demonstration area and promote the industrialization of geothermal. Sun jia[6] proposed the cascade utilization of geothermal resources to maximize the geothermal resource utilization, the application of the injection technology can protect the geothermal resources, improve resource utilization efficiency, reduce environmental pollution, and strengthen the geothermal water thermal insulation and heat exchange technology can effectively increase the geothermal water heat exchange efficiency. Lin Li [38], Zhou Dan [39] thought that increasing injection wells and improving injection rate is the important guarantee to realize sustainable development and utilization of geothermal resources. Zhang su'e [40] put forward to divide mining partition to realize the balanced development of the geothermal resources, and strictly control the exploitation quantity, reasonably plan geothermal well spacing to realize sustainable development and utilization of geothermal resources has important significance.

Besides, the domestic scholars' research on geothermal resources sustainable development index system is quite less. Su yong-qiang [41] chose Hebei hengshui city geothermal resources as the research object, established six subsystems of economy, society, resource, population, environment and science and education, used the Delphi method to determine the resource conditions, the development and utilization efficiency and management level as the three primary evaluation index, and respectively was subdivided into temperature flow, water quality, buried depth, distribution scope; Cascade utilization ratio, injection rate, water temperature, environmental pollution; Degree of exploration, dynamic detection rate, the scientific research level these eleven secondary evaluation index, and then based on Delphi method system, organized experts scoring, then applied fuzzy evaluation method comprehensively evaluated the sustainable development and utilization of geothermal resources in hengshui city, evaluation factors weights were determined by analytic hierarchy process, and according to the evaluation results, put forward the countermeasures of sustainable utilization of geothermal resources in hengshui city. Kong Weizhen [42] put three-e into the DEA model and established the economic evaluation model of geothermal resources.

4 The Geothermal Industry Sustainable Development Research Problems and Research Direction in China

Global geothermal resources within 5 km is about 4.9 trillion tons of standard coal, about one six of the global geothermal resource is in China. Studies have pointed out that, by 2020, Chinese non-fossil energy accounted for primary energy total consumption will increase to 15%, geothermal resources development and utilization will reach 50 million tons of standard coal; in 2030 will reach 100 million tons of standard coal. The development and utilization of geothermal is of great significance for energy structure adjustment, energy conservation and emissions reduction in China. At present, the geothermal utilization in our country is still in its infancy, and the sustainable development of geothermal industry research also exists many problems, such as lacking of thorough research on the geothermal industry sustainable development connotation, theoretical framework, evaluation system and methods of research.

Integrated the current situation and existing problems of the geothermal industry sustainable development in China, the following several aspects can be studied in-depth in the future: (1) define the geothermal industry sustainable development connotation and point out the factors influencing the geothermal industry sustainable development in China; (2) combined with the particularity of the geothermal industry, on the basis of the existing index system of sustainable development, put forward the scientific research methods, build the evaluation index system of sustainable development of geothermal industry development in China, and carry out comprehensive evaluation of the geothermal industry sustainable development ability; (3) put forward the path and countermeasures for the sustainable development of geothermal industry in China .

References

1. H.L. Zhang, Research on the New Energy Development in China, Jilin University, **1** (2014)
2. X. Guan, The situation, measures & suggestions on the exploration and utilization of geothermal resources, CMM, **5**, 78-84 (2010)
3. W.J. You, Y.F. Liu, M.J. Guo, Environmental Issues in the Development and Utilization of Geothermal Resources. Safety and Environmental Engineering, **20**, 24-28 (2013)
4. S.G. Xu, Y.S. Guo. *Geothermics basis* (Science press. 2009)
5. Z.L. Liao, Y.J. Zhang, W.B. Chen,*et al.* Available persist exploitation and utilization of geothermal resources, China Mining Magazine, **15**, 8~11 (2006)
6. G.H.Zhou,R.Huang,P.P.Xie. Component analysis of geothermal industry, Scientific and Technological Management of Land and Resources,**30**,47-53, (2013)
7. J. Sun, Available persist exploitation and utilization of geothermal resources, Resources Economization & Environmental Protection, **4**, 11 (2006)
8. J.H. Zhang, W. Wei, D. Du,*et al.* The Development, Utilization and Sustainable Development of Geothermal Resources, Sino-Global Energy, **18**, 30-35, (2013)
9. X. Guan, Study on method and application of economic evaluation of geothermal resources, CUG, 8-10 (2014)
10. W.J. Lin, Z.M. Liu, W.L.Wang, *et al.*The assessment of geothermal resources potential of China, Geology in China, **40**, 312-321 (2013)
11. J.H. Zhang, Geothermal resource distribution and utilization of our country, Natural Resource Economics of China, **8**, 23-24 (2011)
12. X.J. Kong, Z.T. Sun, L.Z. Yuan, *et al.* Application of Large Aperture DTH Drilling Technology in Dewatering of Subway Engineering, City Geology, **9**, 14-16 (2014)
13. L. Ding, S.J. Song, J. Guan, *et al.*Geothermal resource exploration technique application and prospect, Jilin Geology, **4**, 113-117 (2004)
14. J.X.Wu, The integrated use of various technology in geothermal exploration method, Ground Water, **35**, 105-109 (2013)
15. J.Q. Xu, C.J. Bai, Y.F. Wang, Technique application of geothermal resources exploration in China, *Henan earth science bulletin*, 444-448 (2009)
16. G.Y. Liu, G.H. Li, Y. Zhang, Geothermal resource exploration and development present situation and countermeasure research, Natural Resource Economics of China, **10**, 17-20 (2010)
17. B. Sun, An Opinion on Cognition and Exploration Method of Deep Geothermal Resources in China, Coal Geology of China, **1**, 20-22 (2006)
18. S.Z. Wang, Y.Wang, The basic actuality and the countermove of deep-layer geothermal resources exploring in our country, Journal of Hebei University of Engineering(Natural Science Edition), **3**, 72-74 (2002)
19. X.G. Zhao, G. Wan, Current situation and prospect of China's geothermal resources, Renewable and Sustainable Energy Reviews, **32**, 651-661 (2014)

20. X.C. You, S.Z. Yao, Utilization State and Protection Project in China Geothermal Area, China Mining Magazine, **6**, 1-3 (2007)
21. X.Y. Wang, H.M. Li, The use of geothermal energy analysis and development prospects, Energy Research & Utilization, **3**, 45-58 (2013)
22. Y. Feng, X. Chen, X.F. Xu, Current status and potentials of enhanced geothermal system in China, Renewable and Sustainable Energy Reviews, **33**, 214-223 (2014)
23. Q. Zhan, Y. Cui, The present situation and prospect analysis of geothermal resources development and utilization in China, Theory Monthly, **8**, 170-172 (2010)
24. L.X. Ma, S. Tian, The geothermal energy exploitation and utilization present situation and the development in China, Natural Resource Economics of China, **9**, 19-21 (2006)
25. L. Li, X. Li, Geothermal energy utilization present situation and development prospect of China, Heilongjiang Science and Technology Information, **26**, 11 (2011)
26. W.H. Zhou, The Status Analysis of China's Geothermal Power Generation, Petroleum & Petrochemical Today, **8**, 22-27 (2013)
27. Ministry of Science and Technology of the People's Republic of China.China's geothermal energy utilization, http://www.most.gov.cn/kjbgz/201203/t20120309_93106.htm,03,12,2012.
28. F.G.Liu,D.Hu,M.Wu ,et al. Economic Analysis on Investment in Geothermal Power Generation, SINO-GLOBAL ENERGY,**19**,24-30(2014)
29. J.J. Jiang, Q.F. Tao, J. Hu, The utilization of geothermal resources, existing problems and suggestion in China, *National conference on sustainable development of the geothermal industry* (2005)
30. X. Guan, Chinese Geothermal Development in the Future, City Geology, **7**, 6-8 (2012)
31. X. Guan, Draw lessons from foreign experience in promoting the geothermal industry policy development in China, Hydrogeology & Engineering Geology, **38**, 1-4 (2011)
32. H.Y. Cui, Some thinking for geothermal resources management in Shanxi province, Huabei Land and Resources, **3**, 20-21 (2014)
33. Y.L. Gong, W.B. Ma, Analysis of geothermal energy technology application present situation and the industry in China, High-Technology & Industrialization, **7**, 34-35 (2008)
34. X. Li, The geothermal industry problems and Suggestions, Heating & Refrigeration, **5**, 58 (2012)
35. B.Y. Cheng, Geothermal Resources and Its Sustainable Development in Hubei Province, Resources Environment & Engineering, **21**, 107-110 (2007)
36. Q. Xing, The geothermal industry sustainable development path analysis, Chemical Enterprise Management, **14**, 7-9 (2013)
37. J.X. Xu, China' s Geothermal Resources and Its Sustainable Development and Utilization, China Population,Resources and Environment, **15**, 139-141 (2005)
38. L. Lin, S.M. Zhao, P. Li, *et al.* The countermeasures for the sustainable development and utilization of geothermal resources in Tianjin, Natural Resource Economics of China, **12**, 7-10 (2005)
39. D. Zhou, The Cascaded Exploitation and Sustainable Utiliation of Geotheimal Resource, Science & Technology Information, **19**, 128 (2014)
40. S.E. Zhang, Z.J. Li, X.Z. Zheng, Sustainable development strategy of geotbeanal resource and protection proposals, Journal of Geology, **34**, 220-224 (2010)
41. Y.Q. Su, The geothermal resoUrces continuable development and investination in Hengshui City, Tianjin University, 8-9 (2009)
42. W.Z. Kong, A Dissertation Submitted to ChinaUniversity of Geosciences for the Doctor Degree of Resources Industrial Economics, China University of Geosciences, 10 (2014)

MATEC Web of Conferences **55**, 06009 (2016) DOI: 10.1051/matecconf/20165506009
ACPEE 2016

Multi-Objective Reactive Power Optimization of Distribution Network with Distributed Generation

Hui ZHAO[1,a], Zhaowen LUAN[1], Sixin GUO[2] and Chunpeng HAN[2]

[1]*Electrical Engineering College, Shandong University, Jinan, Shandong Province, China*
[2]*Shandong Electric Power Company, State Grid, Shandong Province, China*

Abstract. Distributed generation (DG) is considered to be a very promising alternative of power generation because of its tremendous environmental, social, and economic benefits. But the randomness and intermittent of DGs brings new problems to the system. This paper analyzes the reactive power optimization problem of distribution network with correlative DGs based on scenario analysis method. A new scenario division rule according to the joint distribution function of wind-PV power outputs is proposed in the paper. Then a multi-objective reactive power optimization model whose objects include the minimum active power losses, the minimum voltage deviation and the maximum static voltage stability margin is established. Non-dominated sorting genetic algorithm-II is used to solve the model. At the last of the paper, the model and the algorithm proposed are verified with an improved IEEE 33-bus system. The results show that the model will be a reference to the reactive power optimization problem in distribution system.

1 Introduction

DG's advantages of being environmental friendly, low-cost and layout flexible have drawn considerable attention in recent years, making them be a very promising alternative of power generation.

But the randomness and intermittent of their outputs bring new problems to the safety and stability of power system, which means the traditional control methods of power system such as the reactive power optimization need some new research[1-4].

This paper studies the reactive power optimization problem of distribution network with correlative DGs. It is organized as follows. In section II, we establish the joint probability model of wind-PV outputs based on Copula theory. A new multi-scene division method based on the joint probability model is proposed in section III. In section IV, we put forward a multi-objective reactive power optimization model and in section V non-dominated sorting genetic algorithm-II is used to solve the model. The simulation results are presented in Section VI and concluded in Section VII.

2 Joint probability model of wind-PV outputs

2.1 Probability model of wind farm

[a] Corresponding author : zhaohui3677@163.com

The output of wind power farm is closely related to the wind speed. It is studied that the wind speed of a certain area usually follows skew normal distribution. Its probability distribution can be described by Weibull distribution [5]:

$$f_W(v) = \frac{k}{c}(\frac{v}{c})^{k-1} \exp\left[-\left(\frac{v}{c}\right)^k\right] \tag{1}$$

where v is the wind speed, k and c are the shape parameter and scale parameter of Weibull distribution.

The relationship of wind farm's output and the wind speed can be expressed as follows:

$$p_w = \begin{cases} 0 & v \le v_{ci}, v \ge v_{co} \\ \dfrac{p_r}{(v_r - v_{ci})} v + \dfrac{p_r v_{ci}}{(v_r - v_{ci})} & v_{ci} \le v \le v_r \\ p_r & v_r \le v \le v_{co} \end{cases} \tag{2}$$

where P_w is the power output of wind farm, P_r is the rated power output, v_{ci} , v_r and v_{co} are the cut-in wind speed, rated wind speed and cut-out wind speed respectively.

The probability density function $f_W(p_w)$ and the probability distribution function $F_W(p_w)$ of wind farm output can be deducted with (1) and (2).

2.2 Probability model of photovoltaic

Similarly, the output of photovoltaic power plants is mainly affected by the light intensity. The light intensity in a certain period of time approximately obeys Beta distribution. So the output of photovoltaic power plants also obeys Beta distribution [6]. Its probability density function can be expressed as (3):

$$f_S(p_s) = \frac{\Gamma(\alpha+\beta)}{\Gamma(\alpha)\cdot\Gamma(\beta)} \cdot \left(\frac{p_s}{p_{s\max}}\right)^{\alpha-1} \cdot \left(1 - \frac{p_s}{p_{s\max}}\right)^{\beta-1} \tag{3}$$

where p_s is the photovoltaic power output, $p_{s\max}$ is the maximum power output; α and β is the shape parameter of Beta distribution.

The probability distribution function of photovoltaic power plants output $F_S(p_s)$ can be obtained by integrating (3).

2.3 Joint probability model of Wind-PV outputs

Under normal circumstances, the output powers of wind farm and photovoltaic power plant in the same area have certain correlation. This paper introduces Copula theory [7] to describe this correlation.

Sklar's theorem [7]: An N-dimensional joint distribution function can be formed by the marginal distributions of the N variables and a Copula function which describes the correlation between the variables. This theorem can be expressed in (4):

$$F(x_1, x_2, \cdots, x_n) = C\left(F_{X_1}(x_1), F_{X_2}(x_2), \cdots, F_{X_n}(x_n)\right) \tag{4}$$

where $F(\cdot)$ is the joint distribution function, is the Copula function, $F_{X_i}(x_i)$ is the marginal distributions of variables X_i .

The common used Copula functions include oval Copula functions and Archimedes Copula functions. In this paper we choose the normal Copula function of the oval Copula functions to form the joint probability distribution function of wind-PV outputs, which can be written as:

$$C(u, v; \rho) = \Phi_\rho\left[\Phi^{-1}(u), \Phi^{-1}(v)\right] \tag{5}$$

DOI: 10.1051/matecconf/20165506009

where $\Phi_\rho(\cdot)$ is the bi-normal probability distribution function with the correlation cofficient ρ, $\Phi^{-1}(\cdot)$ is the inverse function of normal function.

Applying Copula theory to what we are studying, we can obtain the joint probability distribution function of wind-PV outputs:

$$F(p_w, p_s) = \Phi_\rho\left[F_W(p_w), F_S(p_s)\right] \tag{6}$$

3 Multi-scene division method

Scenario analysis method has great advantage in dealing with uncertain problem which can turn the uncertainty into the combination of several deterministic scenarios. We try to adopt scenario analysis method to cope with the uncertainty caused by DGs. How to divide the scenarios is the key of the method. The traditional way is dividing according to the state transition points of wind turbines, as is the case in [4]. But this division method is relatively rough, and the DGs need to be independent from each other..Here in the paper a new multi-scene division method according the joint probability model in the previous section is given.

A. divide

Firstly, divide the wind DG's output interval into N subintervals and similarly the photovoltaic DG's output interval $[0, P_{s\max}]$ into M subintervals.

B. combine

There are N intervals of wind power DG's output and M intervals of photovoltaic DG's output, so in total there are $N \times M$ combinations of power intervals which we denote as $\{S_{11}, S_{12}, S_{13} \cdots S_{nm} \cdots S_{NM}\}$ where S_{nm} is the combination of the *nth* wind DG's output interval and the *mth* photovoltaic DG's output interval.

C. calculate the probability

We define the outputs of the scenario as the average output of the interval:

$$P_W(S_{nm}) = \frac{(2n-1)\cdot P_r}{2N} \tag{7}$$

$$P_S(S_{nm}) = \frac{(2m-1)\cdot P_{s\max}}{2M} \tag{8}$$

And the probability of the scenario is calculated as follows:

$$f(S_{nm}) = \iint_D f(p_w, p_s)\mathrm{d}p_w \mathrm{d}p_s \tag{9}$$

$$D = \{(p_w, p_s) \left| \frac{(n-1)P_r}{N} \le p_w \le \frac{n \cdot P_r}{N}, \frac{(m-1)P_{s\max}}{M} \le p_s \le \frac{m \cdot P_{s\max}}{M} \right.\} \tag{10}$$

4 Multi-objective reactive power optimization model

In the traditional reactive power optimization model, the object function is always the minimum active power losses and the voltage demand acts as the constraint of the model. In this case, the voltage may be close to the voltage limits. In the distribution network with DGs, the fluctuation of the DGs may cause the voltage to go over the limit. So in this paper, we include the voltage demand in the object function and establish a multi-objective reactive power optimization model.

4.1 Objective function

a) minimum active power losses

$$\min f_1 = E\left(P_{loss}\right) = \sum_{n=1}^{N}\sum_{m=1}^{M} f\left(S_{nm}\right) P_{loss\_nm} \tag{11}$$

where $E\left(P_{loss}\right)$ is the expectation of active power losses, P_{loss_nm} is the system active power losses in scenario S_{nm}.

b) minimum voltage deviation

$$\min f_2 = E\left(\Delta u\right) = \sum_{n=1}^{N}\sum_{m=1}^{M} f\left(S_{nm}\right) \cdot \sum_{k=1}^{N_L} \frac{\left|U_k - U_k^{'}\right|}{U_{k\max} - U_{k\min}} \tag{12}$$

where N_L is the number of the load nodes, U_k is the actual voltage of node *k*, $U_k^{'}$ the reference value of voltage, $U_{k\min}$, $U_{k\max}$ is the upper limit and lower limit of voltage on node *k* respectively.

c) maximum static voltage stability margin

$$\min f_3 = E(\frac{1}{\delta}) = \sum_{n=1}^{N}\sum_{m=1}^{M} f\left(S_{nm}\right) \cdot \frac{1}{\delta_{nm}} \tag{13}$$

where δ_{nm} is the minimum singular value of the Jacobian matrix in scenario S_{nm}.

4.2 Constraints

$$\begin{cases} P_{Si} - P_{Li} = V_i \sum_{j\in i}^{N_L} V_j \left(G_{ij}\cos\delta_{ij} + B_{ij}\sin\delta_{ij}\right) \\ Q_{Si} - Q_{Li} + Q_{Ci} = V_i \sum_{j\in i}^{N_L} V_j \left(G_{ij}\sin\delta_{ij} + B_{ij}\cos\delta_{ij}\right) \\ 0 \le Q_{ci} \le Q_{ci\max} \qquad i = 1,2,3\cdots p \\ V_{i\min} \le V_i \le V_{i\max} \qquad i \in N_L \end{cases} \tag{14}$$

where N_L is the node number, P_{Si}, P_{Li} are generator active power output and active power of load at node *i*. Q_{Si}, Q_{Li}, Q_{Ci} are generator reactive power input, load reactive power and capacity of capacitive reactive compensation device at node *i* respectively.

5 Non-dominated sorting genetic algorithm-II

The reactive power optimization problem of distribution network with DGs is multi-objective, non-linear, and multi-constraint. Traditional optimization methods are difficult to obtain global optimal solution. Non-dominated sorting genetic algorithm-II (NSGA-II) algorithm is good at dealing with multi-objective optimization problem.

The main features of the algorithm: 1) Fast non-dominated sorting method based on grading; 2) introducing the concept of crowding distance, which is used to measure the population density of the individuals in the same front after fast non-dominated sorting; 3) introducing elitist mechanism. Parent and offspring together compete for the next generation.

6 Example

In order to validate the model and the algorithm proposed in this paper, a test is done with the improved IEEE33 bus system. The line parameters and the load information of the system is the same with [8]. A 500kW wind power DG and a 100kW photovoltaic DG whose correlation coefficient $\rho = -0.122$ are connected to bus 11 and bus 27 respectively. Both of the wind power and photovoltaic DG work with a constant power factor. The wind power DG's power factor is set to 0.95 and photovoltaic DG's is set to 0.9. Bus 2, 16, and 32 are the candidate places for reactive power compensation. There are 20 groups of parallel capacitors at each place and the compensation capacity of each group is 50kvar.

Fig.1 shows the comparison between the joint probability density function of wind-PV outputs and the probabilities of 120 scenarios calculated using the multi-scene division method given in this paper. From the figure, we can draw the conclusion that the multi-scene division method put forward in this paper has good effect in practice.

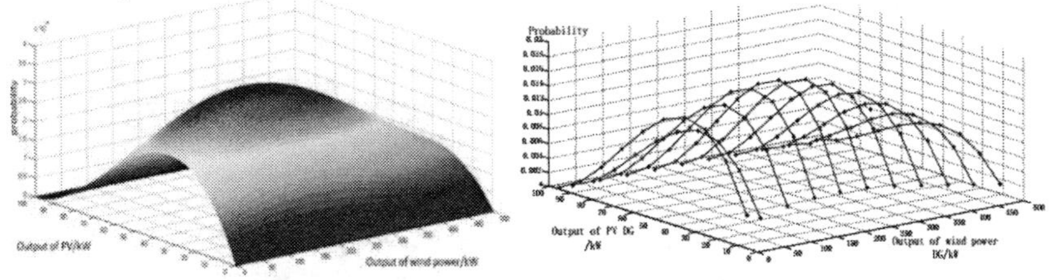

Figure 1. Joint probability density distribution of wind - PV power outputs

We choose 3 solutions from the Pareto and show them in table 1: the minimum power losses one, the minimum voltage deviation one and a comprehensive one.

Table 1. Schemes for reactive power optimization and results

scheme	Q_{c1}(kvar)	Q_{c2}(kvar)	Q_{c3}(kvar)	Q_{c4}(kvar)	Power losses (kW)	voltage deviation	Voltage stability margin
no	0	0	0	0	165.16	15.27	0.1471
scheme 1	550	800	300	700	95.04	9.73	0.1574
scheme 2	950	850	800	950	113.73	6.31	0.1615
scheme 3	550	750	650	1000	104.22	7.04	0.1600

From the table, we can see that compared to the system without optimization, all of the three schemes can reduce the system active power losses, the system voltage deviation and enhance the voltage stability margin tremendously. But scheme 1 reduces the most active power losses and scheme 2 reduces the most voltage deviation. Compared to scheme 1 and 2, none of the three objective functions of scheme 3 is the best, but it makes balance between the three objective functions and seems more moderate.

This is the advantage of multi-objective optimization which provides us more options to choose from, each option laying different emphasis on different objectives.

Fig. 2 shows the voltage expectation curve of the 33 nodes and figure 3 shows the voltage curve of node 18 under 120 scenarios.

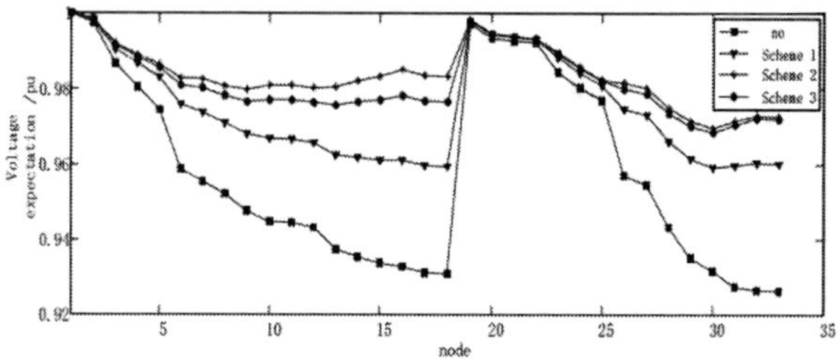

Figure 2. Voltage curves of load bus before and after reactive power optimization

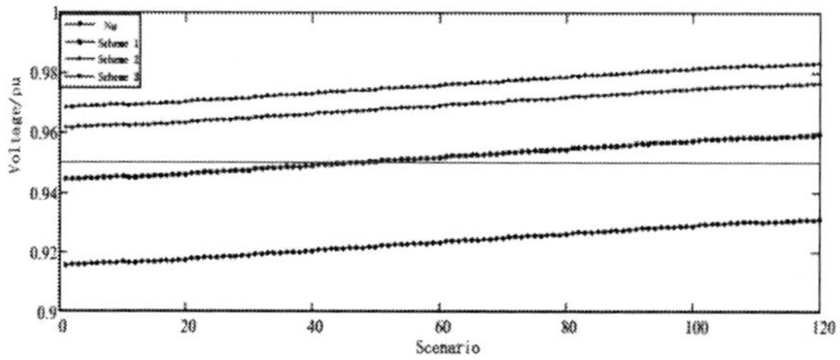

Figure 3. Optimal voltage curve of bus 18 at different scenarios

Once again we can find from the two figures that the three schemes can enormously improve the system voltage. One thing we should point out is that from fig.5,we find that under scheme 1, there are 54 scenarios in which the voltage of bus 18 will be lower than the lower limit 0.95pu.. This give us the lesson that if we are in obsession of reducing the active power losses, there may be threat to the voltage safety. That is just what we have said in the beginning of section IV .

7 Conclusion

In this paper we established a multi-objective reactive power optimization model based on the scenario analysis method and the multi-scene division method. Non-dominated sorting genetic algorithm-II is adopted to calculate Pareto. Simulation of IEEE 33-bus distribution systems shows that the proposed method and model is feasible and effective.

References

1. Su C L. " Stochastic evaluation of voltages in distribution networks with distributed generation using detailed distribution operation models". in IEEE Trans on Power Systems,2010,25 (2) : 786-795.
2. Yanghua L, Zhengqiu W, Youqing T. "A survey on distributed generation and its networking technology". Power System Technology, 2008, 32(15): 71-76.
3. Liang C, Duan X Z. Distributed generation and its impact on power system[J]. Automation of Electric Power Systems, 2001, 25(12): 53-56.
4. Haiyan C,Jinfu C,Xianzhong D . "Reactive power optimization in distribution system with wind power generators". Proceeding of the CSEE,2008,28(7) : 40-45(in Chinese).
5. Yang X, Xiao Y, Chen S. "Wind speed and generated power forecasting in wind farm". Proceedings of the CSEE, 2005,11:1-5.
6. Imad A, Ramakumar R. "Loss of power supply probability of stand-alone photovoltaic systems: a closed form solution approach". IEEE Transactions on Energy Conversion, 1991, 6(1): 1-11.
7. Wu Juan . Copula theory and dependence analysis [D] . Wuhan : Huazhong University of Science and Technology,2009(in Chinese)
8. Aman M M,Jasmon G B,Bakar A H A. A new approach for optimum simultaneous multi-DG distributed generation Units placement and sizing based on maximization of system load ability using HPSO (hybrid particle swarm optimization) algorithm[J]. Energy,2014(66) : 202-215.

MATEC Web of Conferences **55**, 06010 (2016)
ACPEE 2016
DOI: 10.1051/matecconf/20165506010

Emergency Linkage Mode of Power Enterprise

FENG Jie [1,a], YU Zhen [1] and LIU Chao [1]

[1] *State Grid Smart Grid Research Institute, Information &communication research department, Beijing, China*

Abstract. Power emergency disposal needs take full advantage of the power enterprise within the external emergency power and resources. Based on analyzing and summarizing the relevant experience of domestic and foreign emergency linkage, this paper draws the Emergency Linkage subjects, Emergency Linkage contents, Emergency Linkage level, which are three key elements if power enterprise Emergency Linkage. Emergency Linkage subjects are divided into the two types of inner subjects and the external body; Emergency Linkage contents are in accordance with four phases of prevention, preparedness, response and recovery; Emergency Linkage level is divided into three levels of enterprise headquarter, provincial enterprise and incident unite. Binding power enterprise emergency management practice, this paper studies the internal Emergency Linkage modes (including horizontal mode and vertical mode), external Emergency Linkage mode and comprehensive Emergency Linkage Mode of power enterprise based on Fishbone Diagram and Process Management Technology.

1 Introduction

To meet massive power demand caused by the rapid development of national economy, power system has entered the era of ultra-high voltage and interconnected large power system. The safety and stabilization of power system has become an issue that is of great concern to the public. Related research found that power system emergencies feature complicated links, many disaster sources, huge losses and widely affecting [1]. Therefore, it is unable that to deal with the complicated emergencies only relies on the emergency capacity of electricity power.

In modern emergency management, full use of various resources for emergency response has become a trend. Currently, there is less study on the Emergency Linkage of power enterprise. Using the experiences abroad and inside for reference, this paper studies the elements and mode of Emergency Linkage.

2 The Current Analysis of Emergency Linkage Experiences Abroad and Inside

2.1 Foreign Emergency Linkage Experiences

[a] Corresponding author : fengjie_leo@163.com

Through studying the Emergency Linkage experiences of the United States, Japan, Britain, Australia, it is found the following characteristics

(1). In the United States, the main content of Emergency Linkage is that the relevant emergency departments signed a cooperation contract [2];

(2). In Japan, the main content of Emergency Linkage is that government, companies and non-profit organizations played a relevant role in emergency response [3];

(3). In the UK, it is an effective solution to the division of responsibilities by established a "gold, silver, copper," three emergency command mechanism [4];

(4). In Australia, the multi-level emergency response mode provided material and financial for emergency disposal [5].

2.2 Domestic Emergency Linkage Experiences

In China, civil aviation, subway and other industries explored a lot of useful methods in the Emergency Linkage [6,7], such as, the establishment of emergency response mechanisms in line with industry characteristics; the effective integration of internal and external emergency resources; forming corresponding Emergency Linkage processes, methods and technology; explicit emergency response content and approach; combined the Emergency Linkage and emergency management processes.

2.3 Summaries of Emergency Linkage Experiences

Through comprehensive analysis of domestic and foreign Emergency Linkage experiences, we find the following results:

(1) Multiple-subjects Emergency Mechanisms: In emergency management, all kinds the subjects participate in emergency response, mobilized the capacities of each subject get and played their strengths and initiative.

(2) Clear Emergency Linkage Contents: It is clearly defined that procedures and methods of Emergency Linkage, solidified emergency response content, and consolidation Emergency Linkage process;

(3) Scientific Emergency Linkage Grade: For utilizing a limited resource, Emergency Linkage level match to the emergency level.

3 Emergency Linkage Elements of Power Enterprise

By analysing the domestic and foreign Emergency Linkage experiences, the study found that Emergency Linkage has three important elements: Emergency Linkage Subjects; Emergency Linkage Content; Emergency Linkage Level. Based on power enterprise practice, three elements of Emergency Linkage are studied.

3.1 Emergency Linkage Subjects

A lot of Emergency Linkage Subjects involved in responding to emergencies. In order to make emergency response in an orderly manner, Emergency Linkage Subjects are divided into the enterprise interior subjects and the external subjects. The external subjects include related internal departments of power enterprise, and the external subjects are divided into government departments, rescue organizations, monitoring departments, news media, and power consumers in accordance with their function. The classification is shown in Figure 1.

DOI: 10.1051/matecconf/20165506010

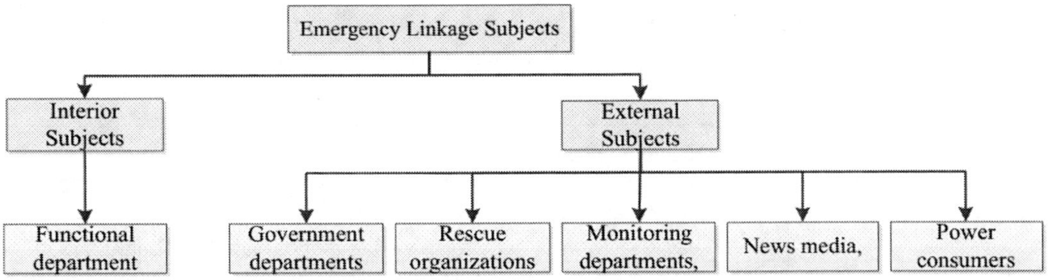

Figure 1. Classification of Emergency Linkage Subjects Emergency Linkage Content

According PPRR theory, emergency management includes prevention, preparation, response and recovery stages [8]. Emergency Linkage Content as an important basis of emergency response work should run through four phases of emergency management. The content involving emergency management four stages is shown in Table 1.

Table 1. Emergency Linkage Related Content

Phase	Content	Phase	Content
prevention	(1)Hazard Management (2)Monitoring and Control (3)Public Education	response	(1)Event Information Sharing (2)Coordination Disposal (3)Tracking and Monitoring (4)Public Opinion Management (5)Resource Allocation (6)Medical Care
preparation	(1)Emergency Linkage Plan (2)Forecasting and early warning (3)Emergency Training (4)Emergency response drills (5)Emergency Supplies Sharing (6)Emergency Response agreements	recovery	(1)Recovery Planning (2)Plan Execution (3)Disposal of Property Damage (4)Incident investigation (5)Summary and Assessment

3.2 Emergency Linkage Level

Power enterprise emergency management is generally divided into three levels: the first level is the enterprise headquarters level; the second level is the provincial enterprises level; the third level is the incident unit level. Each level contains a number of functional departments. In lateral dimension, various functional departments are directed by the emergency leading group. In vertical dimension, lower functional departments are guided by related higher departments. The specific situation is shown in Figure 2

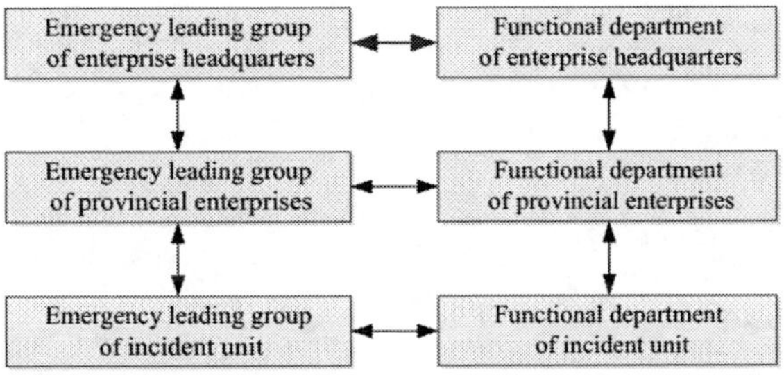

Figure 2. Power Enterprise Emergency Management Organization Relations

Emergency response is a dynamic process. Based on different levels of emergencies, Emergency Linkage Level should be adjusted. When unexpected events affect a large range, it should be taken to a higher Emergency Linkage Level, in order to guarantee the strength and resources of relevant emergency deployment, but also appropriate communication and coordination with external emergency rescue departments. When unexpected events affect a lesser range, it should be taken to a lower Emergency Linkage Level. Specific classification method is shown in Table 2.

Table 2. Classification of Emergency Linkage Level

Emergency Linkage Level	Emergency Disposal	Coordination channels
The first level	Over greater emergencies	The emergency leading group of corporate headquarters or its authorized units are responsible for command and coordination.
The second level	Generic emergencies	The emergency leading group of provincial enterprises is responsible for directing and coordinating.
The third level	Under generic emergencies	The emergency leading group of the incident unit is responsible for directing and coordinating.

4 Emergency Linkage Modes of Power Enterprise

4.1 Power Enterprise Internal Emergency Linkage Mode

4.1.1 Emergency Linkage Mode of Functional Department

Corporate functions around the emergency team, comprehensive support, media disposition, recovery and reconstruction of the four emergency response capabilities [9] establish Internal Emergency Linkage mechanism. This study uses the four emergency response capacities as a starting point to analyse the relevant functional departments involved Internal Emergency Linkage and defines mode of Emergency Linkage mode of functional department by Fishbone Diagram. Emergency Linkage Mode of Functional Department is shown in Figure 3.

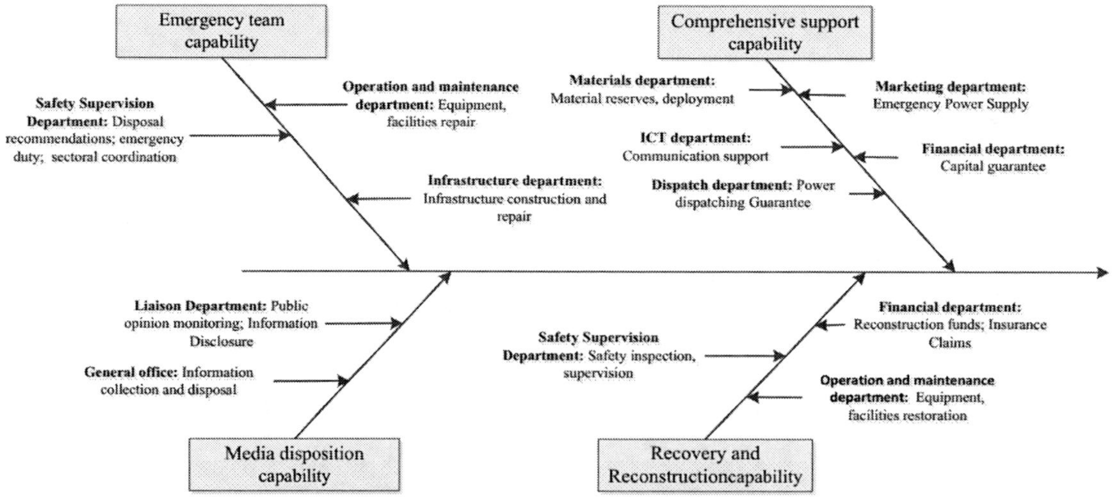

Figure 3. Emergency Linkage Mode of Power Enterprise Functional Department

4.1.2 Emergency Linkage Mode of Interior Units

Emergency Linkage of power enterprise interior units covers prevention, preparedness, response, recovery of the four emergency management links. This study designs the mode between the incident unit and linkage units by Process Management Technology. The mode is shown in Figure 4.

DOI: 10.1051/matecconf/20165506010

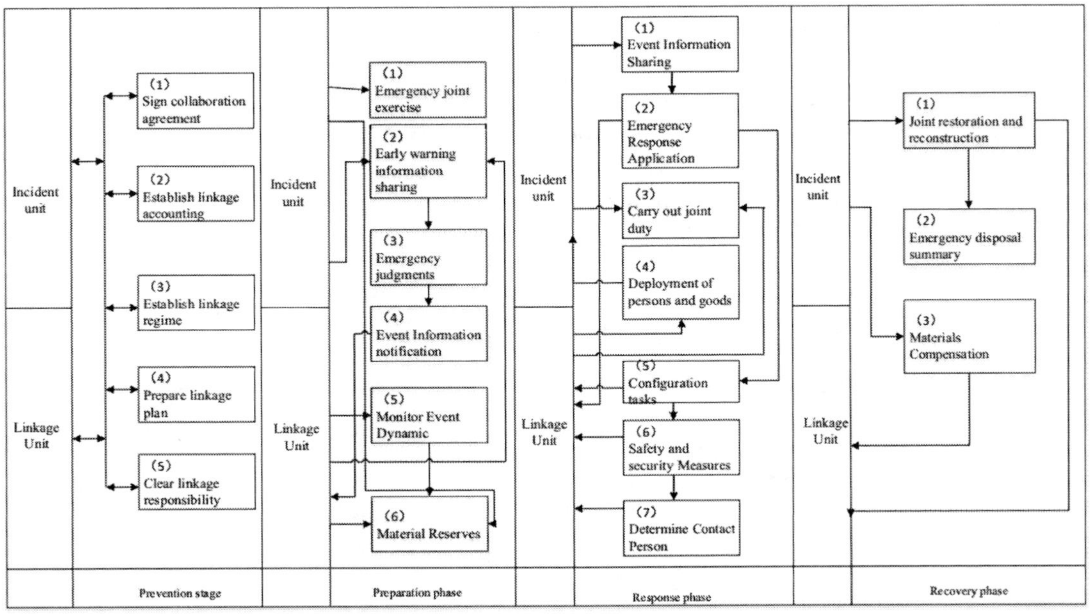

Figure 4. Emergency Linkage Mode of Power Enterprise Interior Units

4.2 External Emergency Linkage Mode

In the external Emergency Linkage, Power Enterprise needs to build connection with relevant external subjects, such as rescue agency, monitoring authorities, news media, power consumers, etc., in order to ensure that the whole process of emergency management emergencies. External Emergency Linkage activities also cover the four links of emergency management. This study designs external Emergency Linkage mode by Process Management Technology. The mode is shown in Figure 5.

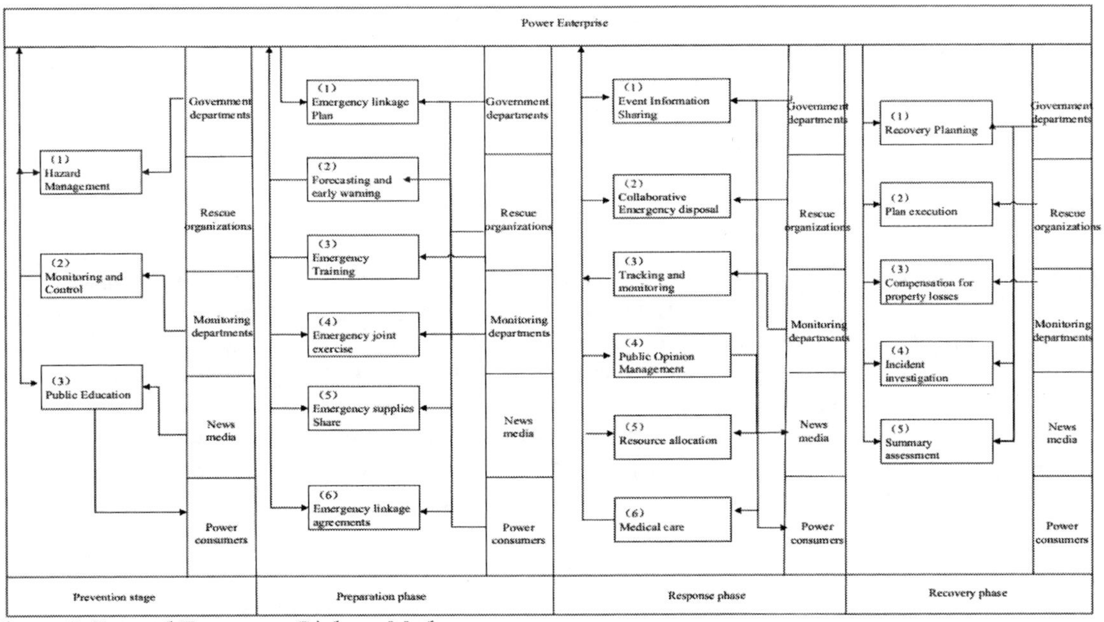

Figure 5. External Emergency Linkage Mode

DOI: 10.1051/matecconf/20165506010

4.3 Comprehensive Emergency Linkage Mode

In emergency management, it is necessary to build linkage mechanism between power enterprise and external departments. Based on the study of power enterprises Emergency Linkage elements, it is designed that comprehensive Emergency Linkage mode by integrating power enterprise Internal Emergency Linkage mode and external Emergency Linkage Mode. The mode is shown in Figure 6.

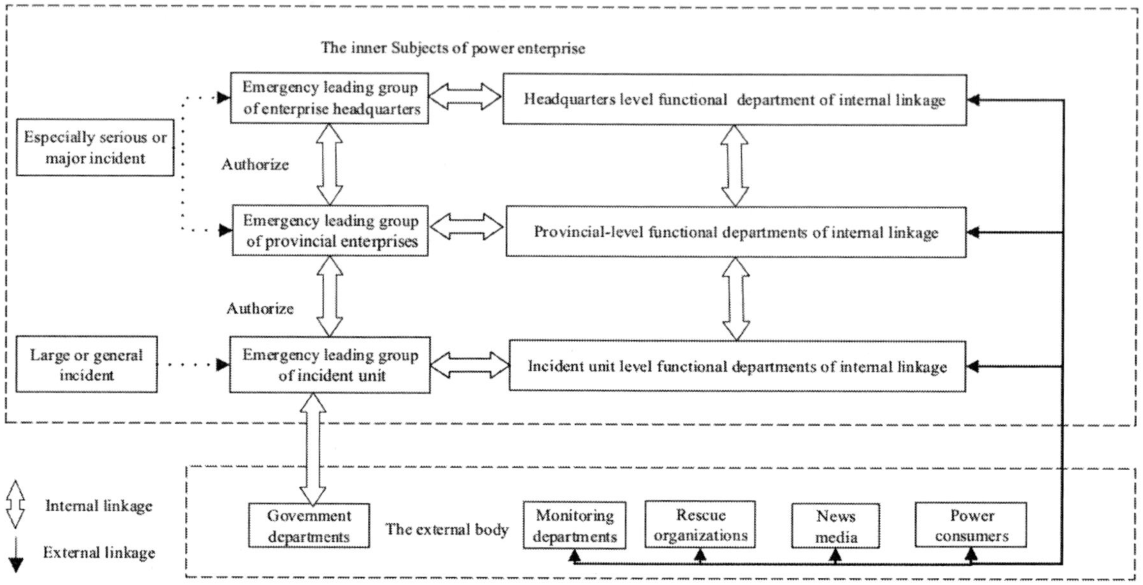

Figure 6. Comprehensive Emergency Linkage Mode

The mode has been applied in the State Grid Corporation of China. The corporation headquarters has established Emergency Linkage mechanism with 27 provincial companies in accordance with Emergency Linkage mode. Each provincial company has signed Emergency Linkage agreements with related units, according to themself characteristics. The State Grid Corporation of China organized 656 Emergency Linkage exercises with all levels of government departments. By applying Emergency Linkage mode, the various types of emergency resources of the State Grid Corporation of China have been scientific integration; external emergency power has been got the full benefits; and the emergency response capacity of corporation has been further enhanced.

5 Conclusion

On the basis of full study at domestic and foreign Emergency Linkage current state, we proposed the three elements of Emergency Linkage: Emergency Linkage subjects, Emergency Linkage content and Emergency Linkage level. They construct the basis for designing the modes of Emergency Linkage. We designed Power Enterprise Internal Emergency Linkage Mode (including Emergency Linkage Mode of Functional Department and Emergency Linkage Mode of Interior Units) and External Emergency Linkage Mode. The modes provide a scientific reference to build Power Enterprise Emergency Linkage mechanism, also help Power Enterprise further enhance emergency response capacity and improve the emergency management level.

References

1. Tian Shiming,Chen Xi,Zhu Chaoyang,et al.Theory of electric power emergency management and its technological countermeasures[J].Power System Technology,2007,31(24): 22-27.

2. William L.Waugh, Jr Kathleen Tierney.Emergency Management Principles and Practice or Local Government[J]. the International City/ County Management Association,2007,58.
3. FU Shichun.Characteristics of Japanese Emergency Management System [J].Party & Government Forum, 2009,4(4),60-62.
4. LI Geqin.On the Emergency and Security Management System in the United Kingdom [J].Journal of International Security Studies,2013,2(3),126-137.
5. GUO Yue. Characteristic of Disasters Management in Australia and Its Enlightenment [J].Journal of Chongqing Teachers College(Natural Science Edition),2005,4(12),59-63.
6. ZHU Xin-yan, LUO Fan.Study on Mechanism Optimization of Joint Emergency Response to Aviation Calamities in China [J].China Safety Science Journal,2007,17(6),9-15.
7. Liu Leyi.On Linkage of Emergency during Metro Operation [J].Modern Urban Transit,2007, 4(8),51-53.
8. ZHU Chaoyang,YU Zhen,LIU Chao.Research on Theory and Technical System for Power Emergency Management [J].Power System Technology,2011,35(2),178-182.
9. LIU Chao, ZHU Chao-yang,GONG Bo-tao,et al.Concepts,Indicators and Methods of Emergency Capability Assessment in City Power Grid Enterprises [J].East China Electric Power,2013,9(9),126-132.

MATEC Web of Conferences **55**, 06011 (2016) DOI: 10.1051/matecconf/20165506011
ACPEE 2016

Cache Management of Big Data in Equipment Condition Assessment

Yan Ma[a], Yufeng Chen, Ying Lin, Chengqi Li, Yujie Geng

State Grid Shandong Electric Power Research Institute, Jinan, China

Abstract. Big data platform for equipment condition assessment is built for comprehensive analysis. The platform has various application demands. According to its response time, its application can be divided into offline, interactive and real-time types. For real-time application, its data processing efficiency is important. In general, data cache is one of the most efficient ways to improve query time. However, big data caching is different from the traditional data caching. In the paper we propose a distributed cache management framework of big data for equipment condition assessment. It consists of three parts: cache structure, cache replacement algorithm and cache placement algorithm. Cache structure is the basis of the latter two algorithms. Based on the framework and algorithms, we make full use of the characteristics of just accessing some valuable data during a period of time, and put relevant data on the neighborhood nodes, which largely reduce network transmission cost. We also validate the performance of our proposed approaches through extensive experiments. It demonstrates that the proposed cache replacement algorithm and cache management framework has higher hit rate or lower query time than LRU algorithm and round-robin algorithm.

1 Introduction

With the rapid development of power grid and the explosion of transmission and transformation equipment, massive and heterogeneous data are generated and separately collected into different business information systems during grid operation and equipment monitoring. In order to make full use of various data from different systems, we build a big data platform for equipment condition assessment. Now it connects production management system (PMS), energy management system (EMS), condition monitoring system of transmission and transformation equipment, geographic information system (GIS) and meteorlogical information system, and is used to research dynamic evaluation of load capacity, fault diagnosis, condition evaluation and risk assessment [1]. For the application with high requirements for response time such as dynamic evaluation of load capacity, users have requirements for processing performance and return time. Therefore, we need to optimize big data processing performance to improve data analyzing efficiency.

Cache is one of the most important means of improving big data processing speed [2]. Data are stored in cache, which largely improve I/O efficiency and then data processing performance. However, cache is more expensive than external storage devices such as disk. Moreover, big data is total sample data. It is uneconomic and infeasible that all the data are stored into cache. In general data accessing is

[a] Corresponding author : yanpony@126.com

frequent and real-time just for a portion of the data. Thus we could put important and frequently accessed data into cache.

Compared with traditional data caching, big data caching has its own characteristics.

1) Data is organized by key-value pair. The caching granularity, caching modality and cache replacement algorithm need discussing to adapt to storage structure of big data.

2) Big data processing relies on cloud computing platform. Data accessed in big data have a certain correlation. It can reduce data transferring cost to put the related data on adjacent positions. For example, a data processing needs data A and B. If data A and B are stored on two nodes, one data should be transferred to the other node. If data A and B are stored on one node, network transmission is avoided and processing efficiency is high. After we get the cached data, a method of placing them on proper nodes needs designing.

2 Related work

Database cache is widely studied to improve database performance [3-5]. Storm et al. [3] gives a self-adjusting cache management mechanism, which builds the model between consumed time and output profits for different users. Malik et al. [4] define hit output rate by unit of byte, which caches the caching objects with high output to reduce network transmission cost. Brown et al. [5] propose fragment fencing to solve cache assignment for different type of applications with QoS requirements.

With the rise of cloud computing, there are shortcomings for cache management of traditional database [6]. On one hand, cache resources are shared and competed by multiple users, which results in low hit rate and poor user experience. On the other hand, there lacks shared cache management strategy for multiple users, which lets users with low query rate have low hit rate. Thus, Yao et al. [6] propose a self-adaptive cache management based on chunk folding architecture, which dynamically generates a series of cache replacement units to consistently achieve users' SLA requirements.

As for cache replacement strategy, the common cache replacement algorithms have LRU, LFU, LRU-K [7]. All the algorithms use pages as replacing unit to minimize page lose rate. LRU algorithm sometimes pollutes cache pool when reading a series of infrequently used pages. The InnoDB [8] improves the above problem based on new and old link tables. However, they do not consider the data structure of big data, which is key-value pairs, which is different from caching the whole page.

3 Cache management framework of big data

In the paper we propose a Distributed Cache Management Framework of big data (DCMF) to accelerate processing speed of big data. The proposed framework includes three parts: cache structure, cache replacement algorithm and cache placement algorithm. Cache structure is the basis of cache replacement algorithm and cache placement algorithm, and it provides the organization format for unit of cache. Cache replacement algorithm selects the cached data, and cache placement algorithm puts the cache data into the cache of cloud nodes.

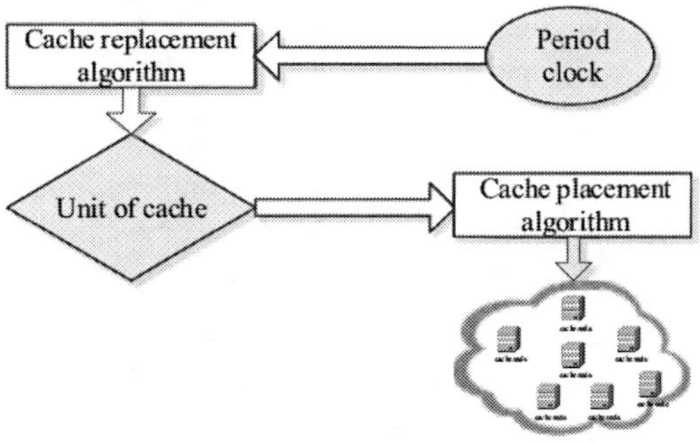

Figure 1. The flow chart of updating data cache

The proposed framework periodically updates big data cache. It first uses cache replacement algorithm to select the cached data from disk, then cache placement algorithm put the cache data on cloud nodes. The flow chart is shown in Fig. 1.

4 Distributed caching method of big data processing

In the section, we introduce the three part of proposed distributed cache management framework: cache structure, cache replacement algorithm and cache placement algorithm.

4.1 Cache structure

In disk, several stored key-value pairs constitute a unit of cache. Therefore, all the data in disk can be divided into certain unit of cache. Data are ordered by Rowkey. Assume the number of key-value pair in a unit of cache is k. We can set k to 100. As shown in Fig. 2, each dotted box is a unit of cache and has k key-value pairs. These units of cache are contiguously stored in disk.

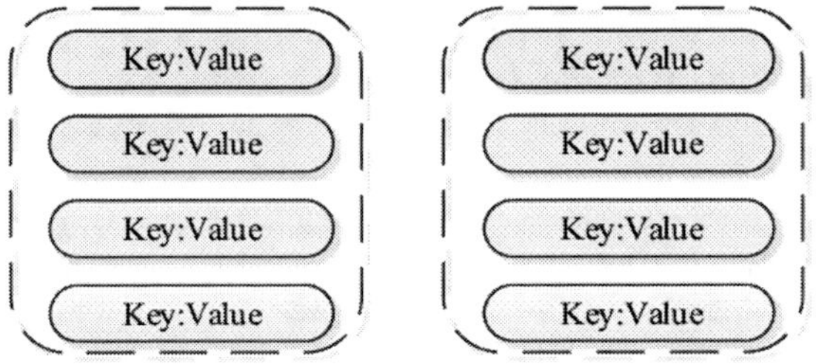

Figure 2. The structure of unit of cache

4.2 Cache replacement algorithm

Cache replacement algorithm [9] gives priority to the unit of cache with high value and accessing frequency and put them into caching set. Caching set is the chosen units of cache from all the units of cache and they will be put into cache on the nodes.

In the section we propose a Cache Replacement algorithm based on periodic Value (CRV). Its concrete steps are as follows.

1)Units of cache in caching set are updated each period.

2)For a period j, we record the accessing frequency of each unit of cache. For a unit of cache c_i, when a data access needs the data in c_i, its number of accesses in period j is computed as $n_i^j = n_i^j + 1$, where n_i^j is initialized as 0.

3)We compute the value of each unit of cache in period j, which is denoted as p 。 For the value of c_i is computed as $p_i^j = \alpha \cdot p_i^{j-1} + (1-\alpha) \cdot n_i^j \cdot \beta$, where p_i^{j-1} is the value of c_i in last period, α is periodic impact factor, β is the data value factor of c_i and $0 < \alpha, \beta < 1$. When c_i is accessed, its return time is tight and the β value is high.

4)We sort all the c_i in the descending order of p_i^j. Assuming that all the cache capacity is *h* and the capacity for a unit of cache is *b*, the number of units of cache that can be put in the cache is computed as $l = h / b$. We select the first *l* units of cache into caching set.

During data access we divide the degree of urgency into real-time, normal and loose according to the requirement on return time. The above three types correspond to different β value and β is selected from 0.9, 0.5 and 0.2.

Once data access is recorded as $<\cup\{c_i\},\beta>$, where $\cup\{c_i\}$ is the units of cache involved into the data access. We calculate the accessed frequency and the degree of urgency for each unit of cache in a period according to data access records.

4.3 Cache placement algorithm

Cache placement algorithm [10] is responsible for putting units of cache in caching set on cloud nodes. In the section we propose a Cache Placement algorithm based on Clustering (CPC). Its concrete steps are as follows.

1)Assuming the number of cloud caching node is n and the caching capacity of each cloud node is m, the total caching capacity is $h = m \cdot n$. The caching mechanism in clouds is organized by Memcache [11].

2)We set a unit of cache as a point. If two units of cache are accessed by a data processing, we add an edge weighted by 1. The weight can be added many times, i.e., if there exists an edgy and there is another data processing, the weight is increased by another 1. All the units of cache in caching set form a connected graph.

3)We divide all the c_i in caching set into n clusters using clustering algorithm. One cluster is denoted as C and the data volume of C is denoted as d_C. In the paper we propose a clustering algorithm called equal-capacity clustering algorithm based on k-means [12]. The algorithm divides all units of cache into the number of nodes (n), and each node has the same data volume.

Next we introduce our proposed equal-capacity clustering algorithm. We put all the l units of cache on all the n cloud nodes with the same caching capacity. Each node has $e=l/n$ units of cache.

1)We compute a weight for each unit of cache. The weight of unit of cache j is computed as $f_j = \sum_{i=1}^{q} w_{ij}$, where w_{ij} is the weight between unit of cache i and unit of cache j.

2)We sort all the units of cache in non-ascending order of f.

3)We first randomly select n units of cache as initial centers.

4)For each unit of cache, we obtain its edge weights with each initial center. The unit of cache is put into the same cluster with the initial center that has the largest weight.

5)When each unit of cache is put into clusters, there appears a situation that the number of units of cache is larger than e. In this case, we choose the unit of cache that has the smallest edge weight to shift out of the cluster, and the removed unit of cache re-chooses another center.

6)The algorithm stops until all the units of cache find their centers.

The proposed CPC algorithm largely reduces network transmission cost during big data access, because data processing only works on one node or several neighboring nodes.

5 Performance evaluation

At present, there are 21 nodes on the big data platform of equipment condition evaluation. Except two name nodes, maximum cache capacity of the 19 data nodes is 76GB. The main applications are searching data in PMS, EMS, GIS, computing the correspondence between equipment ID in EMS and PMS, the Chargeability of surge protectors, calculate the reject rate and time of duration of each attributes. We choose 10 days as a period. The size of each unit of cache is set to 1MB. We use CRV algorithm to choose several units of cache and use CPC algorithm to put clusters on 19 nodes.

In order to validate the performance of CRV and CPC algorithm, we use Least Recently Used algorithm (LRU) and round-robin scheduling algorithm as comparison objects. The performance indexes are hit rate and query time. LRU algorithm is to select the least used unit of cache to replace each time, which is most used in replacement algorithm. It implementation mechanism is simple and it adapts to the high local data access. Round-robin scheduling algorithm is to put units of cache on different nodes in turn, i.e., each time the placed node is computed as $i = (i + 1) \bmod n$, and we choose the ith servers. The algorithm is simple and easy to realize.

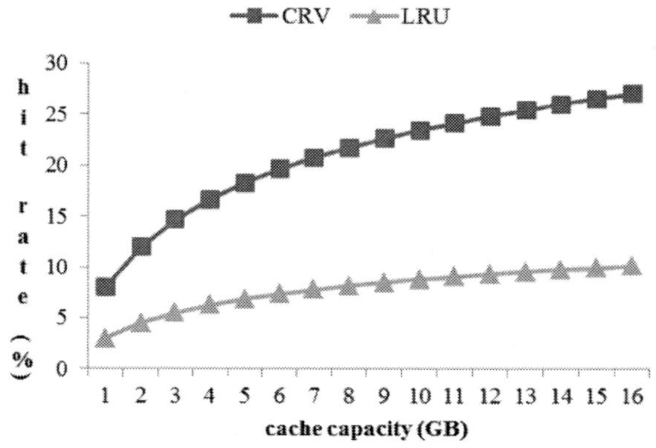

Figure 3. The comparison on hit rate of cache replacement algorithms

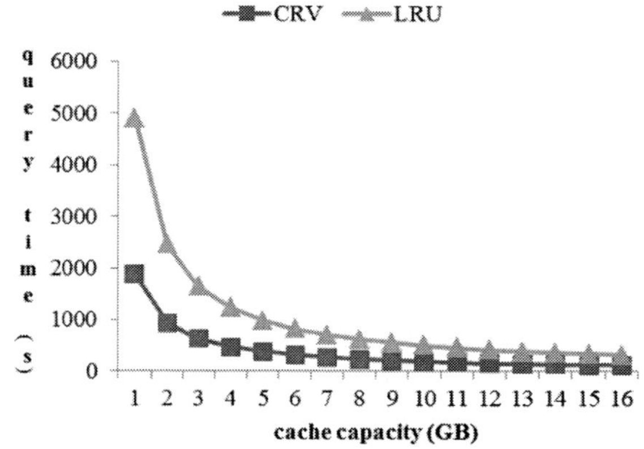

Figure 4. The comparison on query time of cache replacement algorithms

First we just use one node as cache placement node, and use LRU and CRV algorithm to run data accesses in a period. We record the hit rate and query time, as shown in Fig. 3 and Fig. 4. CRV has 13% higher hit rate than LRU algorithm. With the increase of cache capacity, the hit rate also ascends. The query time of CRV is on average 46.5% lower than LRU. Since there is only one cache placement node, the downtrend of query time gradually becomes slow with the increase of cache capacity.

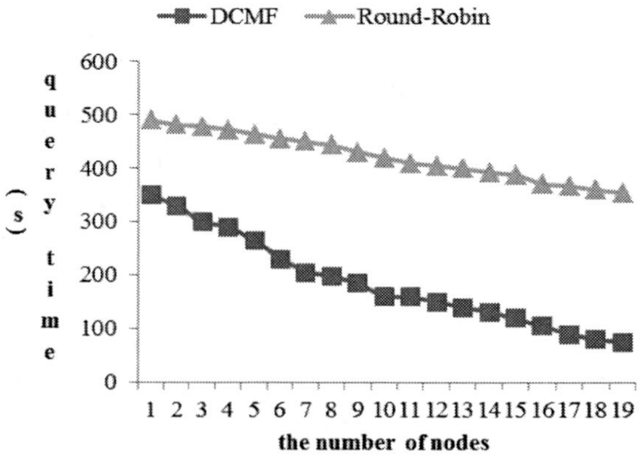

Figure 5. The comparison on query time of cache management frameworks

Second we validate the whole performance of proposed framework. In cache replacement phase we use our CRV algorithm and in cache placement phase we separately use CPC algorithm and round-robin algorithm. We run data access on multiple nodes in a period and record data query time, as shown in Fig. 5. The query time decrease with the increase of the number of caching nodes. Our proposed DCMF framework has 53.3% higher query efficiency than that of round-robin.

6 Conclusions

In the paper we design a distributed cache management Framework of big data for equipment condition assessment, which efficiently improves big data processing performance. Meanwhile, we propose a cache replacement algorithm based on value of units of cache in the last period and cache placement algorithm based on clustering method. All the approaches reduce network transmission cost and data processing time. Our extensive experiments demonstrate that the proposed CRV algorithm has 13% higher hit rate and 46.5% lower query time than LRU algorithm; the whole framework has 53.3% lower query time than round-robin algorithm.

Acknowledgement

This work was supported by the National High Technology Research and Development Program of China (863 Program) under grant no. 2015AA050204.

References

1. X.S. Peng, D.Y Deng, S.J Cheng, J.Y. Wen, Z.H. Li, L. Niu, *Proceedings of the CSEE*, **35**, 1, pp. 503-511 (2015).
2. X.L. Qin, W.B. Zhang, J. Wei, W. Wang, H. Zhong, T. Huang, Journal of Software, **24**,1, pp. 50-66 (2013).
3. Storm A J, Garcia-Arellano C, Lightstone S S, Diao Y, surendra M, *Proceedings of the 32nd International Conference on Very Large Data Bases (VLDB)*, Seoul, Korea, pp. 108-1092 (2006).
4. Malik T, Burns R, Chaudhary A, *Proceedings of the 35th International Conference on Data Engineering (ICDE)*, Tokyo, Japan, pp. 94-105 (2005).
5. Brown K P, Carey M J, Livny M, *Proceedings of the 19th International Conference on Very Large Data Bases (VLDB)*, Dublin, Ireland, pp. 328-341(1993).
6. Jincheng Yao, Shidong Zhang, Shiyu Liang, Liqing Zhong, Chinese Journal of Computers, **34**, 12, pp. 2319-2331(2011).
7. O'Neil Elizabeth J, O'Neil Patrick E, Weikum Gerhard, *Proceedings of the 19th International Conference on Management of Data (SIGMOD)*. Washington, DC, USA, pp. 297-306 (1993).
8. The InnoDB Buffer Pool. http://dev.mysql.com/doc/mysql-monitor/3.0/en/mem-graphs-innodb-buffer-pool.html. (2015-10.08).
9. Chaoyu Wang, The study on cach replacement strategy, harbin Engineering University (2012).
10. Wenzhong Li, Daoxu Chen, Sanglu Lu, Journal of Software, **21**, 7, pp. 1524-1535 (2010).
11. MemCache. http://www.danga.com/memcached, (2015-09-30).
12. Ng R.T., Han J, *Proc. 20th Int. Conf. on Very Large Data Bases*, pp. 144–155(1994).

MATEC Web of Conferences **55**, 06012 (2016) DOI: 10.1051/matecconf/20165506012
ACPEE 2016

A Design of Ultra-wideband Filter with Band-notched Characteristic Based on Left - Handed Materials

LI Shun-li [1,a], ZENG Wen-bo [2]

[1] *School of Electric and Information Engineering,Guangxi University of Science and Technology ,545006,Liuzhou, Guangxi,P. R. China*
[2] *School of Electric and Information Engineering,Guangxi University of Science and Technology ,545006,Liuzhou, Guangxi,P. R. China*

Abstract. Based on the concept of left-handed material (LHM), a novel ultra-wideband (UWB) planar filter with negative permittivity and permeability was designed and proposed in this paper. By etching a cross-coupled structure on the surface of substrate , the filter achieved a frequency pass band of 2.86 - 11.22 GHz, which meets requirement of the UWB applications. Moreover, a notch frequency band of 7.26 - 9.13 GHz also achieved simply by increasing the length of the left hand cross side structure. HFSS software was used to design and optimise the filter .The subtract of the filter is with a relative dielectric constant of 2.55 and the overall size of the filter is $7.6 \times 16 \times 0.8$ mm^3.

1 Introduction

Left - Handed Materials(LHM) refers to those class of materials which electric field **E**, magnetic field **H** and wave vector **K** satisfy left-hand rule in the electromagnetic wave propagation[1]. LHM has a series of bizarre physical phenomena, such as inverse Focus / divergence phenomenon, the phenomenon of negative refraction, inverse Cherenkov radiation effects, inverse Doppler effect, the inverse Gus - Han Sheen displacement and perfect lenses[2]. Because of these unusual electromagnetic properties, LHM has been named backward wave materials (BWM) and negative index of refraction materials (NIP) in the international community. Moreover, because of its negative permittivity ε and permeability μ, LHM was also been called meta-material or double negative material (DNG) [3]. LHM is currently only obtained artificially, but it attracts more and more researcher's attentions for its special properties at present[4].

In recent years, ultra-wideband (UWB) has attracted significant attention in wireless communication, networking, radar detection, and other applications [5].The UWB filter based on left - handed materials is also a hot topic in recent years. The UWB frequency band is from 3.1 to 10.6 GHz. However, there still remains other narrowband wireless communication signals within the UWB spectrum, such as X-band (7.25 - 7.75 GHZ,7.9 - 8.4 GHZ),which may cause several frequency interference to the UWB communication system. In order to avoid and solve the problem caused by this frequency band, UWB filter with band-notched function for UWB applications are required[6].Many UWB filter with band-notched characteristic have been designed and proposed in the past few years, these filter usually with the structures such as adding a single minor sides in the

[a] Corresponding author: ZWB@gxut.edu. cn

feeder, introducing slit structure in the filter ,etc[7].In this paper ,a UWB planar filter based on left-handed material (LHM) concept was designed, furthermore, by increasing the length of the cross structure on the LHM side of the subtract, a frequency notch-band from 7.26 - 9.13 GHz is achieved.

2 Principle of left - handed materials filter

The characteristic of electromagnetic wave propagation in the medium are decided by dielectric constant ε and permeability μ[8].In homogeneous isotropic material medium, single frequency wave phase constant k and angular frequency ω satisfies the relation as

$$k^2 = \frac{\omega^2}{c^2} n^2 \tag{1}$$

where in, n as the refractive index of the material and can be written as

$$n^2 = \varepsilon\mu \tag{2}$$

Usually, permittivity and permeability are positive in the case of right hand material without loss. However ,one can see from above expression, n keeps the same when ε and μ change sign simultaneously ,this means that although the left-handed material does not exist in the nature, it theoretically consistent with the laws of physics.

In this paper, Smith parameter extraction method is used to verify the value of ε and μ. Suppose the filter substrate thickness is h, and the definition of transmission matrix T is expressed as

$$T = \begin{pmatrix} \cos(nkd) & -z/[k\sin(nkd)] \\ k/[z\sin(nkd)] & -z/[k\sin(nkd)] \end{pmatrix} \tag{3}$$

where in, n is the refractive index, z is the wave impedance, k is the free space wave number.

In accordance with the transfer matrix and dispersion relations ,n can be written as

$$n = \frac{1}{kh} \cos^{-1}[\frac{1 - S_{11}^2 + S_{21}^2}{2S_{11}}] \tag{4}$$

$$\mu = nz \tag{5}$$

where in, $z = \sqrt{\dfrac{(1+S_{11})^2 - S_{21}^2}{(1-S_{11})^2 - S_{21}^2}}$

Based on the analysis above, one can find from the expression of ε and μ that the free-space wave number and the thickness of the substrate are higher than zero, so k and h can be ignored while determine the plus-minus of ε and μ. Furthermore, the ε and μ vary from frequency can be obtained based on the simulation value of S_{11} and S_{21} as well as the application of MATLAB software, as shown in FIG 1.

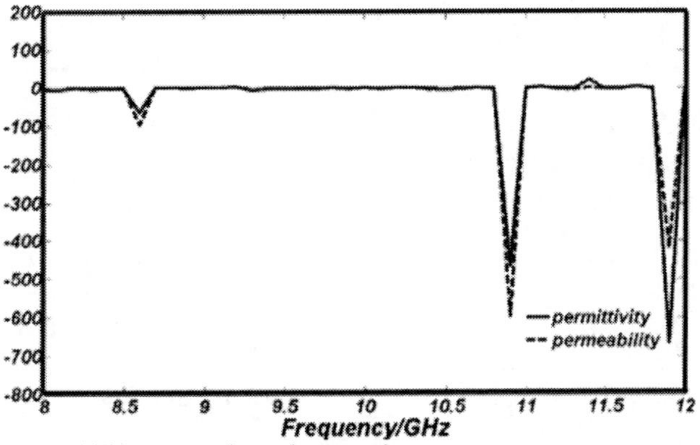

Figure 1. permittivity ε and permeability μ vary from frequencies

One can see from Figure1 that the permittivity ε and permeability μ can be negative simultaneously within a certain frequency range, thus means the designed filter achieve a negative refractivity therefore in accordance with the properties of left-handed materials.

3 Design of the filter

The UWB filter based on the concept of LHM proposed is etched on both sides of a piece of FR4 subtract, and the subtract is with a relative dielectric constant of $\varepsilon_r = 2.55$, dielectric loss tangent tan $\delta = 0.0019$, and a thickness of 0.8 mm. One side of the FR4 subtract is etched as a coplanar waveguide while the other side is etched as a high impedance micro-strip curved lines. Figure 2 shows the structure of the proposed filter .

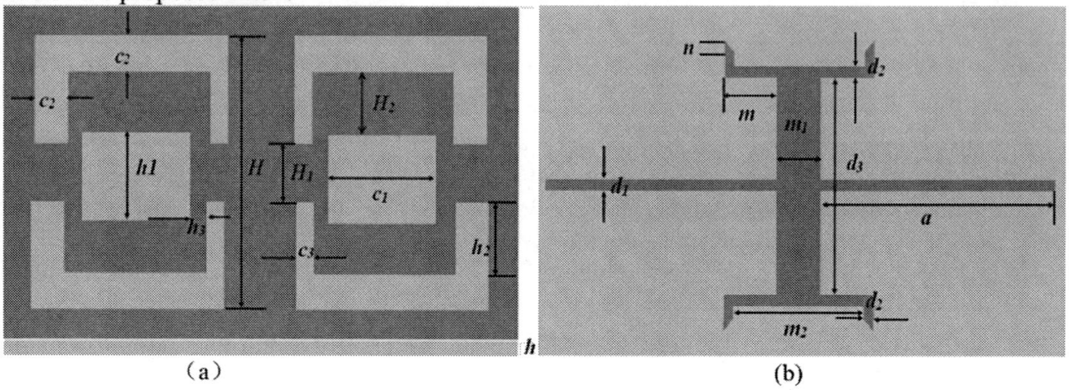

Figure 2. Structure of the proposed UWB filter: (a) top view and (b) bottom view.

The overall size of the filter is 16×7.6 ×0.8 mm^3.HFSS software is used to design and optimise the filter. After intense work, geometrical parameters of the filter can be obtained as: H = 4.94 mm, H_1 = 2 mm, c_1 = 2.2 mm, c_2 = 1 mm, c_3 = 1 mm, h_1 = 1.6 mm, h_2 = 0.47 mm, a = 7.39 mm, d_1 = 0.16 mm, d_2 = 0.22 mm, d_3 = 5.7 mm, m = 1.75 mm, m_1 = 0.5 mm, m_2 = 3.56 mm.

The return loss S_{11} and the insertion loss S_{21}can be obtained by HFSS software, as shown in Fig 3.

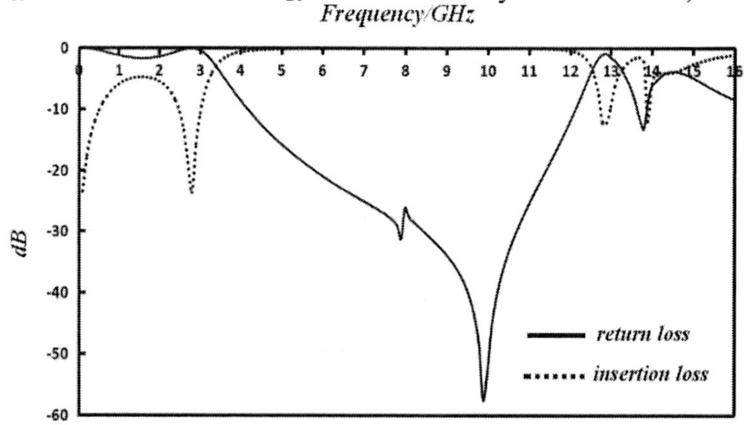

Figure 3. The return loss S_{11} and the insertion loss S_{21} of the proposed filter

To improve the performance of the filter and avoid electromagnetic interference from the X band (7.25 - 7.75GHZ, 7.9 - 8.4GHZ) within the UWB frequency band, a novel structure is introduced. By increasing the left hand cross side length of the material of the planar filter structure, the degree of coupling is increased and the frequency notched-band from 7.26 - 9.13 GHz can be achieved. The band-notched structure is shown as Fig 4 with geometrical parameters of m = 7.25 mm and m_3= 4 mm while other geometrical parameters fixed.

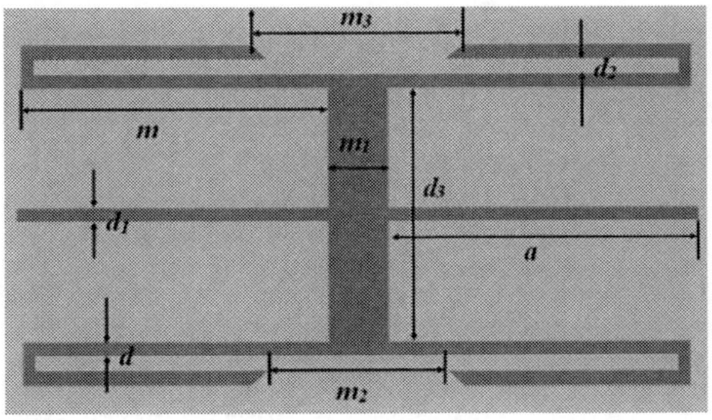

Figure 4. The bottom view of filter with band-notched characteristic

4 Optimization and Analysis

HFSS simulations show that the length of m and m_1 have a great influence on the performance of the filter. In order to optimize the structure of the filter , the insertion loss S_{21} is to be set as the destination function, and the m and m_1 change while other parameters keep fixed. The simulation S_{21} of the proposed filter with different m and m_1 are shown in Figure 5.

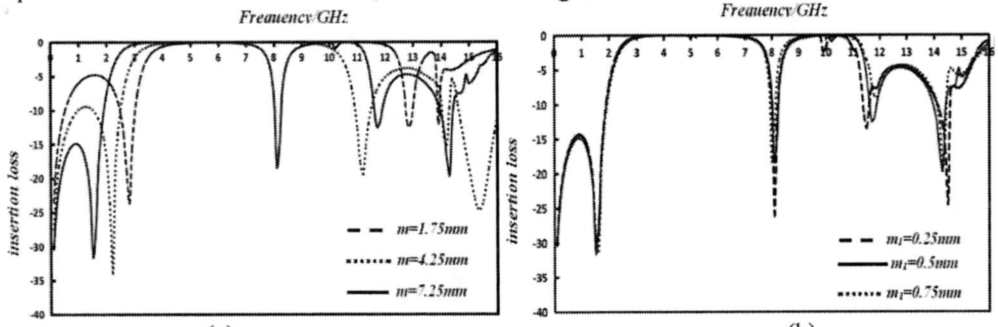

Figure 5. Simulation S_{21} of the proposed filter with different (a) m, (b) m_1

As one can be seen from the figure 5, the notch frequency range varies when m changes, according to the requirement of the size of filter, an ultimately $m = 0.75$mm is adopted .Simulations indicate that m_1 mainly affects the range pass-band of the filter, and the pass-band of the filter gradually widened with an increasing m_1, but when the value of m_1 is greater than 0.5mm, the pass-band trends to narrow, then an optimized $m_1 = 0.5$ mm is adopted. Simulations also indicate that m_1 mainly affects the coupling efficiency of the filter and therefore affects the insertion loss, while m mainly affect impedance of the micro-strip line and therefore affects return loss of the filter[9].

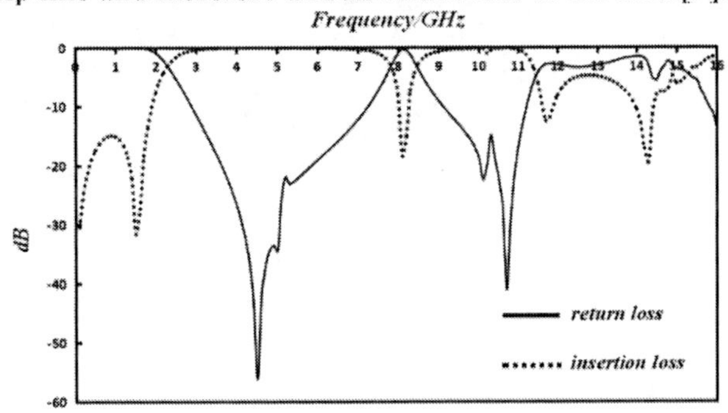

Figure 6. Simulation return loss and insertion loss of the UWB filter with band-notched characteristic based on LHM

The return loss S_{11} and insertion loss S_{21} of the proposed filter is shown in Figure 6, a pass-band from 2.86-11.22 GHz with S_{11} larger than - 10 dB and a notch frequency band of 7.26 - 9.13 GHz with S_{21} less than - 10 dB , are achieved.

5 Conclusion

Based on the concept of left-handed materials ,this paper present a novel design of UWB filter with a notch frequency band from 7.26 - 9.13 GHz , which results the filter a good performance of rejecting the electromagnetic interference caused by the X-band signals. The return loss of the filter on the resonance point achieves -56.21 dB and the gain is close to 1. Moreover, the filter is compact with an overall size of $7.6 \times 16 \times 0.8$ mm^3 ,it has broad development and application prospects in the field of UWB applications.

Acknowledgement

The authors would like to acknowledge the financial support from the projects of Natural Science Fou ndation of Guangxi (NO.2015GXNSFAA139289), and Key Laboratory of Cognitive Radio and Infor mation Processing , Ministry of Education (Guilin University of Electronic Technology) (No. CRKL1 50205)

References

1. J.-Q. Huang,Q.-X. Chu.*COMPACT UWB BAND-PASS FILTER UTILIZING MOD-IFIED COMPOSITE RIGHT/LEFT-HANDED STRUC-TURE WITH CROSS COUPLING*[J].PIER, **8**(2010).
2. Shah Nawaz Burokur, Mohamed Latrach.*Left-Handed Medium effect on the characteristics of a circular patch antenna*.IEEE,**11**(2007).
3 . W.-B. Zeng, J. Zhao, and Q.-Q. Wu, *Compact planar ultra-wideband wide-slot antenna with an assembled band-notched structure*,MICROW OPT TECHN LET,**7**(2012).
4. ZHU Xiao-ming.*Dsihn of miniaturization band-notched UWB antenna*,Journal of Heilongjiang Institute of Technology,**1**(2014).
5. Wu Yiwei,Li Simin,*An microstrip patch antenna based on left-handed material*,Journal of Guilin University of Electronic Technology,**6**(2013).
6. K. Bahadori, Y. Samii, "A miniaturized elliptic-card UWB antenna with WLAN band rejection for wireless communications", IEEE,**11**(2007).
7. C. Yoon, W.-J. Lee, W.-S. Kim, H.-C. Lee, and H.-D. Park, Compact band-notched ultra-wideband printed antenna using inverted L-slit, MICROW OPT TECHN LET,**1**(2012).
8. X.-L. Liu, Y.-Z. Yin, P.-A. Liu, J.-H. Wang, and B. Xu, A CPW-fed dual band-notched UWB antenna with a pair of bended dual-L-shape parasitic branches,PIER,**6**(2013).
9. M. T. Islam, R. Azim, and A. T. Mobashsher, Triple band-notched planar UWB antenna using parasitic strips, PIER,**4**(2012).

MATEC Web of Conferences **55**, 06013 (2016) DOI: 10.1051/matecconf/20165506013
ACPEE 2016

Conceptual Design of a Medium-Sized Combined Smart Photovoltaic - Agriculture System - Case Study in Malaysia

Harijono Djojodihardjo[a]

President, The institute for the Advancement of Aerospace Science and Technology (IAAST), Jakarta 15419, Indonesia

Abstract. With the backdrop of sustainable environment, Photo-Voltaic Power System linked to Climate-Smart Agriculture may offer solutions for Sustainable Energy, Climate Change mitigation and Sustainable Agriculture. An overview of the scope, extent and options of such combined - Co-Located PV Agricultural System appropriate for South East Asian setting, in particular, Malaysia and Indonesia is elaborated, for preliminary insight on steps and choices that have to be taken in undertaking such venture. Possible photovoltaic (PV) system installation and estimate the cost, performance, and site impacts of different PV options are discussed. Technical, financing and procedural aspects that could assist in the implementation of a Co-located PV system at the site should then be studied for decision options. A brief Framework for Conceptual Design of Co-Located PV-Agricultural System Plant is outlined.

1 Introduction

Sustainable growth and the sustainability of the environment, and more pronouncedly, global changes, are not only the concern of governments and international organizations, but for each and every individuals. Energy, food security, poverty and climate change are closely linked sustainable development. Climate-Smart Agriculture is a very significant part of the solution for both Climate Change mitigation and Sustainable Agriculture. Agriculture has much to contribute to a low emissions development strategy. Climate-smart agriculture is essential for building capacity, experience and guiding future choices, as well as smart management of natural resources [1,2]. Climate-smart agriculture is closely related to sustainable energy. Accordingly, to address sustainable energy and climate-smart agriculture relevant for sustainable development on the macro scale, and economical sustainable energy in the micro scale, the present work reviews and summarizes relevant aspects on the state of the overriding considerations and then present a preliminary concepts of Combined Smart /Co-Located Photovoltaic - Agriculture System, appropriate for South East Asian setting, in particular, Malaysia, as a case study. Furthermore, the purpose of this paper is to provide an overview of the scope and extent of such combined - Co-Located PV Agricultural System, for preliminary insight on steps and choices that have to be taken in undertaking such venture. A comprehensive assessment on the site for a possible photovoltaic (PV) system installation and estimate the cost, performance, and site impacts of different PV options would be the follow up of the decision choices. Financing options

[a] Corresponding author : harijono.djojodihardjo@yahoo.com

 DOI: 10.1051/matecconf/20165506013

that could assist in the implementation of a PV system at the site are discussed for decision options. A brief framework for conceptual design of Co-Located POV-Agricultural System Plant is outlined.

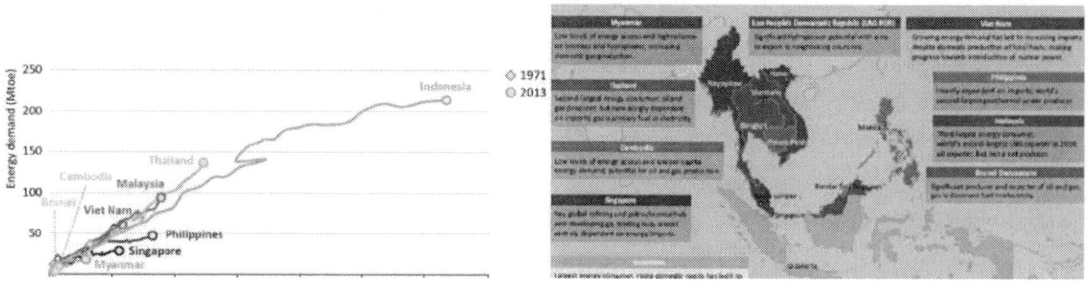

Figure 1: (a) Total primary energy demand and GDP in selected Southeast Asian countries, 1971-2013 [3,4]; (b) Energy demand prospects [4].

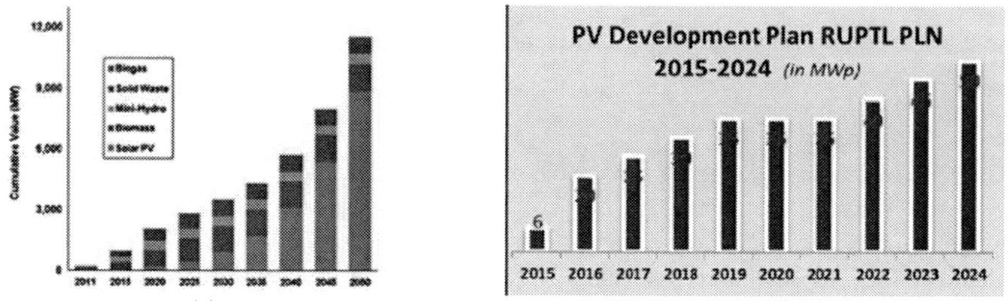

(a) (b)

Figure 2: (a) Potential cumulative values of RE installed capacity in Malaysia for the period (2011–2050)[5]; (b) PV Development Plan, 2015-2024, Indonesia [6].

The focus of the present paper is on the ASEAN region, as one of the fastest developing regions in the world which requires rising energy supplies for their rapid economic development growth. Since Southeast Asia's energy use and economic growth gradually decouple, driven by the adoption of more efficient technologies, increasing saturation of appliances and equipment in some markets, and as the economic structure shifts to less energy-intensive activities [5], this ratio is expected to gradually decline over time. The energy demand evolution is illustrated in Fig. 2, which summarizes the potential cumulative values of Renewable Energy installed capacity in Malaysia for the period (2011–2050)[5,7] and Indonesian PV Development Plan, 2015-2024 [6].

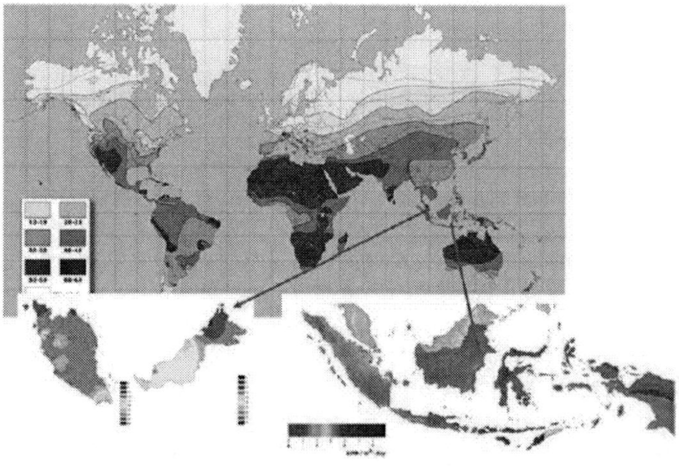

Figure 3: Solar Irradiation in Malaysia and Indonesia in (MJ/m2/day) in the backdrop of Global Radiation in (synthesized from [7] and [8]).

2. Solar Photo Voltaic applications in Malaysia and Indonesia

To illustrate the dedicated application of Photo Voltaic Solar Energy, in 1998, Tenaga Nasional Berhad (TNB) in Malaysia has installed grid-connected PV systems for National power utility [7]. Six pilot grid-connected PV systems were installed with power capacity ranging from 2.8 kWp to 3.8 kWp in 1998. first grid-connected PV system was installed by Tenaga National Research (TNR) Sdn. Bhd. on the rooftop of the College of Engineering, University Tenaga Nasional (UNITEN) with the system capacity of 3.15 kWp [7]. Next two grid-connected photovoltaic systems were installed, one at a BP petrol station with the capacity of 8 kWp by BP Malaysia while the other one was installed at Solar Energy Research Park in University Kebangsaan Malaysia (UKM) with a capacity of 5.5 kWp. One of the objectives of these initiatives is to introduce energy savings in the area of high rise buildings construction and design through air flow improvement. Integrated installation of PV system in a building can improve energy efficiency of a building through meticulous air flow improvement, eventhough the building is tightly sealed to the outside. Prototype Solar House (PSH) has been designed and built in order to emphasize the multifunctional aspects of PV in buildings for living house at urban areas in the country [7]. In 2000, the first Malaysian Building Integrated Photovoltaic (BIPV) system was installed in Port Dickson Malaysia with the capacity of 3.15 kWp. By November 2000, another BIPV system was installed in Shah Alam Malaysia with a capacity of 3.24 kWp.

Figure 4: **Some** Examples of Solar PV Power Plant Installed in Malaysia; (a) 362 kWp Building Integrated Solar Photovoltaic System at PV system Technology Park Malaysia; (b) 646 kWp Building Integrated (Roof-Top) Integrated Solar PV Power Plant at Robert Bosch (M) Sdn Bhd, Penang, Malaysia; (c) 19 MWp Kuala Lumpur International Airport, combined ground-mount, parking canopy and rooftop PV systems.

The installations, being a standalone system, were fixed on the top of existing roof tiles and appeared to be successful since it is functioning without any problem in maintenance of the unit [9,10]. Some PV Power Plants that have been installed in Malaysia recently are exhibited in Fig. 4.

3. Co-Located PV system plants imperatives

In a recent shift towards Grid-Connected Systems, at the end of 2008, cumulative installed PV capacity totaled more than 14 gigawatts (14,000 megawatts) worldwide, according to the International Energy Agency (IEA)[4,11].

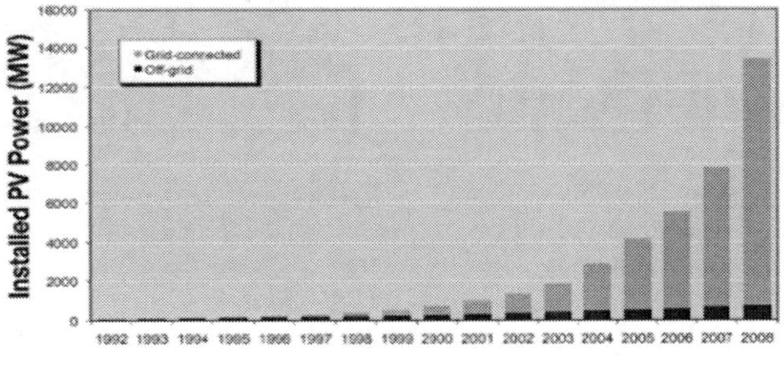

Figure 5: Worldwide Grid-Connected and Off-Grid PV Power [4,11].

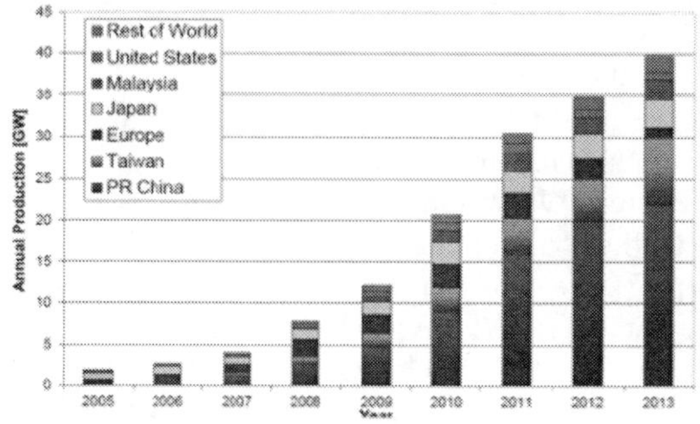

Figure 6: World PV cell/module production from 2005 to 2013 (extracted from [12])

Approximately 5.5 GW of PV was installed in 2008, an increase of about 140% over the previous year. The IEA data clearly indicate tremendous growth in grid-connected PV since the mid 1990s when it was the minority of the market. This trend has continued such that grid-connected applications accounted for about 94% of cumulative installations by 2008. Off-grid applications have continued to grow worldwide, although less vigorously than grid-connected systems.

Fig. 6 illustrates the growth in PV capacity for both off-grid and grid-connected applications. Based on market and policy considerations, grid-connected PV systems for commercial, industrial, and residential uses appear to be entering a period of long-term, accelerated growth after recovery from recession of 2009's. Fig. 5 shows the World PV cell/module production from 2005 to 2013 (extracted from [12]).

3.1 Vegetation-Centric Approaches to Co-Location

Vegetation-centric approaches to co-location of solar energy and vegetation are characterized by actions that serve to maximize biomass production activities and minimize changes to existing vegetation management activities, while also incorporating solar energy production activities. Vegetation-centric approaches may be well suited to areas that are land constrained or that are already developed agricultural areas. The basic premise is that the vegetation productivity of the land being utilized is not sacrificed for the sake of solar generation. An example of Vegetation-Centric Approach to PV System Co-Location is exhibited in Fig. 7

Figure 7: (a) Pasture crops growing under elevated PV panels in Massachusetts. Photo from Stephen Herbert, University of Massachusetts [13].

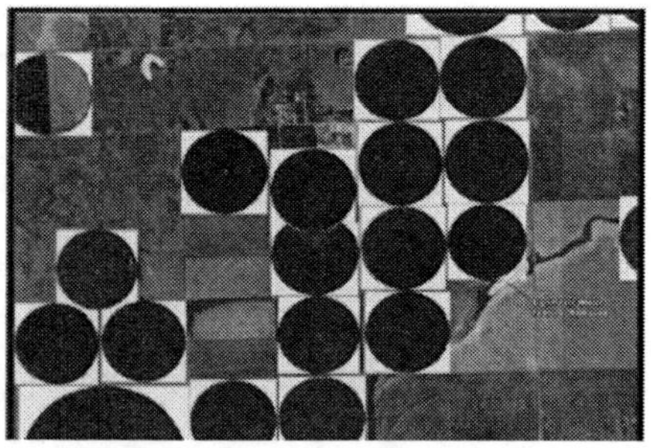

Figure 8: Center-pivot irrigation envelopes with irrigated lands. Yellow areas show unused and non-irrigated lands that could be sites for solar energy technologies[13].

3.2 Energy-Centric Approaches to Co-Location

Energy-centric approaches to co-location of solar energy developments and vegetation are characterized by actions that serve to maximize solar energy output, minimize changes to solar development standard practices, while also promoting vegetation growth under and around the solar installation. Ideally, energy-centric approaches to co-location do not lead to a reduction in expected solar energy output. Energy-centric approaches may be well suited to areas with high quantities of land targeted for solar energy development or areas where solar development has already occurred. An example of Energy-Centric Co-Location of PV panels and vegetation at NREL site is exhibited in Fig. 8. Fig. 9 exhibits some examples of various PV Plants recently installed in Malaysia and Indonesia.

No	Type/ Place	Solar Radiation, kWh/m2	Scale/ Capacity, kW or MW	Investment, RM	Prorated Investment/kW	Return Period for PV Efficiency %		Annual Revenue RM		Annual Output kWh/year	
						15	20	15	20		
1	Rooftop/ Columba	1,571	5.28 kW	52,800	10,000	4.55	3.37	12,133	16,177	9,975*	*based on 350 x daily output; Source: Mun et al,2015
2	Rooftop/Aquila	1,571	.5.78 kW	57,800	10,000	4.84	3.59	12,516	16,688	10,290*	
3	Rooftop/ Jelutong	1,571	4.8kW	48,000	10,000	11.19	8.17	4,768	6,357	3,920*	
4	PV system Technology Park Malaysia		362 kWp	100,000,000 (Razali,2004, based on estimated USD 25 million)	27,624.3						
5	Robert Bosch (M) Sdn Bhd, Penang, Malaysia		646 kWp,								
6	PV power plant at Pajam, Malaysia by Cypark Resources Berhad		8 MW								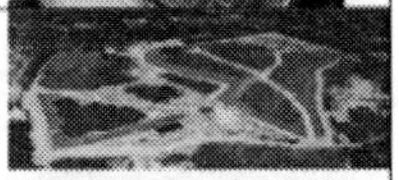
7	Amcorp Power Sdn Bhd, Gemas, Negeri Sembilan, Malaysia		10.25 MW								
8	Kuala Lumpur International Airport, combined ground-mount, parking canopy and rooftop PV systems		19 MW (DC)								
9	Solar PV Power Plant in Bangli, Bali Province, Indonesia Source: Solar PV Feed-in Tariff in Indonesia - Regulation and Framework Conditions, 2014, Ministry of Energy and Mineral Resources		1 MW	~2,000,000.- (IDR 26 billion)							
10	ADB sponsored - Solar energy panels in Sumba island										

Figure 9: Summary of some PV Plants Examples in various categories

4. Preliminary conceptual design of Co-Located PV-Agricultural system

Basic Configuration. An more detailed elaboration of the conceptual design described here should be made, based on in depth study for practical implementation like Small Scale Experiment for Technology and Model Identification, Feasibility Study and Collateral Assurances, Site selection and Land Requirements of Solar Energy Installations, as well as Land-Use and Environmental Impacts Study of Co-Located PV-Agriculture System Projects.

One option of the Co-Located PV Power system that can be proposed is exhibited in Fig. 10, following the configuration of Pasture crops growing under elevated PV panels as example. Its capacity can be chosen between 300KWp to 1 MWp. The circuitry schematics are reproduced on the right of this figure.

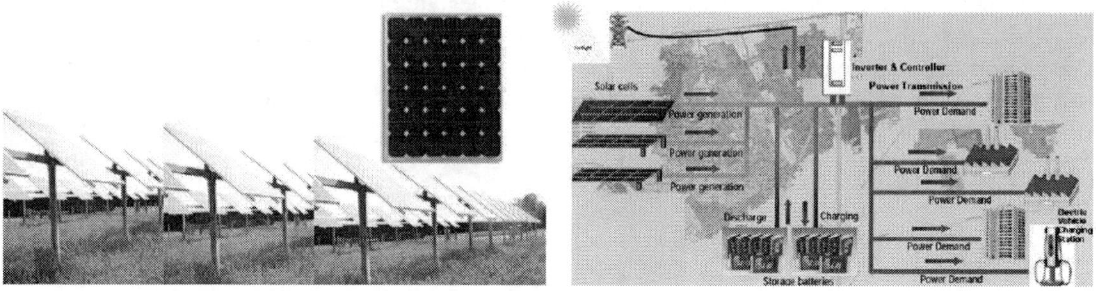

Figure 10: Configuration Option No.1, following Pasture crops growing under elevated PV panels example. Capacity can be chosen between 500KWp to 1 MWp. The circuitry schematics are shown on the right. Illustrated in the inset is the image of monocrystalline solar panels, which can be utilized.

Summary of Sample Design Sizing and the Calculation of the Number of PV Panels. The summary is given in Table.1.

Table 1: Sample Estimate of PV System Preliminary Design

	PV Panels/row	No.Rows/ module	Total No.PV Panels/module	Watts-peak/PV Panel	Total Power (Watts-peak)	Watts. m2	Price/Panel(estimate,MYR)	Price PV Panels (MYR)	
	600	5	3000						
Solar Panel 100Wp Monocrystal line									RM 126.90
Equation 1	Energy Generated (kWh/day)=Annual Solar Radiation x Sun Hours x PV Efficiency x Area								
Equation 2	Annual Revenue = Energy Generated x Days in a Year x FiT rate								
Equation 3	Return of Investment = Initial Investment/ Net Annual Benefits								
Option 3, PV Panels Only, each 100 Wp				100	300,000	pm	127	381,000	1 module, 300,000 Watts-peak
3 Modules, PV Panels Only								1,270,000	~3 modules 1 MW-peak
Initial Investment Cost		PV Module+Inverter+Batteries+Installation cost+contingencies							
Net Annual Benefits		Revenue-Operation & Maintenance Cost							
Revenue		Income from selling power to the utility							
Operation and Maintenance (O&M)		Assumed to be 1~2% of the initial investment cost; can be changed due to situation							
Inverter		SPM3000 MS	3,000MW, Modified Sine Wave	RM1,378.00	US$459.40				
Dee4p Cycle Batteries		SPM100SB	100AH, 12V,AGM	RM772.00	US257.40				
Wiring					US2.45/m				
Support Frames									
Installation Cost									
Initial Investment	RM	4,125,000	Prorated from earlier Studies	Payback period, years	4.258	years	Annual Revenue	Rough Estimate 676,363	

5. Concluding Remarks

Based on an exhaustive review on Photo Voltaic Power System technology and Applications in various parts of the world, and particular applications in Malaysia and Indonesia, a comprehensive insight on conceptual design and technical requirements for an independent venture of Photo Voltaic Power System has been gained. Based on these, a conceptual design and preliminary configuration of a Co-Located Photovoltaic Agriculture system has been proposed, including the probable investment and rate of return.

References

1. Djojodihardjo H. (1995). Perspective of Fifty Mega Watt Peak (50 MWp) Photovoltaic Solar Home System for Rural Electrification in Indonesia. Paper presented at UNESCO-BPPT-LIPI Round Table Conference on Solar Energy Jakarta, 1995.
2. Djojodihardjo, Aerospace Technology Vision for Green Energy, Paper ICSAFEI-189-24 August 2015, resented at the 7th Conference on Sustainable Food, Energy and Industry in Regional and Global Context, 25-27 August 2015
3. IEA, Southeast Asia Energy Outlook - WEO Special Report-IEA, 2013
4. IEA-Southeast Asia Energy Outlook - World Energy Outlook Special Report 2015
5. Ismail, A.M. et al, Progress of solar photovoltaic in ASEAN countries: A review, Renewable and Sustainable Energy Reviews, 2015
6. Abd.-Rosyid, Opportunities and Challenges For Solar PV Rooftop in Indonesia, BPPT, 2015
7. Mekhilef et al, Ismail, A.M. et al, Progress of solar photovoltaic in ASEAN countries: A review, Renewable and Sustainable Energy Reviews, 2015
8. Solar PV Atlas: Solar Power in Harmony with Nature, http://awsassets.panda.org/downloads/solar_pv_atlas_final_screen_version_feb_2013.pdf
9. Erge T, Haw LC. PV in buildings for Malaysia: prototype solar house. In: Proceedings of 3rd world conference on photovoltaic energy conversion, vol. 3. 2003. p. 2358–61.
10. Anonymous, Equatorial sunshine: the Malaysia BIPV programme. Renewable Energy World Magazine 2009.
11. IEA-Trends 2015 in Photovoltaic Applications
12. Waldau, J., PV Status Report-2014, November 2014, European Commission, DG Joint Research Centre,
13. Macknick, Beatty, Hill, NREL, Overview of Opportunities for Co-Location of Solar Energy Technologies and Vegetation, 2013

AUTHOR INDEX

AUTHOR INDEX

EDP Sciences – Web of Conferences
17, Avenue du Hoggar
Parc d'Activité de Courtabouf
BP 112
F-91944 Les Ulis Cedex A
France

ISBN 978-1-5108-2333-4